FIREFIGHTING STRATEGIES
AND TACTICS

FIREFIGHTING STRATEGIES AND TACTICS

Second Edition

James Angle

Michael Gala

David Harlow

William Lombardo

Craig Maciuba

Australia   Brazil   Canada   Mexico   Singapore   Spain   United Kingdom   United States

DELMAR
CENGAGE Learning

**Firefighting Strategies and Tactics,
Second Edition**

**James Angle, Michael Gala, David
Harlow, William Lombardo, Craig
Maciuba**

Vice President, Technology
and Trades ABU:

David Garza

Director of Learning Solutions:

Sandy Clark

Managing Editor:

Larry Main

Acquisitions Editor:

Alison Pase

Product Development Manager:

Janet Maker

Senior Product Manager:

Jennifer A. Starr

Marketing Director:

Deborah S. Yarnell

Marketing Manager:

Erin Coffin

Marketing Coordinator:

Patti Garrison

Director of Production:

Patty Stephan

Production Manager:

Stacy Masucci

Content Project Manager:

Jennifer Hanley

Art Director:

Benjamin Gleeksman

Editorial Assistant:

Maria Conto

Cover Image:

Rick Brady

For product information and technology assistance, contact us at
Cengage Learning Customer & Sales Support, 1-800-354-9706

For permission to use material from this text or product,
submit all requests online at **cengage.com/permissions**
Further permissions questions can be emailed to
permissionrequest@cengage.com

Library of Congress Control Number: 2007024707

ISBN-13: 978-1-4180-4893-8

ISBN-10: 1-4180-4893-3

Delmar Cengage Learning
5 Maxwell Drive
Clifton Park, NY 12065-2919
USA

Cengage Learning products are represented in Canada by Nelson Education, Ltd.

For your lifelong learning solutions, visit **delmar.cengage.com**

Visit our corporate website at **www.cengage.com**

Notice to the Reader

Printed in the United States of America
4 5 6 7 11 10 09

CONTENTS

FOREWORD TO THE SECOND EDITION

It has been almost seven years since the first edition of *Firefighting Strategies and Tactics* was released, and while many of the risks facing our nation's firefighters have remained unchanged, the threat of international terrorists striking on American soil has been realized. Once again fire and emergency medical services personnel find themselves in the front line defense of the citizens they have sworn to protect. But what has changed in an effort to improve the safety of our firefighters? The Commission on *America Burning, Recommissioned* has identified several key findings that should assist the fire service in protecting its own. "Implementation of Loss Prevention Strategies," "Improvements Through Research," and "Codes and Standards for Fire Loss Reduction in the Built Environment" are all key findings that should improve not only the survivability of a buildings occupants, but also the safety of those responding to the emergency needs of those occupants. But advances in building technology and the geographic variances in building sciences has proven to be a double-edged sword.

The building construction industry has made dramatic improvements over the years to meet the changing demands of the market in terms of hazard mitigation, energy conservation, and environmental sensitivity. Their mission has been accomplished through the introduction of new technology, high-tech materials, and engineering that have allowed the building construction industry to capitalize on methods that reach greater heights, obtain longer spans, and increase weight distribution, while still remaining responsive to the financial concerns of the consumer. Although all this engineering and technology has added to the inherent safety of these buildings occupants, the safety of firefighters responding to emergencies inside buildings has additional complications. Firefighters must stay informed and be educated to these new challenges and techniques of building construction, and adjust their firefighting strategies accordingly.

Security threats, both domestic and international, have forced Americans to seek new levels of protection within both new and existing buildings. Interior and exterior doors are no longer a simple transition from one space to another; they are specifically engineered to prevent intruders and protect building occupants through extraordinary design. Firefighters responding with local law enforcement agencies to security threats at incidents inside a structure need to be cognizant of the characteristics and design of the involved building. Lessons learned as a result of The World Trade Center have resulted in new design ideas to further reinforce elevator and stairwell shafts to provide occupants and emergency responders even greater protection during an evacuation.

Natural disasters continue to threaten communities across the nation with devastating results. Prevention strategies such as the Federal Emergency Management Agency's Project Impact: Building Disaster Resistant Communities have encouraged the building construction industry to showcase materials such as impact resistant glass and high density wallboard in target prone areas of the Country. These new materials are better able to withstand the destructive wind damage from hurricanes and tornados. Firefighters need to be able to recognize building construction that utilizes these types of high impact materials, because by design they are extremely difficult to breach without the right tools and training.

Across the nation the incidence of fires continues to decrease each year. This success, however, has resulted in a lack of experience as firefighters respond to fewer fires. It is therefore incumbent that firefighters actively build on their basic knowledge of building construction in order to apply the appropriate strategy and tactics at various emergency incidents. Computer simulations now provide firefighters with access to multiple scenarios utilizing realistic fire and smoke dynamics based on the building construction characteristics. The National Fire Academy's online training Web site provides firefighters with free access to various interactive fire scenarios. The primary benefit of computer simulations is that they provide firefighters the opportunity to repeat complex fire scenarios in order to master the principles of command and control, without placing themselves in harm's way. Firefighters can also review computer simulations that have been re-created after significant incidents such as the 2003 Station Nightclub fire in Rhode Island. Agencies such as National Institute of Standards and Technology (NIST) utilize research and computer modeling to better understand fire development and thermal conditions that existed under actual fire conditions. Utilizing this type of program for training, firefighters can gain insight and understanding into fundamentals of building construction, fire behavior, and potential losses from fire.

Firefighters need to quickly identify critical information such as the building construction, water supply, and fixed fire protection features in buildings within the jurisdiction that they protect. Access to pre-fire plans can quickly provide key information in a standardized format to the Incident Commander. The development and review of these pre-fire plans will provide firefighters an excellent opportunity for training by discussing incident objectives, strategy and tactics, and task assignments for various simulated fire scenarios based on the building characteristics. It is essential for firefighters to understand the structural limitations and expected fire behavior of any building involved in fire.

As in the first edition of *Firefighting Strategies and Tactics,* the authors of this second edition have once again shared their many years of experience and expertise in planning effective strategies to manage today's emergency responses. To assist those firefighting personnel in their duties, a special emphasis has been placed on the science of building construction. Readers are encouraged to use that as a primer to expand their base of knowledge to better understand not only how building components react to the ravages of fire, but how quickly a building's envelope can go from a protective shield to a mechanism for entrapment to all those that have ventured within.

Fire Chief Herman W. Brice
Palm Beach County Fire Rescue
Palm Beach, Florida

FOREWORD TO THE FIRST EDITION

We enter the twenty-first century with the technology and ability to virtually eradicate the nation's fire problem. Yet every reference available reminds us that we suffer from the greatest loss from fire of any industrialized nation in the world. Each year there are more than two million American fires and more than 100 firefighters killed in the line of duty. The original "America Burning—The Report of the National Commission of Fire Prevention and Control" identified America's fire problem more than 25 years ago. This report is considered the single most influential fire service document of the previous century. Unfortunately, "America Burning, Recommissioned" will echo the same challenges for the new millennium.

These challenges are now compounded by the multitude of responsibilities the fire services have inherited during the past 100 years. Emergency medical services, comprehensive code enforcement, hazardous materials, terrorism first response, and local emergency management are part of a long list of examples.

Fire suppression represented almost 100% of the fire services' response workload in 1900. Today, fire suppression represents less than 25% of responses. This phenomenon represents, on one hand, a great demand for diversification and, on the other hand, a great demand for specialization, from the senior fire executive to the recruit firefighter. A firefighter at the beginning of the previous century never envisioned paramedic or specialty team assignments in addition to firefighting responsibilities. Fortunately, the practices and principles presented in *Firefighting Strategies and Tactics* provide a firm discipline for addressing this large spectrum of responsibilities.

As an example, formal firefighting tactics and strategies prepared the fire services for the structure and accountability required to assume emergency medical responsibilities in the 1960s. The next transition was the nation's need to control the hazardous materials threat. The fire services were there again. Hot zones, isolation, protection, and other procedures learned in the control of hazardous materials incidents were a natural for application to the effective first response to terrorism incidents. Practically overnight the nation realized that when New York City's Trade Center and Oklahoma City's federal building were terrorized, 911 was called, not the Federal Bureau of Investigation or the Bureau of Alcohol, Tobacco, and Firearms. This phenomenon is reinforced when the firefighter is the first responder to our nation's numerous natural disasters, ahead of the Federal Emergency Management Agency.

Although the local community expects the fire service to continue to embrace and assume the all-hazards approach to public safety, it must understand that the community must also continue to take a greater role in life and fire safety. We simply cannot provide a firefighter on every street corner.

The National Fire Protection Association estimates there are more than 33,000 fire departments in the United States. These departments, regardless of size and whether they realize it or not, address their fire protection responsibilities through a cyclical performance model. It starts with life and fire safety education, followed by the administration and application of life and fire safety codes. These elements, coupled with code enforcement and preincident planning efforts, represent the fire department's responsibility prior to the fire occurrence.

Once the hostile fire occurs, the fire suppression element is deployed. Completing the fire investigation responsibility follows. The information gleaned here can reinitiate the education element and the cycle. Fire service tactics and strategies can be considered a common thread through all the elements of this fire protection cycle.

The five authors of *Firefighting Strategies and Tactics* represent the fire officers of the twenty-first century. Angle, Gala, Harlow, Lombardo, and Maciuba skillfully present the lessons, methods, and processes necessary to resolve all aspects of fire emergencies. This book is designed to sequentially prepare fire officers to resolve the potential fire scenarios they may encounter on a daily basis. The material has been pulled together in a single resource based on the knowledge, skills, and abilities gleaned from years of experience, training, and education.

Chief John M. Best, Manager
Reedy Creek Emergency Services
Walt Disney World, Florida

PREFACE

ABOUT THIS BOOK

Fireground strategy and tactics are the essence of fire department operations. Study of the methods and processes used to fight fires are a necessary requirement of all those in the fire service, from the newest recruit to the chief fire administrator. This need for knowledge was the basis for writing this text. It is a text designed for a post-secondary course in firefighting strategies and tactics, and is designed to meet the National Fire Academy FESHE course objectives for *Strategy and Tactics*.

Basically, fires burn in the same manner and can cause the same destruction to lives and property, regardless of the size of the town or the type of fire department that protects it. What is different are the procedures and resources available to combat the fire. It is our intent with this book not only to introduce these concepts to first-time learners, but also to guide firefighters, officers, and chiefs to effectively plan a fire attack.

HOW TO USE THIS BOOK

The text begins with a history of firefighting and the reasons that it has evolved throughout the years. It is followed by background information necessary to make effective fireground decisions. The next chapters include information on fire dynamics and extinguishing agents. The following set of chapters deal with concepts common to all incidents, including incident management systems, command and control considerations, firefighter safety, company operations, built-in fire protection, and considerations for after the incident. The book then focuses on the information necessary to apply the common concepts for specific occupancies to include one- and two-family dwellings, multifamily dwellings, commercial occupancies, places of assembly, high-rise structures, vehicle fires, and wildfires, as well as basic information for special fires. The final chapter allows the reader to apply the concepts to written scenarios. Logically organized, each set of chapters builds on the previous ones for a comprehensive understanding of firefighting.

Differences in Procedures and Terminology

The differences in procedures and terminology utilized are important to understand before reading this text any further. It is necessary to realize that there are different ways to approach any fire and achieve the same outcome. The outcome, of course, is met when the actions necessary are taken to achieve fire extinguishment in a safe, effective manner. In this text, we have chosen to present certain ways of doing things that we believe are appropriate to achieve the goals. Whether your department calls your procedures standard operating procedures (SOPs), standard operating guidelines (SOGs), or something else is irrelevant as long as these procedures meet the needs and goals that they are designed to meet. Does it matter whether your town refers to the front of the fire building as side 1 and another department across the state uses side A? Of course not. What is important is that everyone who functions with your department understands the designation and that there is a scene management procedure in place to sector or divide the fireground.

For this text we have applied consistent terminology. For example, standard operating guidelines (SOGs), rapid intervention crew (RIC), and vent-enter-search (VES). As you read the text, do not allow the use of one term or another get in the way of the point of the material. Remember, it is not important that the text uses RIC, and your department uses rapid intervention team (RIT), or firefighter assist team (FAST). What is important is that some process that allows for rapid assistance to firefighters in need is in place to meet the goal of firefighter safety.

FEATURES OF THIS BOOK

This book contains a variety of features to further enhance the understanding and importance of concepts related to strategies and tactics:

- **A new chapter on building construction** provides a review of basic concepts that influence how a fire behaves in a building. This important review provides a foundation and a basic understanding of construction prior to moving onto the complexities of fighting fires in different types of buildings—covered in the chapters that follow.

- **Safety** is emphasized throughout the book, including a separate chapter, "Firefighter Safety." In addition, those chapters focused on developing firefighting strategies for various types of occupancies include a special firefighter safety section. This specific safety information highlights the potential hazards associated with attacking fires in each of these occupancies in order to help mitigate firefighter injury and/or loss.

- **Case studies** highlight actual events that emphasize the importance of developing sound strategies and tactics to effectively fight fires. Each Case Study focuses on a lesson learned—including the value of implementing an Incident Management System (IMS) as well as an accountability system, planning for safety on scene, observing critical clues in fire behavior, and other concepts related to scene operations.

- **Notes** and **key terms** are highlighted in the margins of each chapter—providing emphasis of important concepts and terminology that are critical to effective communications on the job. These features also provide a handy reference for review when studying for promotional exams.

- **Review Questions** and **Activities** accompany each chapter. Readers should complete the reading of each chapter and then complete both the Review Questions and the

Activities. The Review Questions are designed to ensure that readers have gained the knowledge set forth in the objectives. The Activities allow readers to compare the concepts learned to those their own department uses.

- A chapter on **"Putting It All Together"** provides scenarios of different types of buildings and fires, with a hypothetical fire department involved in response. Based on the resources available to this fire department, the text walks you through the goals and objectives of each fire scene, as well as the necessary fireground management concepts, in order to successfully combat the fire.

NEW TO THIS EDITION

Thoroughly revised, this new edition provides the necessary information to keep you up to date with the latest considerations in the fire service:

- **Updates on firefighter PPE and other equipment** provide the knowledge necessary to keep fire departments aware of the latest resources available.
- **Expanded chapter on firefighter safety,** including current statistics and information on the increased efforts to reduce line-of-duty injuries and deaths, provides an awareness of this growing concern in the fire service.
- A **new section on manufactured mobile homes** in the chapter on "One- and Two-Family Dwellings" covers the latest concerns in construction materials.
- A **new section on undeployed air bags and hybrid and tribrid vehicles** in the chapter on "Vehicle Fires" covers some of the safety concerns and other response challenges with these technologies.
- A **new section on electrical sub-stations** in the chapter on "Special Fires" provides the necessary knowledge to fight fires in this specialized scenario.

SUPPLEMENT TO THIS BOOK

In addition to this book, an **e.resource** on CD-ROM is available to instructors to assist in preparing and planning for a course on Strategies and Tactics. Included on the CD are the following components:

- **Instructors' Guide** with Lesson Plans and Answers to Review Questions. The lesson plans for each chapter follow the sequence of topics, highlight important concepts, and are conveniently correlated to the accompanying PowerPoint presentations.
- **PowerPoint** presentations review objectives and the important concepts presented in each chapter. These correlate to the lesson plans in the Instructor's Guide to provide a seamless classroom presentation.

- **Chapter quizzes** in Microsoft Word provide instructors with the necessary basis for evaluating student knowledge of the concepts presented in each chapter. Each chapter quiz is editable, allowing instructors to edit, add questions, or remove questions to fit their course. Questions can also be edited based on the local AHJ's procedures and policies.
- **FESHE correlation grid** outlines the objectives for the National Fire Academy FESHE course *Strategies and Tactics* and correlated each to a chapter/and or pages in the book that contain the corresponding content to meet those objectives.

Order# 1-4180-4894-1

We welcome you to visit us at **www.firescience.com** to review other educational resources available for the fire service!

ABOUT THE AUTHORS

This text was written by a team of five authors. The purpose of this approach was to provide information from different parts of the country. This input is important, as construction types differ throughout the country, as do hazards. The authors have varied backgrounds that are described in the following paragraphs.

James Angle, Fire Chief: James S. Angle is the fire chief of Palm Harbor Fire Rescue in Pinellas County, Florida. The department protects 62,000 people in a 20-square-mile area operating from four fire stations, providing a full range of services including fire prevention, public education, advanced life support, rescue, hazmat, and fire suppression. He is a 33-year emergency service veteran, beginning his career in the Monroeville Fire Department in suburban Pittsburgh, as well as the Pittsburgh EMS Bureau. His background includes employment with five emergency service agencies both small and large, and he worked through the ranks from firefighter/paramedic to his current position as fire chief.

His education includes an associate degree in fire science administration from Broward Community College, a bachelor's degree in fire science and safety engineering from the University of Cincinnati, and a master's degree in business from Nova University. He also holds instructor certification in three areas from the Florida Bureau of Fire Standards and Training, is certified as a paramedic, is a graduate of the Executive Fire Officer program of the National Fire Academy and holds Chief Fire Officer Designation (CFOD) and Certified District Manager (CDM) designations.

He teaches at St. Petersburg Junior College and has delivered seminars to national audiences at numerous Fire/Rescue Conferences. He has also published articles in two national publications.

Michael F. Gala, Jr., Battalion Chief: Gala has significant experience as a fire officer in a large urban fire department and experience in teaching fire service personnel at the New York City Fire Department Training academy. Because of the diversity in assignments, he has developed experience in both engine and truck company functions, as well as operations in multiple story buildings. Mike is currently assigned to one of the busiest ladder companies in New York City where he serves as a Captain.

Mike hold a bachelor's degree in fire science administration and a master's degree in protection management from John Jay College in New York.

David Harlow, Fire Chief: David Harlow is retired after 31 years of active service. Currently David is retired from active fire service after serving two years as a fire chief in south Florida. Before moving to Florida, he served more than 29 years with the Fairborn Fire Department, where he retired as a Division Chief. Most of his career was spent in the field as a battalion chief. The Fairborn Fire Department is a full-time fire department serving more than 40,000 people outside of Dayton, Ohio. The city surrounds Wright-Patt Air Force Base and is home to Wright State University, a state college with more than 17,000 students. The city is a mix of residential, commercial, light industry, and an old downtown area.

Harlow holds an associate of applied science in fire engineering, a bachelor's degree in fire administration and a master's degree in urban administration. He is a graduate of the Executive Fire Officer program at the National Fire Academy, where he was honored nationally as the author of an outstanding research paper on organizational development. Harlow has significant experience in fireground operations, disaster management, and incident command. He is a certified fire instructor for the State of Florida and Ohio. He served as a member of the faculty at Sinclair Community College, Dayton, Ohio, where he was instrumental in the development and teaching of the Hazardous Materials and Fire Officer Development programs. David teaches for private industry and has presented at various seminars and the Fire Department Instructors Conference. He holds the International Association of Fire Chiefs (IAFC) Chief Fire Officer Designation.

William B. Lombardo, Deputy Chief: In 1987 William B. Lombardo started as a career firefighter with South Trail Fire and Rescue in Fort Myers, Florida. He worked his way up the ranks and served as a lieutenant on an engine company for seven years. He was promoted to Training Chief in 1998 and has played a key role in implementing advanced life support engines at his department. In 2006 he was promoted to Deputy Chief.

Fort Myers and the surrounding area have consistently experienced extreme population growth and ongoing changes in demographics. The response area is a mix of commercial and industrial, mid- and high-rise complexes, various dwelling complexes, hospitals, nursing facilities, assembly occupancies (such as the spring home of the Minnesota Twins), and wildland/urban interface areas. The constant changes have given him diverse experience in firefighting and emergency services.

Lombardo's education includes an associate of arts degree in general education, and an associate of science degree in fire science from Edison College. He also holds a bachelor's degree in executive management from International College, and he is a graduate of the National Fire Academy's Executive Fire Officer Program. Lombardo is certified as a paramedic and holds certifications from the Florida Bureau of Fire Standards and Training as a fire inspector, company officer, and Instructor III.

Lombardo has experience teaching recruit firefighting classes and local fire-related seminars. He has presented seminars at the Fire Department Instructor's Conference in Indianapolis, Indiana, and Fire Rescue East in Jacksonville, Florida.

Craig M. Maciuba, Deputy Chief: Maciuba holds an associate's degree in fire science, is a graduate of the National Fire Academy's Executive Fire Officer Program, and is nearing completion of his bachelor's degree. He has significant experience in incident command and initial fire scene operation. He had served as a company officer on one of the busiest engine companies in Pinellas County, a densely populated area of Florida's West Coast. Maciuba has a very strong background in training and education at the fire department and fire academy level. He is currently Deputy Chief of Operations for Palm Harbor Fire Rescue.

NOTES FROM THE AUTHORS

James S. Angle, Fire Chief

It is my hope that the information presented here will help current and future firefighters and officers perform better at fires. By better, I mean safer and with a greater degree of success.

This text is the second that I have written for Delmar. In the preface to the first, *Occupational Safety and Health for the Emergency Services*, I noted that thanking all those who affected me during my career would be too lengthy. I take the same approach in this text. In the interest of brevity, I would like to thank the Board of Commissioners and members of Palm Harbor Fire Rescue for their interest in this project and many for being involved in some of the pictures. I would like to thank all those people with whom I have enjoyed the fire service over the years at the Monroeville, Pennsylvania Fire Department; Hallandale, Florida Fire Rescue; Palm Beach County, Florida Fire Rescue; and South Trail, Fort Myers, Florida Fire Rescue. You have all taught me something that probably was used somewhere in this book.

Once again I both acknowledge and thank my lovely wife, JoAnn, and great children, Anthony and Austin, for yet another round of having their father sit at a computer while the rest of the family was having fun. I cannot describe the support that they have given me throughout this project, and appreciate them more than can be described in words.

I also again thank the staff at Delmar, specifically Jennifer Starr who puts it all together, provides support and coaching, and is always there to lend a hand or just an ear.

Last but certainly not least, I thank the team of authors. When I had the vision for this project, I felt that one of the most important components would be to bring together a team of experienced fire service professionals in order to get different perspectives and to provide a flavor of geographic and fire department diversity. I believe this has worked.

The authors all worked very well together through tight deadlines and through many discussions over procedures and terminology. For being so understanding and willing to take on this project, I thank them. We made a great team.

Michael F. Gala, Jr., Battalion Chief

I would like to thank our team leader, James Angle, for asking me to be part of this project, as well as the coauthors for a great experience. Thanks to my parents, Doris and Mike Gala, who have always encouraged me to reach for my dreams. My father, who served 35 years in the FDNY was always my inspiration and was always ready to share his fire stories with me, as well as take me to work with him from the time I was 12 years old. He showed me what the fire service was all about. Finally, I acknowledge my beautiful wife, Vita, who has always been by my side. Her encouragement, not only through this project, but while I studied for promotional exams and attended college, went well above the call of duty, all while she raised our sons, Robert and Anthony, and our daughter, Bianca. I love you.

David Harlow, Fire Chief

During your career, you may choose to stay on the line as a firefighter or seek to rise through the ranks toward the chief of the department. Regardless of your desires or ambitions, you must still study the concepts and ideas learned from others to survive in this career. As a firefighter, I knew I had to learn what was going on inside a building while I was attacking the fire. What were the smoke and fire telling me? As a company officer, I must look out not only for myself, but also for the members of my crew. I must know how to read the fire's behavior, the smoke conditions, and the building construction. As a battalion chief, my concerns focused on the entire picture of the incident, not just in a fire, but in any type of emergency. I must know and understand the concepts of firefighter safety and survival, fireground command, building construction, and many other facets of this business. No one can learn everything he or she needs from experience alone, especially today, as we are seeing fewer fires. That is why books such as this are valuable. They give us the chance to learn from others' experiences. As you read this book, place yourself in the situations described and see the knowledge gained through another's eyes. We must always be learning.

William Lombardo, Deputy Chief

The wise King Solomon once said that anyone inexperienced puts faith in every word, but the shrewd one considers his steps. The words of this book are the culmination of a great deal of knowledge and experience and it is an excellent learning tool. It is my earnest hope that the reader will take this knowledge and apply it accordingly. Consider your steps and never forget to think. No book can cover every topic or every problem that you may encounter. To be a really great officer, you have to think on your feet and use every bit of knowledge and experience you have learned to guide you in the correct decision and action. This will gain you respect and ensure that you and the people that serve with you go home to their families after the tour is through.

The list of people I would like to acknowledge is too lengthy to mention in full; however, I would like to take this opportunity to thank some key individuals who have made my involvement in this text possible. First, I would like to thank Jim Angle for involving me in this project and for being my mentor. Jim has always put more faith in me than I put in myself. I would like to thank Chief Clifford H. Paxson, the Board of Commissioners, and all the members of South Trail Fire District for their ongoing support. A special thanks to Craig Brotheim and Greg Sutton, who supplied pictures from several personal collections to be included in this book. Thanks to Joe Gugliuzza, who convinced me to take a firefighter survival class in 1988 that forever changed my career goals.

My wife, Jennifer, is the ongoing support that motivates me constantly to improve and reach my goals. Thank you, Jenn, for all you do. I would also like to thank my children, Stacy and Billy, who gave up a little bit of their dad during the writing of this book. The support my family gives me is astonishing.

I thank the staff at Delmar for all their support and for the opportunity to fulfill a dream.

I would like to thank Craig Maciuba, Mike Gala, and Dave Harlow, who worked hard to complete this project and traveled many miles, stayed in below-par motels (sorry, Dave), and had fun doing it.

Finally, I would like to thank my father and mother. Dad, a teacher for 40 years in the public school system, instilled in me the ability and passion to write. Although I hated doing phonics on our back porch in Beach Haven West, New Jersey, when I was 7 years old, it must have worked. Thanks must also go to Mom for convincing Dad that "play time" would help my creative side. It did and still does.

Craig Maciuba, Deputy Chief

I would like to echo the comments of the other authors. So many people have helped and supported me through this project that there are too many to list. I would, however, like to thank my wife, Dawn, and children, Kristine and Kevin, for allowing me the time and space needed to complete this text. I would also like to thank my brother, Keith Maciuba, a Lieutenant with Palm Harbor Fire Rescue for his contributions in some of the drawings, photos, and text of this book.

ACKNOWLEDGMENTS

The authors and staff at Delmar Cengage Learning gratefully acknowledge the time and effort of those who reviewed this project. The comments and suggestions of the review panel were invaluable. Our sincere thanks to:

Michael Arnhart
High Ridge Fire District
High Ridge, MO

Tim Capehart
Bakersfield College
Fire Technology
Bakersfield, CA

Ted Cashell
Township of Princeton
Princeton, NJ

Captain Tony Kostecka
Fairfax County Fire & Rescue
Amisville, VA

Jason Loyd
Weatherford College
Fire Science Technology
Weatherford, TX

Charles McDonald
Trinity Valley Community College
Terrell, TX

David McFadden
Fox Valley Technical College
Neenah, WI

Michael Morrison
Elk Grove Fire Department
Elk Grove, CA

Paul Pendowski
Northeast Wisconsin Tech
Green Bay, WI

John Salka
FDNY
New York, NY

Timothy Sendelbach
Missouri City Fire Department Training
    Division
Missouri City, TX

William Shouldis
Philadelphia Fire Department
Philadelphia, PA

Craig Smith
Hutchinson Community College
Fire Science Training
Hutchinson, KS

Michael L. Smith
District of Columbia Fire & Rescue
Washington, DC

Thomas J. Wutz
NYS Office of Fire Prevention and Control
    and Midway Fire Department
Albany, NY

FIRE AND EMERGENCY SERVICES
HIGHER EDUCATION (FESHE)

In June 2001, The U.S. Fire Administration hosted the third annual Fire and Emergency Services Higher Education Conference, at the National Fire Academy campus, in Emmitsburg, Maryland. Attendees from state and local fire service training agencies, as well as colleges and universities with fire-related degree programs, attended the conference and participated in work groups. Among the significant outcomes of the working groups was the development of standard titles, outcomes, and descriptions for six core associate-level courses for the *model fire science* curriculum that had been developed by the group the previous year. The six core courses are *Fundamentals of Fire Protection, Fire Protection Systems, Fire Behavior and Combustion, Fire Protection Hydraulics and Water Supply, Building Construction for Fire Protection*, and *Fire Prevention*.[1]

In addition, the National Fire Science Curriculum Committee developed similar outlines for other courses commonly offered in fire science programs. These seven non-core courses included *Fire Administration I, Occupational Health and Safety, Legal Aspects of the Emergency Services, Hazardous Materials Chemistry, Strategy and Tactics, Fire Investigation I,* and *Fire Investigation II* .

---

[1] *2001 Fire and Emergency Services Higher Education Conference Final Report,* (Emmitsburg, Maryland: U.S. Fire Administration) 2001, page 12.

FESHE Content Area Comparison The following table provides a comparison of the FESHE *Strategy and Tactics* course outcomes with the content in this text.

FIRE AND EMERGENCY SERVICES HIGHER EDUCATION (FESHE) COURSE CORRELATION GRID

Name:	Strategy and Tactics	
Course Description:	This course provides an in-depth analysis of the principles of fire control through utilization of personnel, equipment, and extinguishing agents on the fire ground.	**Firefighting Strategies and Tactics, Second Edition Chapter Reference**
Prerequisite:	*Principles of Emergency Services*	
Outcomes:	1. Demonstrate (verbally and written) knowledge of fire behavior and the chemistry of fire.	2, 3
	2. Articulate the main components of pre-fire planning and identify steps during a pre-fire plan review.	5
	3. Recall the basics of building construction and how they interrelate to pre-fire planning.	10
	4. Recall major steps taken during size-up and identify the order in which they will take place at an incident.	5
	5. Recognize and articulate the importance of fire ground communications.	5
	6. Identify and define the main functions within the ICS system and how they interrelate during an incident.	4
	7. Given different scenarios, the student will set up an ICS, call for appropriate resources and bring the scenario to a mitigated or controlled conclusion.	11–19
	8. Identify and analyze the major causes involved in line-of-duty firefighter deaths related to health, wellness, fitness, and vehicle operations.	6

Chapter 1

HISTORY OF FIRE SERVICE TACTICS AND STRATEGIES

Learning Objectives

Upon completion of this chapter, you should be able to:

■ Explain significant historical changes in fire service strategies and tactics.

■ List the firefighter's role in contemporary strategy and tactics.

■ Explain why it is important for firefighters to understand strategy and tactics.

1

CASE STUDY

"Chief . . . I'm still stuck back in the right rear of the building, in the closet. We're out of air . . . first-floor, underneath the collapsed ceiling." This was the last transmission made by two firefighters who perished in a Hackensack, New Jersey, Ford dealership at 332 River Street on July 1, 1988. Three other firefighters also were immediately trapped and died when the wooden bowstring truss roof failed due to extensive fire involvement in the attic/truss area below the roof assembly.

The initial call to the Hackensack Fire Department was from a woman approximately one block away who reported seeing flames and smoke. An engine from the Hackensack Fire Department was the first to arrive on the scene of the 40- to 45-year-old building. Engine 304 initially reported finding heavy smoke and a working fire. Initial attempts were made to gain access to the attic area from the interior. Ultimately this access was gained through scuttle holes in the service area of the building. Meanwhile, Engine 301 was establishing a water supply and Truck 307 was laddering the building to vertically ventilate from the roof. As the scuttle holes were opened, the firefighters were met with high heat conditions. Hose streams were directed into the attic from interior positions. Also, Truck 307 stretched a hose line to the roof. The fire grew in magnitude, but unfortunately the interior crews did not realize the growth of the fire and they were making no notable progress toward extinguishment. Truck 307 reported that a vent hole had been cut in the roof and a large amount of fire was coming out of the hole. Additionally, they reported that fire was breaking through the rear of the roof. The battalion chief recognized that there was a problem and ordered all companies to withdraw from the building. However, there was no radio traffic indicating that interior firefighters heard this important message. Approximately 2 minutes after the order was made to evacuate the building, at least one roof truss failed and the collapse occurred.

The collapse trapped five firefighters inside the still-burning automobile dealership. Chief officers knew they had firefighters trapped but no one on the scene knew initially how many firefighters were still in the building. Unfortunately, the heroic rescue attempts were in vain. A 3-inch line from Engine 305 was in the process of being stretched to Truck 307 to supply it as an elevated stream when the collapse occurred. Instead of a master stream, a nozzle was placed on the line and it was directed into the burning building. This line was extremely difficult to handle and was never really effective. Ultimately, handlines were stretched into the service area and a breaching operation took place. The fire was brought under control 1 hour and 37 minutes after the initial telephone call. Five firefighters died in one of the worst tragedies in the history of the Hackensack Fire Department.

This fire was analyzed, scrutinized, and investigated by various agencies including the National Fire Protection Association (NFPA), who performed a 3-day, on-site investigation and the International Association of Firefighters (IAFF), who commissioned Demers Associates, Inc., a consulting firm directed

International Association of Firefighters
labor organization that represents the majority of organized firefighters in the United States and Canada

by David P. Demers. The commission developed an independent "Fire Incident Analysis" for the IAFF that examined the strategic and tactical aspects of the Hackensack Ford fire. The Demers report stated the following:

- "The most significant factor that resulted in the fatalities was the failure of the fireground commanders to recognize the wood bowstring roof truss construction."
- "(There was a) lack of effective communications both on the fireground and between the fireground and fire headquarters (dispatch)."
- There was "the absence of an effective fireground command structure."
- "An important factor that had an impact during the incident was the level of manpower."

The NFPA "Summary Investigation Report," prepared by Director Thomas J. Klem, was requested by Chief Anthony A. Aiellos of the Hackensack Fire Department. The following are some of the significant statements from that important report:

- "Trusses are vulnerable to collapse under fire conditions . . . ."
- "Unfortunately, in this incident the inherent wood truss construction hazard (as shown in **Figure 1-1**) was again a primary contributing factor resulting in firefighter fatalities. This and earlier tragic incidents provide important lessons that must be stressed throughout the fire service to avoid future loss of life."

Figure 1-1 *An exterior view of typical bowstring roof truss construction.*

The Hackensack Ford fire was a catalyst for change in the way fires are fought as risk management has come into the equation. Additionally, strategy, tactics, and training have changed to some degree as a result of major fires such as this one. There have been larger fires and unfortunately more firefighter deaths, but this fire was known and discussed in firehouses nationwide.

INTRODUCTION

■ **Note**
When Ben Franklin organized the first fire company in Philadelphia, Pennsylvania, in 1736, its goals and objectives were clear: to save lives and reduce property loss. Today the fire service exists for the same basic reasons.

When Ben Franklin organized the first fire company in Philadelphia, Pennsylvania, in 1736, its goals and objectives were clear. The company had to save lives and reduce property loss. It did so by putting water on the fire and thus, extinguishing it. Essentially, the fire service exists today for the same basic reasons.

Some of the major changes since Franklin's inception of the fire service can be seen in the way modern fire departments operate. There have been advances in operating methods, including how water is moved to the fire scene, and advances in protective equipment that allow firefighters to enter and venture farther into burning structures in search of victims or for fire attack. New equipment and other advances that are discussed later in the chapter ultimately affect the decisions made on the fireground to meet the challenges at hand.

■ **Note**
When the Philadelphia fire companies arrived on the scene of a structure fire in 1736, they organized a bucket brigade.

When the Philadelphia fire companies arrived on the scene of a structure fire in 1736, they organized a bucket brigade, as shown in **Figure 1-2.** It was time-consuming and labor-intensive; however, at the time it was simply the most

Figure 1-2 *The bucket brigade was the most efficient way to move water to the scene in the early history of firefighting.*

efficient way to move water to the scene. Teamwork was a critical aspect of the brigade, and any breakdown in the human chain that moved the water would seriously affect the outcome. The bucket brigade was the tactical operation utilized to meet the strategic goal of extinguishment. By meeting its goal, the brigade would save lives, reduce the fire losses, and prevent large extensive fires known as *conflagrations.*

A fire department today may have the same overall goal of extinguishment; however, the tactical operations it employs are far different. Modern hose, pumps, and nozzles are used to move the water to the fire (see **Figure 1-3**). Teamwork remains critical. Providing a supply line from a hydrant or drafting site to an engine that, in turn, pumps water to the handline is the modern version of a bucket brigade (see **Figure 1-4**). The fire is extinguished, lives and property are saved, and conflagrations are prevented (see **Figure 1-5**).

A similar comparison can be made between the various types of fire departments that are operating today. There is a large divergence depending considerably on the location of the fire department and the area it serves. This divergence can span from a small-town all-volunteer fire department (see **Figure 1-6**) to a combination volunteer/career, to a large metro-sized department, and everything in between. Of all the department types, none should be considered better or worse than any other. The existence of the various types depends on the demands of the community, the services provided by the fire department, and

■ **Note**
Providing a supply line from a hydrant or drafting site to an engine that pumps water to the handline is the modern version of a bucket brigade.

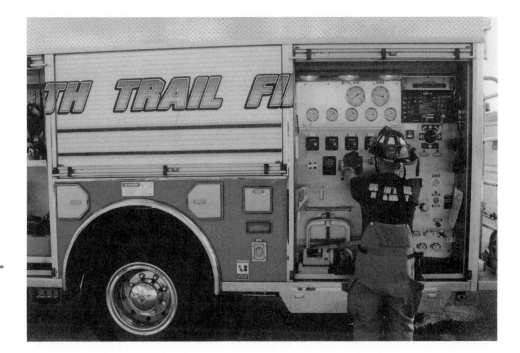

Figure 1-3 *Today's fire service uses hose, pumps, and nozzles to move water to the fire.*

Figure 1-4 *Early fire apparatus significantly improved extinguishing capabilities over the bucket brigade.*

the taxpayers' ultimate decision on the type of fire protection they are willing to support. This issue is mentioned here because, regardless of the resources available, the goals and objectives on the fireground do not change. The methods used to meet the goals and objectives, however, change drastically. Some fire departments, for example, may respond to fires with adequate equipment but may seriously lack personnel and adequate water supply. Conversely, other fire departments may respond with adequate equipment, personnel, and water supply. These issues must be considered and dealt with by each community. Staffing levels are an important concern because they have a direct impact on the effectiveness of the unit. Many career departments have lost numbers of positions over the past several decades. Operational guidelines must take this into account and be adjusted accordingly. Some departments may respond with staffing levels of two personnel per responding unit; others may have up to five. Water supplies vary from fixed to shuttles, depending on the infrastructure of the area they protect. Do the goals and objectives change for any of these different types of departments? The answer clearly is no. The mission of every fire department is to save lives and reduce property loss.

Figure 1-5 *Modern fire apparatus has certainly evolved.*

Figure 1-6 *An example of a small-town fire department.*

The comparison made between the assorted types of modern day fire departments reveals that the strategic goals and objectives are the same as that of the first Philadelphia Fire Department, but the manner in which the tactical operations are employed to meet these goals has changed. These tactical operations vary drastically, depending on the availability of equipment, personnel, extinguishing agents, and training.

It is under this context that critical topics are covered in this text, such as fire dynamics, various extinguishing agents, incident management, command and control, firefighter safety, company operations, fire protection systems, incident termination, structural firefighting in one- and two-family dwellings, multifamily dwellings, commercial fires, fires in places of assembly, high-rises, vehicle fires, wildfires, and various other fires and special situations.

REASONS FOR CHANGE

■ **Note**
There have been changes in building structural materials and the fire load contained within them. For example, there are variations and different types of engineering in building construction and significant changes in the flammable materials commonly found in homes and businesses.

If the objectives on the fireground are the same today as they were 250 years ago, then what events or issues have created changes in the manner in which fires are extinguished and lives are saved? Several factors prompted changes in the fire service. Some guiding factors are positive, such as advances in equipment and personal protective clothing, substantially better educational opportunities for firefighters, stronger codes and standards, and private fire protection systems. Some regulations and standards that have had a direct effect on fireground strategies and tactics are explained further in Chapter 6. Other significant factors, such as firefighter and civilian deaths, have been negative. There have been changes in building structural materials and the fire load contained within them. For example, there are variations on and different types of engineering in building construction and significant changes in the flammable materials commonly found in homes and businesses (see **Figure 1-7**). Additionally, hazards have been created through architectural design, wildland/urban interface, equipment in automobiles such as air bags, and the ever-increasing amount of hazardous materials. Finally, some changes in the way fire departments operate come from the activities and design of the community, state, and country in which they operate. For example, the threat of terrorism in the United States has created changes in response and operational policies. Nonetheless, changes are inevitable and come as a result of the never-ending desire and critical need for improvement.

■ **Note**
The threat of terrorism in the United States has created changes in response and operational policies.

SPECIFIC STEERING AGENTS FOR CHANGE: CHANGE MANAGEMENT

How, specifically, have the various steering agents created tactical and strategic changes?

Figure 1-7 *Changes in building construction and materials have increased the hazards.*

Equipment Advances

Personal protective equipment (PPE) has improved immensely over the years. Generally, it is unheard of today to fight a fire with a metal helmet and rubber coat (see **Figure 1-8**). Today's firefighter is equipped with the latest fire-resistive materials such as Nomex and Kevlar. Personal protective equipment now has thermal protection that actually "breathes" and reduces the firefighter's chances

Figure 1-8 *Typical personal protective equipment of the fire service 20 years ago.*

personal alert device
device such as the
Personal Alert Safety
System (PASS) that
emits a loud alert or
warning that the
wearer is motionless

**National Emergency
Training Center**
location in
Emmitsburg,
Maryland, for the
United States Fire
Administration, the
National Fire Academy,
and the Emergency
Management Institute

**National Fire
Academy (NFA)**
agency under the
USFA that develops
and delivers training
and educational
programs for
firefighters and fire
officers

**United States Fire
Administration**
agency under the
Department of
Homeland Security
that directs and
produces fire
programs, research,
and education

of stress from overheating (see **Figure 1-9**). An additional safety device added to the protective coat is the firefighter safety strap, which is sewn into the coat and reinforced under the arms to provide a pulling strap for downed firefighters (see **Figure 1-10**).

Manufacturers have made advances with lightweight, composite helmets. Numerous state and federal standards, such as the Occupational Safety and Health Administration (OSHA) standard 1910.134, mandate breathing apparatus to be worn in any environment that is immediately dangerous to life and health (IDLH). Also, **personal alert devices,** audible warning devices that sound at a high decibel level when a firefighter stops moving for 20 to 30 seconds, are no longer optional.

Other equipment advances include improved forcible entry equipment, such as handheld hydraulic tools that have reduced property loss and expedited the entry. Infrared helmet-mounted or handheld thermal cameras increase the survivability rate of potential victims and help find hot spots in hidden locations such as ceilings and walls, which can ultimately reduce property loss.

These advances in gear and equipment change the strategic goals and tactical considerations by allowing firefighters to perform their tasks more effectively.

Figure 1-9   *Personal protective equipment has changed immensely over the years.*

Figure 1-10 *New equipment included with protective gear is the firefighter rescue strap, which is built into the coat.*

The United States Department of Homeland Security established under the National Strategy for Homeland Security and the Homeland Security Act of 2002, the mission of the DHS is to, "lead the unified national effort to secure America. [To] prevent and deter terrorist attacks and protect against and respond to threats and hazards to the nation. [To] ensure safe and secure borders, welcome lawful immigrants and visitors, and promote the free-flow of commerce" (http://www.dhs.gov, 2006)

Firefighters can go deeper into structures in search of victims, they can force doors and enter burning buildings quicker, and they can be more efficient at applying water. However, the tactics must be modified to utilize the equipment to its fullest potential. Ultimately, the incident priorities of life safety, incident stabilization, and property conservation are met more quickly.

Advanced and Increased Educational Opportunities

The National Emergency Training Center, shown in **Figure 1-11,** specifically the National Fire Academy (NFA) in Emmitsburg, Maryland, has been a catalyst for advanced formal education in the fire service.

The National Fire Academy is funded by the federal government and is part of the United States Fire Administration (USFA). Interestingly, it was the terrorist attacks that occurred on September 11, 2001, that spurred an organization change within the federal government. In 2001, President George W. Bush established the Office of Homeland Security (OHS) and later (in 2002) created the Department of Homeland Security (DHS). Subsequently (in January 2003), the U.S. Fire Administration (previously part of the Federal Emergency Management Agency [FEMA]) was reorganized under the DHS. Another change came in October 2005, when the USFA was separated from FEMA and became part of the Department of Homeland Security's Preparedness Directorate.

The National Fire Academy (NFA) was originally created in 1974 as a result of the Federal Fire Prevention and Control Act of 1974. This act was the outcome of the landmark document "America Burning: The Report of the National Commission on Fire Prevention and Control" (available through the USFA).

Figure 1-11 *The National Fire Academy has been a catalyst for higher education in the fire service.*

America Burning: The Report of the National Commission on Fire Prevention and Control
a 1973 report to the President of the United States describing the nation's fire problem

Executive Fire Officer Program (EFOP)
an initiative of the United States Fire Administration/ National Fire Academy designed to provide senior officers and others in key leadership roles with . . . enhanced executive-level knowledge, skills, and abilities necessary to lead . . . transforma-tions, conduct research, and engage in lifelong learning (http://www. usfa.dhs.gov/training/ nfa/efop/, 2006)

Although the NFA has experienced many organizational changes over the years, it has remained true to its core values and services. The National Fire Academy has brought together firefighters from all over the nation and the world to be educated in a myriad of subjects. According to the USFA Web site (http:// www.usfa.dhs.gov/about/orgchart/nfa.shtm), one of the primary functions of the NFA is to: "Provide educational opportunities through resident and distance de-liveries in management technology, executive development, emergency medical services technology, incident management, planning and information manage-ment, hazardous materials, emergency response to terrorism, arson prevention/ detection, fire prevention, public education, counterterrorism and instructor de-velopment." The NFA has also moved forward in providing for advanced-level education such as the **Executive Fire Officer Program**, and various baccalaureate programs such as the Degrees at a Distance Program (DDP), an online degree pro-gram that has agreements with seven colleges and universities throughout the country (see **Figure 1-12**).

State-level fire colleges have also provided many advances in education for firefighters. Most have established state certification programs for subject areas such as fire officer and fire inspector. State fire colleges also provide quality hands-on training in critical areas that ultimately affect the tactics employed

Figure 1-12
Firefighters have numerous educational opportunities available to them, such as the National Fire Academy's Degrees at a Distance program.

on the scene, such as extrication, high-angle rescue, pumper operations and hydraulics, urban search and rescue, and farm medic classes (see **Figure 1-13**).

Further advances in education come from local community colleges and technical schools. Many of these institutions provide ongoing training and education such as hazardous materials and confined space rescue. Most community colleges are offering associate degree programs in fire science or emergency management, and many states and individual departments will pay an educational incentive for those that acquire a degree.

Firefighters have available to them numerous educational opportunities from basic skills to master's degree and doctorate degree programs. Commanders of any incident, small or large, must know the abilities and limitations of equipment and the personnel working on the scene. The decisions made can be

■ Note

Firefighters have available to them numerous educational opportunities, from basic skills to master's degree and doctorate degree programs.

Figure 1-13 *Training towers provide quality hands-on training.*

instituted with confidence when working with well-educated, experienced firefighters and officers. For example, the decisions would be different when working entirely with a group of new firefighters. An analogy can be made to the manager of a baseball team. The manager places the best hitters at the top of the lineup and the best fielders are placed in the infield. The manager is aware of the talent available on the team. This tactic provides for the best opportunity to win. Imagine the waste of talent if a good pitcher were placed in right field to catch pop flies. Similarly, the prudent incident commander knows the education and talents of the responding personnel and uses them accordingly. If a firefighter has knowledge in a specific area such as hazardous materials, then the abilities of that firefighter should be considered in the tactics applied.

Codes and Standards

Codes and standards have had a major impact on fire department tactics. Examples of some of the more major items that are a result of improved building codes and standards are required sprinkler systems, standpipes, fire alarm systems, and exit requirements in various buildings (see **Figure 1-14**).

Figure 1-14
Sprinkler systems are an important ally of the fire service.

Sprinkler systems are provided in various types of structures. The most important function that can be accomplished when fighting a fire in a sprinklered building actually occurs long before the fire starts. In order to apply the right tactics for rescue and extinguishment, the incident commander and firefighter must first be aware of the specific type of sprinkler system that is in the building. This can only reasonably be accomplished through good preincident planning and knowledge of the particular buildings in the incident commander's response zone. Sprinkler systems have been a factor in tactics chosen simply because they now exist and are a superb ally to the firefighter. Proper use of the sprinkler system allows for the accomplishment of several of the strategic goals discussed in Chapter 5.

Standpipes, as shown in **Figure 1-15,** are another major improvement in multistory and large area buildings. As with sprinkler systems, the incident commander and firefighter must be aware of the type of standpipe system in the building. This knowledge ultimately directs the tactical decisions. A standpipe allows firefighters to carry less hose and thus be less fatigued. Additionally,

■ Note

A standpipe allows firefighters to carry less hose and thus be less fatigued.

Figure 1-15
Standpipes are a major improvement in firefighting capabilities within multistory buildings.

siamese

device to connect two hose lines into one line with either a clapper valve or gate valve to prevent loss of water when only one line is connected

the amount of time needed to lay the hose to reach the needed areas is vastly reduced through the use of standpipes. It is usually easier to take a standpipe pack, made up of 150 feet of hose, a reducer or gated wye, and a spanner wrench to the standpipe, and attach it to begin flowing water than it is to drag 300 to 400 feet of $2\frac{1}{2}$- or 3-inch hose up several flights of stairs. These factors must be considered when determining the tactics to use.

Apparatus placement to augment or provide water flow to both the sprinkler and standpipe systems must be considered early in the incident. In some buildings, both systems are supplied by the same siamese. However, in large malls, high-rises, and multiwing buildings, several siamese may supply various portions of the fire protection systems in the structure. It is critical that the fire officer know in advance which siamese supplies which wing or floors of a given building. The only way to have this advance knowledge is though building tours and preincident plans. The tactics chosen will be based on the previous knowledge of the building, its components, and the availability of equipment and personnel.

Exit requirements (see **Figure 1-16**) are another standard that has guided tactics and strategies. Firefighters know that when they respond to residential fires in fairly new structures, they will find some form of an exit out of every sleeping

Figure 1-16
Standards and codes continually address minimal exit requirements. (Photo courtesy of Amy Mozes.)

Safety
When a fire department regularly responds to false alarms in specific buildings in its district, there is a tendency to become complacent. Complacency can be deadly and cannot be tolerated.

area. In addition, various types of commercial occupancies have a wide variety of exits and minimum exit widths. These exit requirements have been provided through codes and standards, which have created major changes in tactics and provided for increased life safety. Firefighters must be cautious, however, and consider when establishing strategies and tactics that exits may be blocked, as shown in **Figure 1-17,** or locked.

Fire alarm systems (**Figure 1-18**) allow early recognition of fires and early response, which affect the strategies and tactics employed. Alarm systems have dropped in cost and are thus in many more businesses and private homes. The problems encountered are those of false alarms. When a fire department regularly responds to false alarms in specific buildings in its district, there is a tendency to become complacent. Complacency can be deadly and cannot be tolerated.

Codes and standards have provided positive changes in the fire service. Many more factors are probably taken for granted but have been provided through countless codes. Some examples are the size of steps on a staircase, the width of disability ramps, and the material that can be used as decorations.

Figure 1-17 *When establishing strategies and tactics, consideration must be given to the potential for blocked exits.*

Figure 1-18 *Fire alarm systems allow for early detection and notification of fires.*

Firefighter and Civilian Deaths

In the event of a firefighter death, the incident must be investigated and the lessons utilized as a learning experience. The goal of this type of research is to prevent deaths and injuries in the future. Firefighter and civilian deaths have

Figure 1-19
Firefighter deaths have an impact on decisions made at the fire scene. However, approximately 100 firefighters still lose their lives each year.

National Institute of Occupational Safety and Health (NIOSH)
a government agency dealing with workplace safety and health issues. NIOSH began a project in 1997 in which it will investigate firefighter line-of-duty deaths

played a major role in changes and improvements in regard to tactics. Firefighter deaths (see **Figure 1-19**) have influenced the way fires are fought, the decisions made on the fire scenes, and, in some cases, code changes. For example, after a Philadelphia high-rise office building fire claimed the lives of three firefighters, the code was changed to require sprinklers in all high-rise office buildings in Philadelphia. Civilian deaths have also motivated changes in building codes. For example, multiple deaths in places of assembly have changed requirements for exits. All fire deaths affect individual firefighters in a very personal way (see **Figure 1-20**). Firefighters who have dealt with a fire death tend to look inside themselves to see whether they can personally improve their skills and abilities.

The National Institute of Occupational Safety and Health (NIOSH) investigates every firefighter death in the United States. Further, the NFA has printed a firefighter death autopsy protocol for medical examiners to use. In addition to actual autopsy details, this protocol describes the procedure for securing the personal protective clothing of the deceased firefighter.

News media and coworkers will want to know that the firefighters who perished did not do so in vain. Some agencies and people want to make sure that the firefighters were provided with the proper equipment and training, while others want to ensure that the same mistakes were not made that have killed firefighters in the past.

Figure 1-20 *Every firefighter is affected by firefighter deaths in his or her own personal way.*

The Hackensack, New Jersey, fire is a prime example of firefighter deaths that resulted in changes in tactical considerations. Before that fire, the term *bowstring truss* was not nationally known. In fact, most of the country was not even aware of other dangers, such as lightweight truss roofs. At the 1988 fire in Hackensack, no one thought the roof would fail with such consequences. Since this and other deadly fires, the fire service looks at these types of incidents differently and as a result tactics have changed. Sadly, though, firefighters continue to die in the line of duty at the rate of approximately 100 deaths per year. In spite of the lessons learned at the Hackensack fire, we still have incidents in which firefighters are killed in the line of duty saving unoccupied buildings with known hazards, such as the bowstring trusses. Actual fire service deaths are detailed annually by various agencies such as the International Association of Firefighters (IAFF), the National Fire Protection Association (NFPA), and the United States Fire Administration (USFA). (For further information on this topic, see Chapter 6.) A review of these statistics shows that the fire service continues to make the same mistakes year after year. We still have firefighters dying in apparatus crashes because they did not wear seat belts, we still have firefighters dying in buildings that have no occupants, we still have firefighters dying in wildfires protecting trees. Most unfortunate, of the approximate 100 deaths per year, about half of these are consistently due to firefighter stress, yet there is still not a national mandatory physical fitness program for career or volunteer firefighters. The fire service talks about fitness, and we have fitness initiatives, even a standard called *NFPA 1500 Fire Department Occupational Safety and Health Programs* that says we must have a fitness program. Yet, we still do nothing real to prevent these deaths. Fitness must be mandatory; officers must inculcate in their firefighters the need to work hard at it and be dedicated to it. Fitness programs will save firefighters' lives and the lives of those in their communities. If we do not learn from the mistakes of the past, we are destined to repeat them.

Change is gradually happening, though, and we are seeing some changes. For example, the Hackensack Ford fire has caused incident commanders to examine their operations with regard to interior fire attacks and when to switch to exterior attacks, especially in situations in which the life safety of building occupants is not an issue.

Firefighter deaths have also resulted in many other changes. Some have had national influence, whereas others have had only local influence. Examples of recent change include rapid intervention teams to save firefighters in trouble, personnel accountability, the two-in, two-out ruling, and even physical fitness programs. Firefighter deaths, their investigations, and the media attention that surround them will always be a stimulus for change; however, often this change is only short term. Firefighters must embrace these incidents, learn from them, and create long-term change.

How have civilian deaths influenced changes in the fire service? Most changes that have come as a result of civilian deaths have been in the form of codes or increased enforcement of existing codes. One of the best examples of this is the Coconut Grove Fire in Boston, Massachusetts, in 1942. Four hundred ninety-two people died due to three primary reasons: flammable decorations, inadequate exits, and overcrowding. An interesting point about this fire is that many of the hazards found were covered by the 1942 edition of *The Building Exits Code*. Much discussion followed this fire regarding enforcement of the codes and applicable laws. Further, some changes in the code were recommended in 1945 at the NFPA's annual meeting and incorporated into the code in 1946. These changes included methods of measuring exits, clarification of the need for stairway enclosure, collapsible revolving doors, changes in lighting and signs, and a special note on interior finishes.

The changes that are made to codes as a result of civilian fire deaths also affect strategies and tactics. For example, some of the code changes that resulted from the Coconut Grove fire brought interior finish to the attention of firefighters in the 1940s. Firefighters had to understand fire behavior and consider how fast the fire would move through the structure in order to apply tactics correctly.

Building Materials and Building Contents

Most people enjoy spending quiet time in front of a fireplace or around a campfire. The wood burning is delightful and, although the smoke sometimes gets in your eyes, it does not cause any property damage or life hazard. However, place Styrofoam cups and plastic forks in the fire and the smell of the smoke is rancid and dangerous. This illustration demonstrates what has happened to building materials and contents of buildings in recent history (see **Figure 1-21**).

■ **Note**
Furniture now can be made of plastics, foams, and synthetic fibers. Homes have wall-to-wall synthetic carpets and cabinets made of pressboard with laminate. All these factors change the dynamics and behavior of the fire: They create toxic fumes and dense smoke, and the fire burns much hotter.

Figure 1-21 *Types of building contents can drastically affect strategy and tactics.*

Homes in the 1920s and 1930s were generally furnished with wood, leather, cotton, and wool, with structures built mostly of wood. It was commonplace to find hardwood floors made of oak and ceilings made of tongue-and-groove pine.

Changes in building contents and materials have definitely changed and created problems for firefighters. Furniture now can be made of plastics, foams, and synthetic fibers. Homes have wall-to-wall synthetic carpets and cabinets made of pressboard with laminate. As a result, the more modern building materials and contents typically burn twice as hot as the older-style homes made and furnished predominantly of wood and natural fibers. All these factors change the dynamics and behavior of the fire: They create toxic fumes, dense smoke, and, as noted, the fire burns much hotter. Firefighters must wear breathing apparatus and protective gear. As a general rule, firefighters should not aggressively fight interior structure fires for more than the time it takes to expend two 30-minute self-contained breathing apparatus (SCBA) cylinders without rehabilitation. In a Philadelphia Fire Department test of 750 members, the average time of work from a 30-minute SCBA cylinder was 12 minutes 50 seconds. Firefighters should call for relief early if in offensive mode and the fire attack is going to be of an extended time. This time changes drastically depending upon weather conditions, heat of the fire, specific job functions, and the physical ability of the person doing the work. The stamina of the firefighter, the availability of backup personnel, and available incident rehabilitation must enter into tactical decisions. Although these items may have always been tactical considerations, the changes in building materials have resulted in needed and continued modifications.

self-contained breathing apparatus type of respiratory protection in which a self-contained air supply and related equipment are worn or attached to the user; fire service SCBA are required to be of the positive pressure type

Architectural Designs and Engineering

The Hackensack Ford fire was another example for the fire service to show the dangers of certain types of engineering, specifically, wooden bowstring truss roofs. Firefighters learned that trusses lost their integrity quickly under fire conditions. Buildings are designed and engineered to stay erect against the forces of gravity and wind. Under normal conditions, trusses perform as designed; however, when weakened and affected by the heat of fire, the materials that make up the trusses can fail (see **Figure 1-22**). Effects of fire and heat are not a consideration in the design of truss members.

Figure 1-22
Lightweight wood trusses are common in new construction. (Photo courtesy of Craig Brotheim.)

■ **Note**
Proper strategy and tactics can be better employed if the firefighter knows the construction types, building materials used, architectural designs, and general engineering principles of the specific structures in the response zone.

Lightweight truss roofs are now common in most communities and are especially prevalent in newer construction. Lightweight trusses are usually made of 2-by-4-inch pieces of lumber engineered into a series of triangles, chord and web, which create a very strong member. The two-by-fours are held together with small metal plates that have $\frac{1}{4}$-inch protruding spikes, called gusset plates, stamped in them to attach to the wood. Several trusses are lined up on bearing walls 16 to 24 inches apart to create the skeleton of the roof. The gusset plates work well during normal conditions; however, when exposed to the heat of fire, gusset plates pull out from the wood and have been known to fail. Although no specific research has been performed to determine when the trusses will fail, firefighters must be aware that they can fail early in an incident. The result is a complete or partial roof collapse. Further, the sheer amount of lumber used in truss design can create a situation in which there is literally a lumberyard full of wood perched above the working firefighters. The advent of lightweight truss engineering has dramatically changed strategies and tactics of firefighting. An attic fire must be located, confined, and extinguished quickly or there is a very real threat of roof collapse. Trusses used in lightweight metal construction are used to span large areas in buildings such as supermarkets and discount stores. These can fail early from the heat of fire or through the efforts of firefighters cutting through a chord while ventilating a roof.

Changes in the roof design of single-family housing also have had a negative impact on the fire service. The pitch of newer contemporary style homes is very steep, which creates additional hazards for firefighters performing vertical ventilation operations. Void spaces, or unseen areas in buildings, such as cocklofts and areas between floors and walls, are present in many occupancies and can allow fire to burn undetected, further weakening structural components. These void spaces allow an avenue of fire extension as well. Firefighters must be aware of the presence of void spaces and the associated hazards.

Other architectural changes have been very helpful to the fire service. Balloon-frame construction was a building style in which no fire stops were located between floors or attic spaces. If a fire started on the first floor, it would run to the attic unchecked. This construction now requires fire stopping. However, many balloon-frame homes are still fully occupied in this country (see **Figure 1-23**).

The key to adaptation of strategy and tactics to the various types of building materials and architectural and engineering changes is preincident planning. A firefighter must be aware of the various building features to which he or she will respond and make the strategic and tactical changes that are necessary to mitigate the incident safely. Proper strategy and tactics can be better employed if the firefighter knows the construction types, building materials used, architectural designs, and general engineering principles of the specific structures in the response zone.

Figure 1-23
Firefighters must consider the possibility of balloon-frame construction in older wood-frame structures.

Wildland/Urban Interface

Forest fires have traditionally been fought by the various state and federal divisions of forestry, whereas structure fires have traditionally been fought by the local fire department. How, then, can wildland fires influence changes in tactics for structural firefighting forces? Many homes are built in and around forest and brush areas, which creates a strategic and tactical challenge when a forest fire is moving toward occupied homes (see **Figure 1-24**). This phenomena is known as *wildland/urban interface*. This problem has escalated over the years as people continue to move out of the cities and into suburban and rural areas. The population has longed for the peace and shade of the forest. As this trend continues, so does the problem (see **Figure 1-25**).

New Equipment and Materials in Vehicles

Airbags, fiberglass, catalytic converters, shock-absorbing bumpers, sealed air-conditioning units, alternate forms of power such as propane, hydrogen, and electric, the use of plastics and lightweight metals, and other advances in vehicles have had powerful influences on tactics for vehicle fires. Like many of the other motivators for change that have been mentioned, this area continues to grow and diversify.

Figure 1-24 *It has become very popular to build homes in and around wildland areas and forests. (Photo courtesy of JoAnn Angle.)*

Figure 1-25 *Large wildland fire in progress.*

Hazardous Chemicals

Hazardous materials are probably in every structure firefighters respond to. They are in garages, factories, retail stores, homes, sheds, and businesses. Hazardous materials are transported over the roads, rails, and seaways of virtually every community in the developed world. Hazardous materials have affected the ways in which fires are fought. Even though a typical house fire may not be considered a hazardous material fire, homes usually have some sort of pesticides, herbicides, gasoline, chlorine, and other chemicals in them. These items have had an impact on procedures such as SCBA usage. Hazardous chemicals can affect the behavior of the fire; as the fire behavior changes, so must the firefighting tactics.

Ongoing Changes

The influences that affect strategy and tactics will never stop. Some of the influences come gradually and over many years, whereas others, such as OSHA rulings and standards, impact the fire service immediately. Firefighters must adapt to the changes in order to best meet the most basic goals and objectives—to save lives and reduce property loss. Ben Franklin did not have as many factors to consider as today's firefighters do, but the goals and objectives stay the same. The procedures used to meet these goals must change as the influences change or they will go unmet.

THE FIREFIGHTER'S ROLE

■ Note
Regardless of rank, every firefighter must clearly understand strategy and tactics.

The role of the firefighter with regard to strategy and tactics is diverse. Regardless of rank, every firefighter must clearly understand strategy and tactics. This understanding customarily does not occur in basic recruit school. The probationary firefighter is usually precise in the action needed to meet the objective given by the company officer. For example, when told to force open a door, the probationary firefighter has no trouble meeting that objective. It takes experience and education, however, before the probationary firefighter clearly understands the overall methods and techniques used to bring the entire incident under control. Nevertheless, all firefighters have the responsibility to be an asset to the operation. The firefighter actively working at an incident is the eyes and ears of the incident commander. He or she should have a basic understanding of many of the concepts discussed later in this book, such as fire dynamics. An understanding of how a fire will behave and where it will go can greatly assist in extinguishment. The firefighter should also know the extinguishing agents available and be expert at application. Further, an understanding of command and control, safety, and various strategies, tactical objectives, and methods are important aspects of a firefighter's role.

SUMMARY

The fire service and the major influences upon it are constantly changing. The strategies and tactics that are chosen for a given incident must be adaptable to the changes. In order to accomplish this well, the firefighter must have a basic understanding of the history of strategy and tactics and how the fire service has used them to its advantage. Firefighters capable of making the right decisions are the ones who understand how the changes in items such as building materials and furnishings, architectural design, personal protective clothing, building codes, and hazardous materials impact the need for particular fire strategies and tactics. The prudent firefighter must understand that this premise holds true for every aspect of every strategy and tactic chosen.

The fire service is steeped in tradition and history and, as a result, changes usually don't come easily or quickly. Some of the changes may result from the many influences that have already been discussed. These changes force the fire service to constantly move forward and reevaluate itself. If the prudent fire officer does not also move forward, he or she will be left behind.

REVIEW QUESTIONS

1. Describe the different types of fire departments discussed in this chapter.

2. True or False? The National Fire Academy was established in 1974 as a result of the Federal Fire Prevention and Control Act of 1974.

3. Briefly describe improvements in personal protective clothing.

4. Name the landmark document that helped to create the Federal Fire Prevention and Control Act of 1974.

5. True or False? NIOSH investigates every firefighter death in the United States.

6. The Hackensack Ford fire brought to the attention of firefighters nationwide the dangers of which of the following?

   A. Bowstring trusses

   B. Flashover

   C. Backdraft

   D. Parapet walls

7. Briefly explain the structural and building materials differences between homes built in the 1920s and homes built today.

8. The metal plates on lightweight truss roofs that fail under fire conditions are known as what?

9. List three items that a firefighter should have a basic understanding of to be an asset to the incident commander.

10. Why is it important for a firefighter to understand the major catalysts that have changed and molded strategy and tactics?

ACTIVITIES

1. Review any incident that has occurred in the last year that resulted in a firefighter death. What were the contributing causes of the death? Were there issues brought to the attention of the fire service that may influence strategies and tactics? Explain.

2. Discuss new policies or equipment purchases that your department has adopted or put into service that will affect strategies and tactics.

Chapter

2

FIRE DYNAMICS

Learning Objectives

Upon completion of this chapter, you should be able to:

- Compare the components of the fire triangle and the components of the fire tetrahedron.
- Define the five classes of fires.
- Describe the stages of fire growth.
- Describe the four methods of heat transfer.
- Define the terms *flashover* and *backdraft*.
- Describe the concept of smoke behavior.
- Describe the relationship between fire dynamics and the application of strategies and tactics.

CASE STUDY

In the early morning hours, the first-due engine on a structure fire response finds fire engulfing a vacant three-story wood-frame, wood-sided building. The fire building is part of a block-long row house. The first exposure is on the west side of the building and is labeled side B. This building is a three-story wood-frame apartment with brick exterior. The fire is extending by radiant heat to this occupied tenement on exposure side B. The exposure on the east side, side D, is the remainder of the row house. There are no exposures on the north and south sides of the building. The fire is spreading rapidly by convection currents through the cockloft of the row house of origin on side D. Based on the speed of the flame spread and intensity of the fire, a second alarm is sounded. As units deploy to contain the fire, extra help is obviously necessary because of the speed of the fire spread. Minutes later a third, then a fourth alarm is sounded. Finally, an hour into the incident a fifth alarm is sounded and the units working bring the fire under control.

The fire department was able to use video imaging provided by local news helicopters as a tool for critique of this fire. By use of normal and thermal imaging modes from both the helicopters and ground crews, they could track the rapid spread of the fire and develop lessons learned. The incident commander was faced with large amounts of heat from a free-burning structure. In a fire like this, master streams are required to reduce the heat production and cool the exposures. The fire rapidly extended to exposure B through intense radiant heat exposure. As it extended up the side and rear of the building's combustible exterior siding, the fire produced significant radiant heat to affect exposure B. The fire extended horizontally through exposure side D by convection currents forced under pressure. This required extensive vertical ventilation as close as possible to the main body of the fire to release the trapped heat, gases, and pressure. The thermal imaging cameras provided insight into the advancing plume of heat in the cockloft. The fire also spread through the structure's void spaces. It traveled through the brick nogging of the walls and roof structure. Ultimately, this fire was extinguished with no loss of life and only the vacant part of the row house destroyed.

Source: This case study taken in part from *With the New York Firefighter, 3rd ed.* [1998].

■ **Note**
Incident commanders should know their enemy, whether it be a fire, a hazardous chemical release, or a building collapse.

INTRODUCTION

When the United States entered the Gulf War, it had every reason to know it would be victorious. The battles were quick and decisive because the military leaders knew their enemy, just as incident commanders should know their enemy, whether it be a fire, a hazardous chemical release, or a building collapse.

The military leaders created their battle plans based on their knowledge of how their enemy was armed, how it prepared for battle, and how it fought. A fireground commander must know how a fire grows and expands, how it spreads, and what methods are necessary to stop it. Firefighters know that fire is a chemical process where heat, fuel, and oxygen come together in an uninhibited chain reaction. Fire produces an **exothermic reaction** releasing heat, light, smoke, toxic gases, and other products of incomplete combustion. The military leaders knew how to overcome their enemy. They knew the enemy's strengths and weaknesses. They knew their reserves and they knew how to get ahead of and cut off the enemy. The same must be true for the effective fireground commander. On the fireground, the firefighter must be able to read the fire and determine the fire's potential course and harm. Firefighters must be able to determine how much fuel is ahead of the fire and the speed with which it will grow. Furthermore, knowing the type of building helps the firefighter to determine how this fire will move through the building. Will the building add to the fire load as fuel, such as a wood-frame building, or will the building contain the fire the way a concrete-block building does? Will the building allow extensive smoke to travel to the upper levels or contain it in compartments? Essentially, knowing what a fire is, how it grows and spreads, and what it takes to stop its progress is critical knowledge for any incident commander.

exothermic reaction
a chemical reaction that releases heat

■ Note
Knowing what a fire is, how it grows and spreads, and what it takes to stop its progress is critical knowledge for any incident commander.

THE CHEMISTRY OF FIRE

How can the oxidation process known as fire be both a friend and a foe? Fire provides heat and light for keeping us safe and warm or it provides heat and flame that destroys and kills. Fire is a combustion process that can be a self-sustaining reaction. This chemical process is exothermic in that one of the products of fire is heat. The fuel of the fire can be in any form of the three physical properties of matter: solid, liquid, or gas (see **Figure 2-1**).

Fire Triangle

When looking at the components necessary for fire to occur, the most common approach is to examine the **fire triangle** (**Figure 2-2**), which basically explains the three components that must be brought together for a fire to occur. The first component of the triangle is fuel. The fuel must be a combustible material. It is probably composed of carbon and hydrogen atoms that can be oxidized. The second component is oxygen, the most common oxidizing material in the atmosphere. Oxygen is found in normal breathing air in the atmosphere at a volume of 21 percent. The rest of the air is mostly nitrogen, an inert chemical. The third component of the triangle is heat energy.

fire triangle
three-sided figure showing the heat, fuel, and oxygen necessary for combustion

FUELS OF THE FIRE

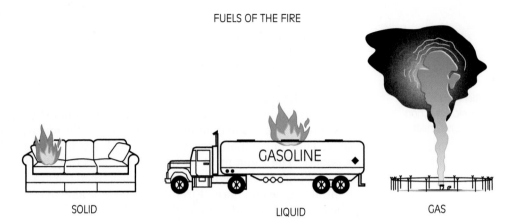

SOLID   LIQUID   GAS

Figure 2-1   *The fuels of a fire can be in any form of solid, liquid, or gas.*

When the fuel combines the oxidizing agent, oxygen, with the heat of ignition in a chemical process, it releases heat, light, and other chemical products. This heat and light is released in an energy form we refer to as a flame or fire. As the heat energy is released, it continues to vaporize the liquid or solid fuel, which ensures the continuation of the chemical reaction, or fire.

If any of the three components of the triangle are removed, the fire will go out. Now visualize a small fire. If you could reach inside it and remove the fuel, it would go out. If you remove the oxygen, such as covering a candle with a jar, the fire will go out. If the third component, the heat, is removed by an extinguishing agent, the fire will go out. Water, therefore, is an effective firefighting agent because it absorbs heat, thus cooling the fuel, and because it is readily available. It can also be used to smother flames by submerging some fuels. In an Ohio train derailment involving an air-reactive chemical, the fire was extinguished by covering the fuel with the smothering effects of water. By excluding the oxygen, the fire's chemical reactive process stopped.

Fire Tetrahedron

fire tetrahedron
four-sided pyramid-like figure showing the heat, fuel, oxygen, and chemical reaction necessary for combustion

A more complete explanation of the fire process involves the fire tetrahedron (**Figure 2-3**). This concept was developed in the early 1950s by Walter Haessler, who was interested in finding out why the dry chemical ammonium phosphate

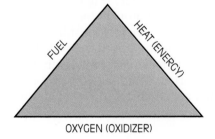

Figure 2-2   *The new fire triangle.*

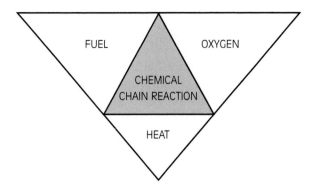

Figure 2-3 *Fire tetrahedron.*

was so effective as a fire-extinguishing agent. His belief was that the current fire triangle theory was inadequate as an explanation. In this four equal-sided configuration, the three components just described are included along with a fourth—an uninhibited chain reaction. In the fire process known as flaming combustion, the tetrahedron better describes the reaction taking place. The fourth component deals with free chain reactions. In this process, there is a transfer of electrons in the molecular combustion reaction. This transfer of electrons at the molecular level is caused by the oxidizing agent adopting electrons from the fuel. The fuel is known as a reducing agent. When the oxidizing agent and the reducing agent combine and transfer their electrons, they emit heat and light as an energy source. In a large reaction, this energy source presents itself as a flame.

Classes of Fires

Historically there were four classes of fire: A, B, C (see **Figure 2-4**), and D. A new class, K, was added in 1998. Made up of the common fuel sources such as wood and untreated paper products, class A fires are referred to as fires involving ordinary combustibles. Class B fires involve fuels made up of flammable petroleum products. These fuels can be liquids, solids, or gases. Class C fires involve

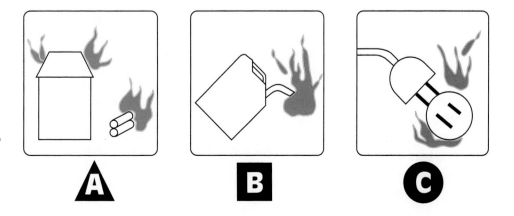

Figure 2-4
Classifications of fire symbols for A, B, and C classes.

electrically energized equipment. In a class C fire, the fuel itself can be either class A, B, or D. The electrical energy hazard denotes this category as being a class C fire. Note that the use of water, foam, or other product that can conduct electricity must not be used until all power has been eliminated. Once the electrical energy is eliminated, the appropriate extinguishing agent for the class of material burning can be used. Class D fires involve combustible metals such as sodium, magnesium, and titanium. These metals can burn with intense heat. The firefighter must avoid bringing water into contact with these products and use extinguishing agents specific to the material. Class K, the new classification of fire, involves fires in combustible cooking fuels, such as vegetable or animal oils and fats. These fuels are similar to class B fuels but involve very high temperatures and require the use of wet chemicals for extinguishment. Wet chemicals are described further in Chapter 8.

It is important to know the classification of fuels in order to choose the appropriate extinguishing agent and select the proper application. The proper extinguishing method must be used to extinguish these materials safely.

■ **Note**

It is important to know the classification of fuels in order to choose the appropriate extinguishing agent and select the proper application. The proper extinguishing method must be used to extinguish these materials safely.

Smoke

A final issue with the chemistry of fire is the products from incomplete combustion. In uncontrolled fires, the combustion process may not be complete. When class A fires burn, they produce carbon dioxide (CO_2), carbon monoxide (CO), and water. When class B fires burn, they produce heavy, black, sooty smoke because of the carbon that remains unburned (**Figure 2-5**). Plastics are also considered to be hydrocarbon-based. They produce heavy black smoke and high levels of carbon monoxide (CO), carbon dioxide (CO_2), hydrogen cyanide (HCN), and hydrogen chloride (HCL). The smoke produced by the incomplete combustion process creates a toxic atmosphere that is dangerous to both the occupants of the building and the firefighters entering it. The smoke will severely obscure vision and irritate the eyes and respiratory tract (**Figure 2-6**). The gases are toxic and are the greatest cause of deaths in fires. Most fire deaths result from the toxicity of the smoke, not from actual contact with the flames. The smoke is heated, and under the right conditions, it can ignite because of the buildup of carbon monoxide.

The most abundant gas produced at any fire is carbon monoxide. It is the gas that has killed most people; therefore, it is the major threat in most fire atmospheres. In a fire, the low oxygen levels cause the fire to burn inefficiently, releasing great quantities of carbon monoxide. The heat and smoke carry this gas in convection currents throughout the structure and to upper floors, incapacitating the occupants before they can escape. Carbon monoxide poisoning is the cause of death in more than 50% of all fire fatalities. When fires involve natural and synthetic materials that contain nitrogen, such as wool, silk, acrylonitrile, polyurethane, and nylons, they release hydrogen cyanide. Simple fires such as those involving automobile interiors have sickened firefighters due to large amounts of gases released from the polyurethane foam rubber seats and coverings. Acrolein, a very potent irritant, is released from burning polyethylene.

■ **Note**

Carbon monoxide poisoning is the cause of death in more than 50% of all fire fatalities.

Figure 2-5 *This tire fire would be classified as a class B fire. Note the heavy black smoke.*

Figure 2-6 *Heavy smoke can cause limited visibility and an abundance of carbon monoxide in the structure.*

Hydrogen chloride, a deadly gas released from polyvinyl chloride (PVC), is irritating to the eyes and upper respiratory tract. The lack of oxygen, or oxygen deficiency from the combustion process itself, is also deadly. The greater the supply of oxygen, the less gases produced as more complete combustion occurs. In a smoldering fire, tremendous amounts of toxic gases can be produced because the oxygen levels are diminished.

THE PHYSICS OF FIRE

When thinking about the physics of fire from the vantage point of the incident commander, the concerns are for fire stages and growth (see **Figure 2-7**). How the fire grows, spreads, and moves through a building must be considered by the incident commander and firefighters. They also must consider the amount of fuel available ahead of the fire. It is important for an incident commander to be able to visualize the concept of the fire phenomena and its progression and growth. The following are all important aspects to consider in determining the stage of the fire's growth:

- The shape and color of the flames
- The density and color of the smoke
- The pressure behind the smoke movement

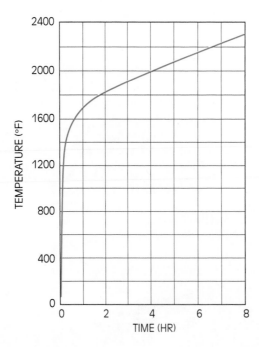

Figure 2-7   *The standard time–temperature curve showing the temperature increase of fire over time.*

Fire Growth

A fire starts when the three legs of the fire triangle are brought together with an uninhibited chemical chain reaction. A combustible material begins to burn, producing heat, smoke, gases, and other products of incomplete combustion. The fire's progress depends on the fuel load, the combustibility of the fuel, and the intensity of the heat produced. Depending on these factors, the fire can develop quickly or smolder for hours until finally reaching flame stage.

Fires can go through any one of or all three phases: incipient or growth stage, flaming or fully developed stage, and smoldering or decay stage. In the first two stages, there is plenty of oxygen and the fire can grow as long as there is sufficient fuel. The growth of the fire depends on:

- The type of fuel (combustible or flammable?)
- The physical property of the fuel (solid, liquid, vapor, or gas?)
- The surface-to-mass ratio (Is it a four-by-four piece of wood, wood shavings, or dust?)
- The arrangement of the fuel (Is it tightly packed or stacked with spaces in between?)
- Adequate ventilation (natural ventilation or an air-tight room?)

Fires can progress through these three specific phases. The key to using this information is to understand that not all fires, especially in buildings, go through distinct phases. Some begin by smoldering and remain that way throughout their life spans. Others begin as a free-burning flame and continue to grow larger or burn out quickly. Still others begin as an incipient fire and rapidly develop into a free-burning phase. This concept can be understood by following the development of a fire within a room.

The incipient phase is the actual ignition stage. The fire involves only the original material ignited and produces some heat and smoke. If there is adequate oxygen in the air, the flame can continue to grow as long as fuel and air are available. Depending on the fuel, the flame itself may produce a flame temperature of 800–1,000°F, but the room temperature may not change significantly.

The fire then progresses into the free-burning phase (see **Figure 2-8**). In this phase, the fire produces significant smoke and heat. It preheats the surrounding materials and the growth rate increases. The fire spreads vertically as a thermal convection current. As the thermal column meets the resistance of the ceiling, it begins to spread out horizontally. The room is heated from the top down as the heated gases continue to fill the room The fire also spreads in all directions through radiant heat waves. As the materials in the room continue to absorb the heat, they begin to reach their ignition temperatures. As long as the fire keeps getting air and the fuel supply does not diminish, the room will build to flashover conditions.

After the room has reached flashover and has consumed the oxygen in the room, it enters the smoldering phase. The flame diminishes and is reduced to

Figure 2-8   *Fire in the free-burning phase. (Photo courtesy of Fire Marshal James Fletcher.)*

convection
a method of heat transfer by which the air currents are the means of travel

radiation
a method of heat transfer through light waves, much like the sun warms the Earth

conduction
a method of heat transfer through a medium, such as a piece of metal

direct flame impingement
a method of heat transfer by which there is a direct contact to the object by the open flame

■ **Note**
Direct flame impingement is sometimes actually considered a form of radiation.

glowing embers. As the flames diminish, the fire emits dense smoke and other products of incomplete combustion. There is not enough oxygen in the area for complete combustion, so the smoldering phase is heavily laden with combustible and flammable gases. At this point, the room temperature can be 1,000°F or more. The increased pressure from the expansion of heated gases forces heat and smoke through small openings, which results in the puffing, or drawing back, of smoke out from small cracks and openings in buildings associated with a potential back-draft condition. If oxygen enters the room improperly, a backdraft can occur. If the room remains completely void of air, the fire will likely burn itself out through oxygen starvation.

Heat Transfer

If the fire grows, it can spread by four principal means: convection, radiation, conduction, and direct flame impingement. Direct flame impingement is some-times actually considered a form of radiation. It is presented in this text as a fourth method so that the reader is clear about the strategic and tactical implica-tions of this method of heat transfer (see **Figure 2-9**).

Convection   As the fire burns, it produces heat, smoke, and products of incomplete combustion, such as tars. These heated products are lighter than air and move

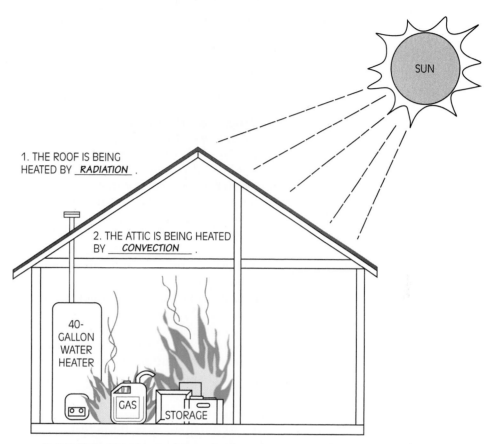

1. THE ROOF IS BEING HEATED BY *RADIATION* .

2. THE ATTIC IS BEING HEATED BY *CONVECTION* .

40-GALLON WATER HEATER

GAS

STORAGE

3. THE FIRE IS TRAVELING THROUGH THE STORED COMBUSTIBLES BY *CONDUCTION* AND *DIRECT FLAME INPINGEMENT* .

Figure 2-9 *Examples of the four main means of heat transfer.*

■ **Note**

Just as water and electricity seek the path of least resistance, so does fire.

freely upward until they meet an obstacle such as a ceiling. As the ceiling tries to contain the heat and gases, they begin to spread outward and move horizontally. Just as water and electricity seek the path of least resistance, so does fire. As the smoke and heat move out of the room, they move up stairs and through corridors and other compartments. In the structure fire presented in the case study in this chapter, convection currents spread the fire quickly throughout the enclosed cockloft of exposure D (the geographic label of a building's sides at a fire scene).

The travel of convection currents can be affected by air currents and ventilation systems. As the fire and its gases move through the building, they ignite other items in their path. The faster the fire grows and the more intense it becomes, the quicker these gases spread through the building.

The fire gases can spread faster than the flames. As they begin to cool, the gases stratify, the lighter gases staying near the top floors of the structure while the flames and heavy gases spread throughout the lower areas of the structure. These gases are toxic and can quickly kill the occupants on any upper floors. This phenomenon is what took the lives of many guests at the MGM Grand Hotel fire in Las Vegas (Clark County), Nevada. Because of the rapid smoke and fire spread, the greatest dangers exist on the fire floor, the floor immediately above the fire, and the top floor of the building. The priority of search should be set in this order.

Radiation The flames from a fire produce heat waves, similar to light waves, that move in a straight line from the fire. Heat waves are an invisible movement of heat that can penetrate windows and unprotected openings in buildings. Radiant heat moves equally in all directions from the fire, thus heating everything it can reach. It is an electromagnetic energy that does not become thermal energy or heat until it strikes an object and excites the molecules.

The radiant heat can travel downward from the heated gases as they move through a room. It radiates in all directions from the flames of rollover and the fire itself. The greater the intensity of the fire, the greater the radiant heat. In the fire incident in the case study in this chapter, radiant heat played a major role in the ignition of the exposure.

Because radiant heat is really electromagnetic heat waves, it can move through glass and water. That movement is the reason that water curtains are of little use in stopping radiant heat. In order to protect an exposure, the water must be applied to the exposure to reduce the temperature of the exposure itself. The water is used to remove any buildup of heat from the radiant heat waves. **Figure 2-10** and **Figure 2-11** show an example of successful exposure protection.

■ **Note**
Water curtains are of little use in stopping radiant heat. In order to protect an exposure, the water must be applied to the exposure to reduce the temperature of the exposure itself.

Conduction In the 1800s, Joseph Fourier defined the law of heat of conduction. It is the heat transfer in a stationary medium (solids, liquids, and gases). The rate of heat flow is directly proportional to temperature difference. The heat transfers from the point of heat outward toward the cooler areas. The handle of a spoon left in a pan of hot soup will become hot. This occurs because the metal spoon has conducted heat from the soup. The heat has traveled from the hot liquid up the spoon to the handle. In a fire, heat can be conducted through pipes, ceilings, and walls. In old buildings with tin ceilings, heat transfer is a common concern.

When examining materials for heat conduction, the most important physical properties are their thermal conductivity, density, and specific heat. Materials such as metals are good heat conductors. Products like bricks, blocks, and concrete are poor heat conductors. A firefighter must remember that the transmission of heat cannot be completely stopped by any heat-insulating material.

The conduction of heat through air and other gases can occur, independent of pressure in the typical range of pressures. There is no heat conduction through a perfect vacuum.

Direct Flame Impingement The spread of fire through direct flame impingement is similar to that of radiant heat spread (see **Figure 2-12**). The flames are in actual

Figure 2-10 *The exposure building on the right was protected from radiant heat by direct water application.*

Figure 2-11 *A closer view of the building in Figure 2-10 shows no damage to the exposure building, because of building construction and good placement of fire streams.*

Figure 2-12   *An example of direct flame impingement.*

contact with the exposed surface. This concern is common when buildings are built so closely together that their exteriors are only several feet apart. The flames from one structure actually contact the surface of the other.

In dealing with hazardous materials, the concern lies in the flames from one container touching another, which has been the cause of many explosions. For example, if the flames from an overturned railroad car are forcefully being expelled under pressure against another tank car lying alongside, this creates direct flame contact and may rapidly cause overpressure in the exposed car, creating the potential for an explosion.

Flashover

A critical point in the development of a room fire occurs when it reaches flashover, the point in the progression of a room fire when all the combustibles in the room have ignited. The fire and smoke begin to push out of the room under pressure, spreading throughout the structure.

In the early or incipient stage of fire, the fire moves slowly across the object of origin, producing smoke and radiant heat. The radiant heat begins to raise the temperature of the items in its path to their ignition temperatures. As this occurs, the fire on the item of origin continues to grow. The heat and smoke produced begin to be carried to the ceiling and across the room. This action continues to heat all items within the compartment. The fire produces smoke and products of incomplete combustion. The temperature in the room can now exceed more than

Figure 2-13 *Rollover is indicative of the early stages of flashover. Ventilation and cooling are required to prevent flashover. (Photo courtesy of Greg Sutton.)*

rollover
the rolling of flame under the ceiling as a fire progresses to the flashover stage

flashover
an event that occurs when all of the contents of a compartment reach their respective ignition temperatures in a very short period of time

1,000°F, and the fire and smoke begin to push out of the room under pressure, spreading throughout the structure. The smoke fills the room at the ceiling level and begins to bank down the walls toward the floor. The combination of radiant and convected heat is now bringing all things in the room to their ignition temperatures. Smoke begins to exit the room and move down the corridors. The smoke movement is slow at first. Then as the fire grows bigger, the pressure in the room from expanding gases pushes the smoke harder and faster down the corridors, through cracks in walls and through other openings. Other objects in the room begin to smoke as they reach their ignition temperatures. Fingers of fire begin to roll across the ceiling, reaching out to other fuel within the room. This rollover (**Figure 2-13**) of fire across the ceiling is one of the first signs that flashover is imminent. Unless water can be quickly applied to cool the ceiling, flashover will occur. As flashover occurs, all the contents of the room have ignited. The fire load in the room determines the fire temperature and the pressure generated by the expanding heat and gases of the self-sustaining chemical process. The fire pushes out of the room, down corridors, up stairways, and through any openings it can find.

An incident commander must be able to understand this fire development and know when and where to commit the companies and crews. As the fire

■ **Note**

The key to recognizing flashover is the smoke movement and fire growth.

■ **Note**

Ceilings, which can be nearly 20 feet high, mask the amount of heat that is actually present at ceiling level.

thermal radiation feedback
as heat is transferred in a compartment, the walls and furnishings in that compartment heat; this heat then feeds back and further heats the compartment

continues to grow, it pushes through the structure, consuming everything in its path. It will eventually vent out of the structure and involve the building itself.

The key to recognizing flashover is the smoke movement and fire growth. Is the fire in a small or large compartment? What is the height of the ceiling? Is the smoke thick and dark black? Is it being pushed out under pressure or moving lazily? Are there fingers of fire, rollover, moving through the smoke?

Inside a building look for a sudden buildup of heat. Watch for flames moving through smoke above your head. Cool down the ceiling in front of you as you move toward the fire. If the smoke is banked down almost to the floor and the heat is escalating, flashover may be imminent. With today's encapsulating fire gear you may not feel the heat until flashover has occurred.

Do not be fooled in structures with high ceilings (**Figure 2-14**). In these buildings, the ceilings, which can be nearly 20 feet high, mask the amount of heat that is actually present. This height of the smoke and heat allows for good visibility across the warehouse and little heat buildup. The high heat at ceiling level can flash over and move so quickly that it will outrun the firefighters.

Definition of Flashover Sudden ignition of the superheated smoke, gases, and all contents of a room. It occurs at the end of the growth stage when the entire room becomes involved in a fully developed fire. The heating process is caused by **thermal radiation feedback**.

Figure 2-14 *High ceilings such as these are typical in warehouses, masking the signs of flashover due to the low heat at the floor level.*

Indicators of Flashover

- Room size
- Extreme, unbearable heat buildup that drives you to the floor
- Rollover of fingerlike flames rolling out through the smoke

Survival Strategies

- Watch smoke movement, flame movement above.
- Know your way out of room and building.
- Use safe search procedures; follow wall in one direction.
- Use safety search rope or hose line for guidance out.
- Vent buildup of heat.
- Do not enter rooms when flashover conditions are apparent.
- Close door to keep heat and smoke in room and out of halls.
- Wear *all* your protective clothing and self-contained breathing apparatus (SCBA) properly.
- Remember that once flashover occurs, the chance of survival in the compartment is almost zero.

Flashover Dangers

- Flashover occurs when everything in the room has reached its ignition temperature.
- The room will become fully involved in fire from floor to ceiling.
- Temperatures can exceed 1,000°F.
- Visibility will be diminished.
- Five feet into the room where flashover has occurred is commonly known as *the point of no return*.
- It means death or critical thermal burn injury to anyone trapped in the room.

Backdraft

Safety

If the fire is not vertically ventilated, the admission of air allows the gases to ignite and explode.

When an oxygen-starved fire in an enclosed compartment suddenly gets a fresh supply of air, it will ignite with explosive force. This ignition is known as a *backdraft*. The fire has enough heat and fuel but lacks the air to burn. When air is introduced into the room, the explosive gases ignite and the room explodes. This explosion has been known to blow out windows and collapse walls.

Today's buildings are being constructed with energy efficiency in mind, with fewer natural drafts and less fresh air moving through cracks and window frames. The use of insulation and thermal pane windows has severely limited natural air flow in modern buildings.

In a room where the fire has grown to the free-burning stage, the fire may use all of the available oxygen. The fire has developed heat, smoke, and products of incomplete combustion. The fire gases in the smoke, such as carbon monoxide, are now at their ignition temperatures. The contents of the room are smoking and ready to ignite. Without adequate oxygen to burn, the fire begins to smolder. Without the oxygen necessary to burn, the fire continues to smolder and produce greater amounts of smoke and other gases, forming an explosive atmosphere. If the fire is not vertically ventilated, the admission of air allows the gases to ignite and explode.

An oxygen-starved fire builds up heat and pressure in the room, forcing smoke out of cracks and crevices in the building. When this pressure is released, the fire then sucks fresh air back in. The fire grows until this air is used and the pressure increases, and then begins to die as the pressure is once again released. This process continues until the fire finally dies out or an unrestricted supply of air is introduced. If a firefighter ventilates this structure at the level of or below the level of the fire, air will rush in and allow all the gases to ignite instantaneously and the room will explode. The oxygen and carbon monoxide mixture ignites at approximately 1,100°F.

The fire does not necessarily have to smolder for a lengthy time. There have been documented cases in which smoke is forced into a compartmented area during a free-burning fire. After that compartment becomes pressurized with the heated gases and smoke, it finally ignites and explodes. This has been a fairly common concern when battling fires in **taxpayers** and **strip shopping centers** with common cocklofts. It can also occur in **row houses** and **garden apartments.** With today's environmental consciousness and energy-saving ideas, it is not uncommon to have a backdraft in single-family houses.

Firefighters must take the actions necessary to prevent backdraft. Commercial occupancies are common locations for backdraft. If you arrive on the scene of a fire in a commercial occupancy after it has been closed for hours, you may face a backdraft. Look for smoke pushing from nearly every crack and crevice of the building. Look around doors and windows for smoke pushing out and being sucked back in. Little or no flame will be visible from outside, because the building has not ventilated itself. The windows will be stained black and be hot to the touch. There may be condensation streaks running down the inside of the windows. These are clues to a backdraft situation. When hose lines are charged and in place, vertical ventilation must take place prior to firefighters entering the structure.

The U.S. Navy has a training center where it can simulate backdraft explosions. In tests it has concluded that the fire must draw air into it for a few seconds before ignition occurs. If water is thrown into the building compartment, hence cooling the environment, while it is drawing this air in, the balance of heat, fuel, and air can be disrupted and backdraft potential can be eliminated.

taxpayers
a term of building, more common on the East Coast, in which a mercantile occupancy is on the first floor and living areas occupy the floors above

strip shopping centers
rows of attached mercantile occupancies with a common look and roof line

row houses
homes attached with common walls and roofs

garden apartments
a two- or three-story apartment building with common entry ways and floor layouts, often with porches, patios, and greenery around the building

❗Safety
When hose lines are charged and in place, vertical ventilation must take place prior to firefighters entering the structure.

facade
an artificial face or
front to a building

Backdrafts can occur in void areas, such as double ceilings or double floors, in cocklofts and attics, and also in facades. Facades can be found on the front and sides of garden apartments and remodeled buildings. They hold smoke and heat in them until they either burn through or, as in some cases, backdraft. Backdraft occurs with little or no warning.

Definition of Backdraft   A forceful explosion can occur in the smoldering stage of a fire. Visible flames have disappeared but high heat and smoke conditions remain. The explosion occurs when oxygen is introduced and gases enter their flammable range, causing them to ignite. This phenomenon can occur from premature horizontal ventilation. These explosions have blown out windows and have blown down walls, causing structural collapses.

Indicators of Backdraft

- Thick smoke is pushing out windows, doors, and openings under pressure.
- Heavy smoke appears with no visible fire in well-sealed spaces.
- Dark yellowish-brown smoke is seeping from a tightly closed building.
- Smoke is pushed out of building and then drawn back in.
- Windows are stained black and are hot.
- There are little or no visible flames.
- The building is tightly closed or contained.

Survival Strategies

- Anticipate this sudden explosion based on indicators.
- Stay away from windows and doors.
- Introduce vertical ventilation.
- Prevent buildup of smoke and heat.
- Drive water deep into the superheated smoke with hose streams in order to cool the environment.
- Wear *all* your personal protective clothing and SCBA properly.
- Stay low on the floor.

!Safety
One of the greatest fire hazards faced by firefighters and civilians is smoke. It is a killer because of the gases contained in the process of incomplete combustion.

SMOKE BEHAVIOR

One of the greatest fire hazards faced by firefighters and civilians is smoke. It is a killer because of the gases contained in the process of incomplete combustion. These gases can be toxic, suffocating, and explosive. Smoke is produced from the incomplete combustion of the fuel. It contains particles of tar, water, and multiple gases. These gases move quickly through a structure and can cause

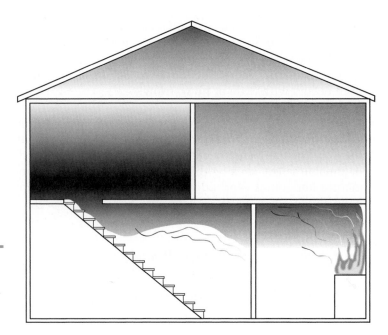

Figure 2-15 *Smoke travels quickly through a structure, following the path of least resistance.*

death before the fire grows large enough to be discovered (see **Figure 2-15**). In mid-rise and high-rise fires, the smoke can travel rapidly to the top floors and kill people far from the actual fire area.

Smoke was the main killer in the MGM Grand Hotel fire, the Beverly Hills Supper Club, the Wicoff Hotel, and many other fires throughout history. Many of the deaths occur when people try to escape (see **Figure 2-16**). As they leave their rooms, they succumb in the hallways, stairways, and elevator cars from the smoke as it moves to the upper floors.

As a fire begins to grow, it produces large amounts of smoke and gases. The less air provided to the fire, the greater the smoke production. The smoke initially moves upward until it is stopped by the ceiling. It then begins to move horizontally, seeking the path of least resistance. It will move out of the room of origin and move down hallways, corridors, and vents. It goes upward whenever it can find an opening, such as a stairway, elevator shaft, or utility shafts (see **Figure 2-17**).

As long as the smoke is heated and being pushed by a flame front, it will move quickly. As it hits the cooler air outside the room, it will begin to stratify and slow in its movements. The smoke will rise upward through stairways and shafts and it can be moved throughout the structure by the heating, ventilation, and air-conditioning system (HVAC).

In fighting a structure fire, the firefighters must locate the base of the fire to extinguish it. Smoke can make finding the fire very difficult. A tightly closed structure causes the fire to produce large amounts of smoke that hides the fire. In some cases, the smoke obscures the entire fire scene. Again, once hose lines are in position, ventilation must occur.

Figure 2-16 *Heavy smoke produced by a high-rise fire. Smoke spread will outpace firefighters' efforts to control it. (Photo courtesy of Gene Blevins.)*

Figure 2-17 *Smoke and flames can spread rapidly in strip malls with common cocklofts.*

Figure 2-18 *Smoke begins to obscure firefighters working outside the structure. (Photo courtesy of South Trail Fire Department.)*

One way an incident commander can help in locating the fire is to get a 360° view around the fire building to look for visible fire, the heaviest concentration of smoke, and other signs indicating where the fire may be. This inspection may be accomplished by the incident command or by firefighters reporting their findings to the incident commander; however, it may only be appropriate if the building is small, such as a detached single-family dwelling. At the very least, the incident commander should obtain a view of at least three sides. A survey around the structure gives the firefighter an idea of where the fire may be located by observing the smoke concentration and movement in the building (see **Figure 2-18**).

If the smoke is very heavy and being pushed forcefully out the windows, the firefighter can expect a fire that is growing quickly with adequate fuel and air. If the smoke is lazily pushing out the windows, the fire growth is slower. Dark black smoke is indicative of hydrocarbon-based fires, such as plastics and foams. Dirty brown smoke indicates an oxygen-starved fire. Lighter smoke is more indicative of class A fires. In very cold weather, any of the smokes can look white because of the water condensation due to the heat of the fire. Be aware that smoke looks different at night than in the daylight (see **Figure 2-19**).

Until the smoke is controlled in a structure, the fire will be difficult to locate and extinguish (see **Figure 2-20**). Fire crews need to ventilate the structure to

Figure 2-19 *Note how smoke can look different at night.*

Figure 2-20 *Single-family dwelling before fire ventilation.*

Figure 2-21 *Single-family dwelling after fire ventilation.*

remove the smoke and create conditions that allow the crews to move inside. Rescues are compromised by heavy smoke conditions, and firefighters have been lost and died in these conditions due to disorientation. Whether the incident commander decides to use vertical, horizontal, or positive pressure ventilation depends on the type of structure and the extent of the fire (see **Figure 2-21**). This lesson is learned more easily in training than on the fireground.

SUMMARY

Firefighters must be able to read a fire and think in a logical process to get ahead of the fire. By knowing the manner in which a fire grows and spreads through a structure, the fire officer can determine the strategy necessary to get ahead and stop the fire. Knowledge of the dangers of smoke allows for a safer fireground. The ability to read the smoke helps in determining the growth rate of the fire. The color of the smoke helps in determining the

extent of the fire and the stage of the fire's growth. As a firefighter, you must understand fire dynamics to make decisions on how and where fire will spread through a building. The initial strategy and tactics are determined by the size-up from the conditions inside and outside the building. Observing construction, occupancy, fire load, smoke, and fire can give you clues as to what strategy and tactics you will need to implement.

REVIEW QUESTIONS

1. List the three phases of fire growth.
2. What warning signs might be present before a flashover occurs?
3. True or False? A backdraft is the sudden ignition of an oxygen-starved fire.
4. Write an explanation of the interrelationship of the three components of the fire triangle.
5. Describe how and why the removal of one side of the fire triangle extinguishes a fire.
6. What is the most abundant toxic fire gas found in a structure fire?
7. Relating your answer to fire growth, describe convection heat currents.
8. Relating your answer to fire growth, describe radiant heat currents.
9. Describe how smoke moves through a structure.
10. List the five classes of fires.

ACTIVITIES

1. Using one of your department's preincident (fire) plans, review the plan and determine routes of smoke travel and fire spread.
2. From this chapter's case study, discuss how the fire grew so rapidly and traveled in different directions.
3. Using pictures of actual fires, discuss the stage of the fire and what can be expected due to the smoke and fire conditions visible.

Chapter

3

EXTINGUISHING AGENTS

Learning Objectives

Upon completion of this chapter, you should be able to:

- Discuss the various properties and use of water as an extinguishing agent.
- Describe the various formulas to determine fireground flow requirements.
- Demonstrate various types of nozzles and explain the advantages and disadvantages of each.
- Perform basic hydraulic calculations.
- Describe the use of foam as an extinguishing agent.
- Discuss the use of dry chemicals and dry powders.

CASE STUDY

On February 23, 1991, at approximately 2040 hours, the Philadelphia Fire Department responded to a reported structure fire on the twenty-second floor of the thirty-eight-story Meridian Bank Building, also known as One Meridian Plaza. The Meridian Bank Building was sprinklered only on certain floors, a fact that is significant to the final outcome. This fire eventually became one of the largest high-rise office building fires in modern American history. Eight floors were consumed and the fire burned for more than 19 hours. The fire went to twelve alarms which responded fifty-one engine companies, fifteen ladder companies, eleven specialized units, and approximately 316 firefighters. Tragically, three firefighters lost their lives and twenty-four firefighters were injured in the blaze.

The first-arriving units found fire venting from one window on the twenty-second floor with heavy smoke already in the stairway and on floors immediately above.

The Philadelphia Fire Department initially were met with three major problems that hampered the firefighting efforts:

1. *Water supply problems.* These problems were related primarily to the standpipe systems. Engine companies proceeded to connect to appropriate standpipes with $1\frac{3}{4}$-inch hose lines and automatic fog nozzles, which required 100 psi to provide effective fire streams of variable gallons per minute. The standpipes, however, had pressure-reducing devices installed that reduced pressures to less than 60 psi. This pressure was not sufficient to create effective streams.

2. *Loss of electrical power.* By the time one of the first-arriving battalion chiefs reached the eleventh floor with a fire attack team, there was a total loss of electrical power in the building. This power failure forced firefighters to perform all suppression operations in complete darkness. The only light available was from handheld battery-operated lights. The electrical system failure actually caused the third major problem that hampered firefighting efforts.

3. *Out-of-service elevators.* The power loss made the operation of the elevators impossible. Firefighters were required to climb at least twenty flights of stairs to the staging area, carrying all suppression equipment, including self-contained breathing apparatus (SCBA), hose, and extra air bottles.

Incident commanders on the scene made every possible attempt to remedy these problems. To overcome the water supply problems, the initial decision was made to stretch 5-inch lines up a stairwell to provide adequate water supply for the handlines. While this operation was being performed, a sprinkler contractor arrived on the scene and manually adjusted the pressure-reducing valves. This action greatly improved the fire streams, but unfortunately by the time the valves were adjusted, the fire involved several floors and could not be contained with manual fire streams.

During the fire fight, Engine 11, staffed with one captain and two firefighters, was assigned to ventilate the center stair tower. The Engine 11 crew radioed that they had left the stairway and were disoriented in smoke on the thirtieth floor. Attempts were made to direct them to safety via the radio and search teams were sent to the thirtieth floor. The search teams were not able to find the missing firefighters. They were found on the twenty-eighth floor 3 to 4 hours after they initially reported they were in trouble. The search efforts were heroic, but tragically all three members of Engine 11 succumbed to smoke inhalation and died.

Firefighting forces found evidence of structural damage, and the command post consulted with a structural engineer, who believed that there was the possibility of a pancake collapse of the fire-damaged floors. Therefore, the fire operations were shifted to a defensive mode, and the building began to be evacuated at 0700 hours on February 24, nearly 11 hours after the initial call.

After the evacuation of the building, portable master streams were placed in operation from adjacent buildings. The fire was ultimately brought under control by ten sprinkler heads that operated when the fire reached the thirtieth floor. The fire was declared under control at 3:01 P.M., February 24, 1991.

Source: This case study was taken in part from the United States Fire Administration Technical Report Series, report number 049, "High-Rise Office Building Fire, One Meridian Plaza, Philadelphia, Pennsylvania (February 23, 1991)" by J. Gordon Routley, Charles Jenning, and Mark Chubb. Further reading on this incident is encouraged and can be found in this publication.

INTRODUCTION

■ **Note**
Extinguishing agents are the materials used to extinguish a fire.

■ **Note**
Different fires and scenarios call for various types of extinguishing agents.

■ **Note**
The prudent incident commander is keenly aware of the materials burning and understands the agent best suited to extinguish the fire.

Water, dry chemical, foam, and dry powder are some of the many forms of extinguishing agents available to the fire service. Extinguishing agents (**Figure 3-1**) are the materials used to extinguish the fire. Sand that is used to smother a camp fire is a good extinguishing agent. It is inexpensive, readily available, and works well if applied correctly. These factors are important considerations when choosing extinguishing agents to be utilized by the fire service. Different fires and scenarios call for various types of extinguishing agents. In order to be proficient at extinguishment, a firefighter must understand the benefits of the agents used. As discussed in Chapter 2, it is important to understand the behavior of fire because understanding where the fire will go and how it will get there is critical to a positive outcome to the incident. But just as important is to understand the properties of the extinguishing agent and how, why, and when a particular agent should be used. These two topics are closely related. The prudent incident commander is keenly aware of the materials burning and understands the agent best suited to extinguish the fire. Additionally, he or she understands the effects of the extinguishing agent. For example, if the incident commander chooses to use water to extinguish a fire, he or she must understand

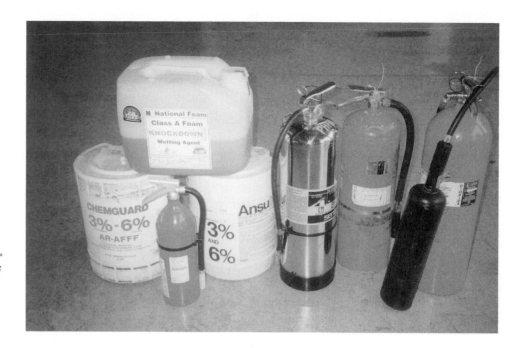

Figure 3-1 *Examples of various extinguishing agents available to the fire service.*

basic facts, such as how many personnel and how much equipment is required to move water from point A to point B. This concept is simple, but other aspects and properties of the various agents available to the incident commander must also be understood.

WATER

■ **Note**
Water is the most prevalent extinguishing agent in use.

Certainly, water is the most prevalent extinguishing agent in use. The reasons for water's popularity in the fire service are obvious to the firefighter. Usually water meets the needs that were discussed earlier.

1. It is inexpensive. (It is customarily free to fire departments; however, considering the costs associated with the infrastructure of the water system, is it really free?)
2. It is commonly available, either through a municipal or private water distribution system and hydrants or by fire departments moving it to the desired location via equipment and trained firefighters.
3. Generally it is efficient at extinguishment when applied properly.

Water almost always meets the objective of extinguishment. In order to be effective with water, though, one must have a complete understanding of its properties including its advantages and limitations (see **Figure 3-2**).

Figure 3-2 *Hydrants help to make water readily available for fire attack. Lack of water supply affects strategies and tactics negatively.*

normal operating pressure
the pressure on a water distribution system during normal demand

Water cannot be compressed. Why is comprehension of this concept important to extinguishment? Because water cannot be compressed, it can be forced to move by applying pressure to it. In simpler terms, firefighters can make water move to the place they need it most. The pressure to make the water move comes from various sources. Each municipal water system has a certain amount of pressure on the system. When a faucet is turned on, the water flows because of this pressure, known as normal operating pressure. If a firefighter hooked a hose line directly to a hydrant and flowed water through a nozzle, he or she would be operating the nozzle at the normal operating pressure on that given system. However, the fire service usually needs more pressure than the local water department can provide, so the fire service utilizes the normal operating pressure to supply a pumper. The pumper, in turn, boosts that pressure through the use of a centrifugal pump that provides the pressure required for an effective fire stream (see **Figure 3-3**).

Water takes the shape of the container it is in. For this reason it can be forced, by pressure through a hose line. As the water enters the nozzle, it again takes on the shape of the container. In this case the container is the nozzle and allows the water to form a usable stream.

Water can be found in three physical states: solid, liquid, and gas. As a solid, water is ice. At 32°F (0°C), water begins to freeze. In the temperature range from 32°F (0°C) to 212°F (100°C) water is in the form of a liquid and from 212°F (100°C) and up water is converted into steam, a gas. Water as ice is important to understand from a safety standpoint, but from a fire extinguishment viewpoint, water and steam are the most important states. When water is heated to the

Figure 3-3 *The fire pump is used to increase water pressure taken from municipal water supply.*

boiling point, it converts to steam and expands. At 212°F it expands 1,700 times its original volume. Further, the hotter the temperature, the greater the expansion rate. It is important to understand this concept because as steam expands, it changes the fire and it changes the atmosphere in the burning building. In order to completely understand the capabilities of steam expansion, you must first understand how water is used as an extinguishing agent. In other words, how does water extinguish a fire (see **Figure 3-4**)?

Water extinguishes fire through three basic methods:

1. Cools the burning material
2. Smothers the fire
3. Separates the fuel from the heat

Generally speaking, structural firefighting forces cool the fire and burning materials below their ignition temperature and the fire goes out. Additionally, when water converts to steam, it can smother the fire by displacing the oxygen in the

Figure 3-4 *Water is the most common extinguishing agent because of its properties. Note in this case the poor stream and poor line staffing. The stream is being used for fire confinement.*

room. However, cooling the fire remains the main reason water extinguishes fire.

Flow

A minimum amount of water is needed to cool and smother a fire. A firefighter cannot just throw a bucket of water at a house fire. It would not do anything to mitigate the incident. The amount of water needed is known as the *flow requirement*. There are three nationally accepted formulas for figuring flow requirement: the National Fire Academy (NFA) formula, the Iowa State University (Iowa) formula, and the Insurance Service Office (ISO, New York City) formula. The NFA and Iowa formulas are simple, easy to understand, and can be done quickly on the fireground if needed. The ISO formula was not designed to be used on the fire scene and would be very difficult to do quickly. It can be utilized during preincident planning. Further information about the ISO formula can be obtained from the Insurance Service Office.

The National Fire Academy formula is

$$NFF = \frac{L \times W}{3}$$

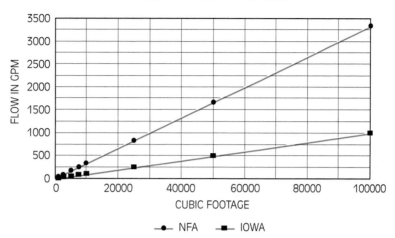

FLOW FORMULA COMPARISON
ASSUMING A CONSTANT 10' HEIGHT

Figure 3-5   The proper amount of water must be applied to be effective. Here is an example of the National Fire Academy formula versus the Iowa State formula for various cubic footages.

This formula is reasonably simple to understand and use. *NFF* = needed fire flow in gallons per minute, *L* = the length of the burning object, and *W* = the width of the burning object. Multiply the two and divide by 3. Three is the constant of the formula and never changes.

The Iowa formula is

$$NFF = V/100$$

V stands for the volume of the container burning, which is found by multiplying length by width by height. In other words, how big is the area that is burning? Divide this number by 100. The 100 is a constant and never changes. This gives you the amount of gallons per minute (gpm) or fire flow. See **Figure 3-5** for a comparison of the flow requirements of these two formulas.

Effective Fire Streams

The required flow is only part of the formula for extinguishment. Application of extinguishing agents is a critical step in extinguishment (**Figure 3-6**). Effective fire streams are necessary. If the forward velocity of the water, which is created by the pressure applied to it by the pump, is sufficient, the water will pass through the nozzle and it will be shaped into a viable fire stream. Effective fire streams have been defined in various ways but simply stated, the stream meets its objective. For example, a room and content fire might require 200 gpm of water flowed at 100 pounds per square inch (psi) through a fog nozzle to be effective. This same stream, though, might not be as effective on a large outdoor rubbish fire with 15 knot winds. Additionally, 30 gpm at 50 psi might be all that is needed to do overhaul on a smoldering couch (**Figure 3-7**). In this case, the

Figure 3-6 *Proper application of master streams are necessary for extinguishment in the defense mode.*

Figure 3-7 *Overhaul typically does not require large amounts of water or pressure.*

30 gpm stream is an effective fire stream because it is doing what the firefighter wants it to do: It is meeting its objective.

Nozzles

In order to meet the objective, the water must be forced through a nozzle. Without a nozzle, the stream would not be able to be directed. Additionally, the type of nozzle chosen is a guiding factor in extinguishment (through cooling, smothering, or both). There are many types of nozzles to choose from and each one creates a water stream differently.

Fog Nozzles Probably the most widely used nozzle in the fire service today is the fog nozzle. Many variations of fog nozzles, including adjustable gallonage, automatic gallonage, and fixed gallonage, are available to the fire service, and some examples are shown in **Figure 3-8.** All fog nozzles, though, break down solid streams of water into thousands of droplets of water that can be adjusted into various patterns. The patterns range from straight stream to wide fog pattern, making the fog nozzle very versatile. Because it has an adjustable pattern, it can be utilized for many different functions. Wide fog patterns may be used for protection of firefighters and straight streams may be used for reach or penetration. Additionally, because the fog nozzle is breaking down a solid stream of water, it can develop enough mechanical agitation to create foam blankets with some types of foam. This is discussed in detail later in this chapter in the section about foam. Finally, fog nozzles have better cooling capabilities than any other nozzle. The cooling comes from the tiny droplets of water. Each droplet of water absorbs heat. A comparison can be made to ice. A large block of ice absorbs less heat than

Figure 3-8 *Various fog nozzles.*

surface area
the exposed exterior
surface of an object

crushed ice; that is why crushed ice melts faster. Consider the fog nozzle as the crushed ice. The more pieces of ice there are, the more surface area of the ice and the more heat is absorbed. The more droplets of water there are, the more surface area, the more heat from the fire is absorbed. For this reason, fog nozzles have a tremendous amount of cooling capability.

Some of the problems with fog nozzles are that they tend to be affected by wind and weather (see **Figure 3-9**). Because of the droplets of water, there is more surface area than a solid stream of water. Thus, wind can make a fog stream ineffective. Further, fog nozzles can disturb the natural thermal layers of a fire if applied incorrectly.

When the fog nozzle is applied to a fire inside a structure, it creates steam because of the expansion properties of water discussed earlier in the chapter. As the steam fills the room, it forces the hot gases and products of combustion down toward the floor where firefighters are operating. Therefore, proper ventilation is critical to the proper use of fog nozzles. Another problem with fog nozzles is that many require 100 psi at the nozzle to create an effective stream. This can be a problem in areas of low water pressure as was the case in the Meridian One Bank

Figure 3-9 *Fog nozzles can be ineffective when affected by wind.*

Building in the case study. In recent years, however, manufactures have recognized this problem and now many new fog nozzles are capable of operating effectively at pressures less than 100 psi. Nozzles are also available with emergency low-pressure switches at the front of the nozzle. When a firefighter needs a stream of water and does not have adequate psi for an effective stream, the firefighter can switch to low-pressure mode. The firefighter then has an effective stream at 50 psi, which is the same pressure required to operate a handline smooth bore nozzle.

Note

The smooth bore nozzle creates a solid stream of water.

Smooth Bore Nozzles The smooth bore nozzle creates a solid stream of water. There are many advantages to the solid stream. It is best suited for direct application to the burning material but can be broken down into uneven droplets of water by ricocheting off objects such as a ceiling or wall. The uneven droplets created by the ricochet do not have the cooling capabilities of the fog nozzle but, if applied correctly, can be extremely effective. Perhaps one of the best benefits of a smooth bore nozzle is that it creates a stream that does not generally disturb the thermal layers of the fire (see **Figure 3-10**), which typically results in fewer injuries to firefighters due to steam burns. Solid streams also tend to have better reach and penetration. Wind does not affect the stream as dramatically as the fog stream adjusted to a straight pattern.

Figure 3-10 *There are many advantages to the solid stream of water created by a smooth bore nozzle.*

Broken Stream Nozzles Broken stream nozzles (**Figure 3-11**) create a stream of water discharged by small orifices of various sizes and shapes. Most broken stream nozzles have a special type of application. For example, a piercing nozzle (**Figure 3-12**) is a form of a broken stream nozzle. Another example is a cellar nozzle. In both of these nozzles, the application of water is done blindly. In other words, the firefighter may not be absolutely sure where the base of the fire is or cannot access the base of the fire. In the case of the piercing nozzle, the nozzle is driven into a wall or ceiling and the water is applied. The cellar nozzle is usually

Figure 3-11
Examples of broken stream nozzles.

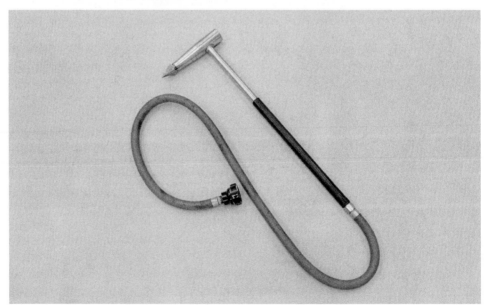

Figure 3-12 *An example of one style of piercing nozzle.*

Figure 3-13 *Various fog and smooth bore nozzles.*

lowered through a hole in the floor and the water is applied. The broken stream pattern is most effective for these applications.

Choosing the proper nozzle is crucial. Each nozzle has its advantages and disadvantages (**Figure 3-13**) and the prudent firefighter must be aware of them. However, nozzle choice, as shown in **Figure 3-14,** is not all that is required: The water being forced through the nozzle must be of sufficient pressure and gallonage to create an effective fire stream. Fireground hydraulics are used to ensure that the proper amount of water and pressure arrive at the nozzle.

Friction Loss and Basic Hydraulics

As the water is being moved through the hose in an attempt to become an effective fire stream, it is being slowed down by friction. The friction is created by forcing the water through the hose, nozzles, pump, piping and couplings, and around bends and up stairs or ladders. The pressure that is lost is known as friction loss (**Figure 3-15**).

For example, if a 50-foot garden hose is laid out and supplied with water at 30 psi, by the time the water comes out the other end, there may only be 20 psi left. The 10 psi that are missing were lost due to the friction created by forcing the water through the hose. The 10 psi loss is the friction loss. Too much friction and the pump will not be able to create enough pressure to overcome it and the

Figure 3-14   *Each fire department must choose the most appropriate nozzle based on local needs.*

Figure 3-15   *Size and length of hose and the type of nozzle all influence friction loss.*

streams will not be effective. The following simple steps can be taken on the fireground to reduce friction:

- Reduce hose length.
- Use larger hose.
- Reduce the amount of couplings and appliances used.
- Reduce the amount of bends and kinks in the hose.

The friction loss that cannot be reduced on the fireground must be overcome with pump discharge pressure. The following three simple steps can be used to figure the pump discharge pressure needed:

1. *Calculate the constants.* For example, the nozzle pressure is always known. It is the amount of pressure needed at the nozzle to produce the fire stream. Fog nozzles generally require 100 psi. Smooth bore nozzles require 80 psi for master streams and 50 psi for handlines. Appliances are also always known. Some departments use a standard number for appliances. For example, 25 psi friction loss for deck guns and 10 psi friction loss for gated wyes is an industry standard that many departments use. Each department, though, should actually perform flow tests on its appliances to be familiar with the accurate friction loss of its equipment.

2. *Calculate the elevation.* If on ground level, this figure will be zero. Otherwise, figure $\frac{1}{2}$ pound per foot of elevation (5 per 10-foot floor).

3. *Calculate the friction loss in the hose.* This number is figured differently for each size of hose. The larger the hose, the less friction loss there will be. The incident commander must understand the relationship between available water supply, required flow, and friction loss to be sure the required amount of extinguishing agent is getting to the fire. Many different friction loss formulas are available to the firefighter. Rule of thumb, fireground formulas, and various combinations of formulas can be utilized. All of the formulas figure friction loss in 100 feet of hose. Therefore, the friction loss must be multiplied by the amount of hose used. A simple friction loss determination formula is $FL = Q(Q + Q + 1)$, where FL is friction loss and Q is gallons per minute/100. This formula is only applicable to $2\frac{1}{2}$-inch hose flowing greater than 100 gpm. To figure other sizes of hose, the formula must be multiplied by a factor (or by using a different formula). If the gallons per minute is under 100, the 1 in the formula becomes 0.5, or $\frac{1}{2}$. The gallons per minute are generally known by being aware of the type of nozzles being used. This information must be known prior to fireground use. The formula for determining GPM is shown in **Figure 3-16.**

■ **Note**
A simple friction loss determination formula is
$$FL = Q(Q + Q + 1).$$

Table 3-1 illustrates the factors and formulas required for the various hose sizes. **Table 3-2** illustrates the condensed Q formula where the factor does not have to be figured and the formula is uncomplicated. Many departments use the $Q(Q + Q + 1)$ formula for $2\frac{1}{2}$-inch hose and smaller and move to the condensed

Figure 3-16
*Formula to
determine gallons-
per-minute (gpm)
flow.*

$$29.7(d)^2\sqrt{NP}$$

In the formula, d = the diameter of the tip and NP = nozzle
pressure. In simpler terms, the formula reads 29.7 × the tip
diameter squared × the square root of the pressure.

Q formula for 3-inch hose or greater. Once all the numbers are known, they can
simply be added together to determine the required pump discharge pressure
(PDP) (see **Table 3-3**).

When everything comes together properly, the result is an amazing combination of efforts that extinguish the blaze. The incident commander understands

Table 3-1   *Friction loss factor for various hose sizes.*

Hose Size	Friction Loss Formula	Multiply by the Factor
1½	$Q(Q + Q + 1)$	13.5
1¾	$Q(Q + Q + 1)$	6.0
3	$Q(Q + Q + 1)$	.4
4	$Q(Q + Q + 1)$	.09
5	$Q(Q + Q + 1)$	.031

Table 3-2   *The condensed Q formula.*

Hose Size (inches)	Condensed Formula	Factor
3	Q^2	No factor
4	$Q^2 \div 5$	No factor
5	$Q^2 \div 15$	No factor

Table 3-3   *Pump discharge pressure.*

PDP = Pump discharge pressure is equal to	
NP	nozzle pressure (given) +
APPL	the appliances (given) +
ELEV (+ or −)	the elevation (given) +
FL	friction loss (figure) +

the required flow and chooses water as the extinguishing agent, the pump oper-ator takes the water from the system and adds the required pressure needed to overcome friction loss, the firefighter adds the correct nozzle, and the officer makes sure the stream is applied correctly. This teamwork is needed to effec-tively utilize water as an extinguishing agent. But what happens when water is not enough?

FOAM

Sometimes water alone is insufficient to mitigate a fire. Sometimes chemicals such as foam must be added to water (**Figure 3-17**). Foams are available to the fire service in many different forms and are created for different types of appli-cations, but generally they have been created for use on various hazardous materials fires and class A combustibles. The hazardous materials foams can be extremely specific, such as hazardous material vapor-mitigating foam or film-forming fluoroprotein foam, which can be applied at the base of a tank and allowed to float to the top of the product to extinguish the fire. This technique is known as subsurface injection. Foam is commonly used for two basic types

subsurface injection
the tactic of introducing foam at the bottom of a bulk storage tank, allowing the foam to float to the top and blanket the product

Figure 3-17 *At times, foam may be needed to extin-guish a fire or remove a hazard.*

hydrocarbons
any of numerous organic compounds such as benzene and methane that contain only carbon and hydrogen

polar solvents
water soluble compounds such as alcohol; require foams that are alcohol-type concentrates (ATC)

of hazardous materials: hydrocarbons and polar solvents. Hydrocarbons are petroleum-based products such as gasoline, benzene, and toluene, whereas polar solvents are alcohol-based products such as acids, ketones, and lacquer thinners.

Foam Components

Foam is created by using mechanical agitation to mix three ingredients together: air, water, and foam concentrate (see **Figure 3-18**). Water is the primary ingredient in foam: In fact, 90% or more of the foam is water. Foam concentrate is the raw foam product before water and air are added by mechanical agitation. Mechanical agitation, the method used to mix the ingredients of foam, comes from nozzles and foam makers. If any one of the components is removed, foam cannot be made. A simple demonstration of foam can be done in any kitchen. Fill a gallon jug half full of water and add liquid detergent. Three of the ingredients of foam—air, water, and foam concentrate—are together, but there is no foam because there has been no mechanical agitation. Place a cap on the jug and shake the jug. Now a foam blanket can be seen.

This simple demonstration does not address a critical factor when applying foam: proportioning the proper amount of foam concentrate for the specific type of fire. Foam is made up of 90% or more water; the other 10% or less is the foam concentrate. The specific amount is dependent on the use of the foam blanket. For example, most manufacturers of foam require that they be proportioned at 6% for polar solvents and 3% for hydrocarbons. This simply means that a polar solvent fire should be extinguished with a foam blanket made up of air, 94% water, and 6% foam concentrate, agitated to an effective foam blanket and applied correctly. If the product burning were a hydrocarbon, the foam blanket would be made up of air, 97% water, and 3% foam concentrate.

■ **Note**

Foam is created by using mechanical agitation to mix three components together: air, water, and foam concentrate.

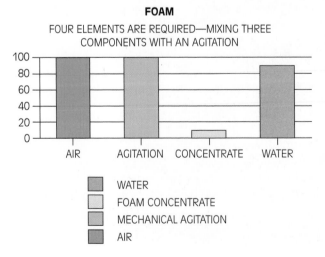

Figure 3-18   *The four components of foam.*

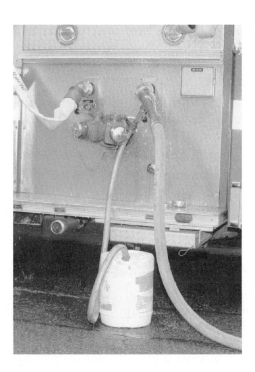

Figure 3-19 *In-line foam eductor.*

Numerous types of proportioners are available to the fire service. Some are in-line proportioners (**Figure 3-19**), such as those that are set up directly in the hose line and the proportioner used to suck foam out of a bucket. This is done by creating a vacuum at the proportioner through a venturi. The venturi is generated by the design of the proportioner and the water rushing by the pickup tube. Most manufacturers of proportioners establish a minimum psi required to create this venturi and thus vacuum the foam out of the bucket. Other proportioners include those built into the pump on the engine such as around-the-pump proportioners and built-in in-line proportioners. Each firefighter must be aware of the type of foam system in use in his or her department and be able to proportion and apply foam correctly.

Application of Foam

Application of foam depends on the type of foam used, the type of incident, the topographics, and the equipment used for the application. For foam to be effective as an extinguishing agent, it must be applied correctly. Hands-on training is critical to proper application. Training should be done regularly at the local level with the actual equipment that would be used in an incident. Additionally, application rates are important. Formulas are available to calculate proper application rates and to determine if the correct amount of foam is on hand to mitigate the incident. The following steps have been used with success in the fire service (on ignited spills):

1. Determine the size of the area burning ($L \times W$).
2. Multiply by the application rate of the foam being used. Aqueous film-forming foam (AFFF) and film-forming fluoroprotein foam (FFFP) is generally 0.10 for hydrocarbons and 0.24 for polar solvents. Check with the manufacturer for specific rates.
3. Multiply by the proportioned percentage (3% or 6%).
4. The result is the number of gallons of foam concentrate needed for 1 minute.
5. Multiply by the number of minutes required by standard operating guidelines.

Major Types of Foam

Many different types of foams are available for firefighting purposes: protein foams, film-forming fluoroprotein foam, class A foams, gold foam, aqueous film-forming foam, and others (see **Figure 3-20**). Many fire departments have moved to AFFF 3% and 6% foam, which is also referred to as AFFF alcohol-type concentrate (AFFF ATC), because it is applicable on both hydrocarbons (at 3%) and polar solvents (at 6%). The dual use of the foam makes it a very cost-effective product. Further, AFFF can be used on class A fires if proportioned very low such as 1% (check with the specific manufacturer for application rates).

Recently introduced to the fire service, gold foam acts as an AFFF ATC type foam but can be used on both polar solvents and hydrocarbons at 3%. This obviously allows the foam concentrate to last longer on polar solvent fires because not as much of the product is needed to create an effective foam blanket.

Figure 3-20 *Foam concentrate.*

Class A foams have seen huge growth in recent years. Their use has been primarily on wildland fires, but many departments have begun to use class A foam for interior structural firefighting. Class A foams work by breaking down the surface tension of water, thus allowing it to soak into class A combustibles better. Essentially, class A foam makes water wetter, so it is sometimes referred to as a wetting agent. Further, class A foam has a very low proportion rate, usually 1% or less.

A fire department does not need to purchase special equipment to use class A foam. The product can be added to the booster tank, mixed with water in a drop tank, or educted from drums. Standard nozzles can be used for the application of the class A foam. Different nozzles and application equipment can be purchased for special applications.

compressed air foam system
a foam system in which compressed air is injected into the foam solution prior to entering any hose lines

The use of a compressed air foam system (CAFS) produces high-quality foam bubbles that are small and consistent in size and density. These bubbles allow this foam to stay together, and provide a 25% better drain time for the foam, allowing the foam to work better and be more persistent. The foam system requires an air compressor to inject air into the foam solution as it is leaving the pump piping. This creates a turbulence inside the hose line that creates a better and more consistent foam. The CAFS allows the operator to control the consistency of the foam from a runny, wet solution for quick knockdowns to a thick, dry foam for vertical exposure applications.

DRY CHEMICALS

Dry chemical is another form of extinguishing agent available to the fire service. Generally used on flammable liquid fires such as grease and gasoline, it is very common in fire extinguishers and hood systems. Dry chemical extinguishers (**Figure 3-21**), can be made for use as A-B-C extinguishers (multipurpose) or B-C extinguishers (ordinary purpose.) **Table 3-4** shows the common chemicals used in the two types of dry chemical extinguishers.

■ Note
Generally used on flammable liquid fires such as grease and gasoline, dry chemicals are very common in fire extinguishers and hood systems.

In general, dry chemicals extinguish class B and C fires by interrupting the chemical chain reaction. On class A fires (if applicable), they coat the material burning and extinguish the fire through smothering.

DRY POWDERS

Dry powders are special purpose extinguishing agents used for specific applications such as combustible metals, class D fires. Combustible metals are not extinguished with normal extinguishing agents. Dry powders have been developed to extinguish them and are available in portable extinguishers and in raw

Figure 3-21 *A dry chemical extinguisher.*

Table 3-4 *Common dry chemicals.*

Ordinary Purpose (B-C) Extinguisher	Multipurpose (A-B-C) Extinguisher
Sodium bicarbonate	Monoammonium phosphate
Potassium bicarbonate	Barium sulphate
Potassium chloride	Ammonium phosphate

powdered form, which is applied by shoveling the agent on the burning material. For exact application rates and proper use, one should refer to manufacturer recommendations. Further description of dry powder agents and their application is found in Chapter 8.

OTHER EXTINGUISHING AGENTS

Carbon Dioxide

Carbon dioxide (CO_2) extinguishers are another type of extinguishing agent currently available for use to fight fires. They are used primarily on class B and C fires and come in the form of portable or wheeled extinguishers. The extinguishment is accomplished when gas discharged from the extinguisher smothers the fire.

Halon

Halon was available as an extinguishing agent in the 1980s and early 1990s but has been discontinued due to the potential damage to the atmosphere upon discharge. Halon systems, however, are still in existence and can stay in use as long as they are not discharged. Once discharged they must be replaced with a different type of agent.

Two types of halons were used and may still be seen in the field: Halon 1211 (bromochlorodifluoromethane), a liquid form of halon, and Halon 1301 (bromotrifluoromethane), a gaseous form. Halon 1301 was primarily used for protecting sensitive electronic equipment, such as computers and communication equipment, because it did not leave a residue.

All of the previously described extinguishing agents are presented in further detail in Chapter 8.

SUMMARY

An integral part of strategy and tactics is knowledge of the equipment available to firefighting forces. The prudent incident commander must be very knowledgeable of the extinguishing agents available and understand how and why they meet the objective of extinguishment.

Water is the primary tool of the fire service and most of the time it works to extinguish fires flawlessly. However, incident commanders need to understand why water sometimes does not work

and when the decision should be made to move to a different agent. Sometimes that decision comes prior to the incident. For example, when preplanning a business that uses combustible metal in its manufacturing process, the decision may be made early not to apply water to the products burning, but to use dry powders. At a minimum, the incident commander needs to understand that copious amounts of water are required for extinguishment of this type of material. Using an incorrect agent can cause additional property loss and, more important, loss of life or injury to firefighters.

A complete appreciation of the strengths and limitations of the various extinguishing agents is also important. Additionally, understanding the personnel and equipment required to move the agent to the desired location to extinguish the fire is an important part of setting up strategic goals and tactical considerations. An incident commander cannot simply list "extinguish the fire" as a strategic goal if he or she does not understand that a great deal of effort may go into the process. If it takes too much time, energy, and equipment to meet the objective, it may never be met. If he or she understands the process, then there can be a reasonable expectation for success. Understanding the process includes fundamental knowledge of extinguishing agents.

REVIEW QUESTIONS

1. Explain how extinguishing agents and fire behavior are closely related.

2. List three reasons why water is the primary extinguishing agent in the fire service.

3. True or False? Water is not compressible and will not take the shape of the container it is in.

4. List the three flow formulas and describe why two are better suited for fireground use.

5. True or False? A straight stream of water is created by a smooth bore nozzle.

6. True or False? A solid stream of water is created by a smooth bore nozzle.

7. Explain friction loss and why it is important to overcome this loss in the fire service.

8. Generally speaking, what psi do fog nozzles require at the nozzle? Smooth bore master stream nozzles? Smooth bore handline nozzles?

9. List the components that are required to create foam.

10. AFFF ATC is proportioned at what percentage for hydrocarbons? For polar solvents?

ACTIVITIES

1. Do a foam demonstration: Fill a gallon jug half full of water. Add liquid detergent. Mechanically agitate by shaking the jug. Look at the foam blanket. This demonstrates the production of foam through agitation of the ingredients of foam.

2. Calculate the flow requirement for your fire station (or any building) using all three formulas. Compare the results.

3. Calculate the pump discharge pressure in the following scenario: 300 feet of $2\frac{1}{2}$-inch hose with a 200 gpm fog nozzle.

4. Calculate the friction loss in 200 feet of 3-inch hose with a 300 gpm fog nozzle using the formula $Q(Q + Q + 1)$. Figure the friction loss in the same scenario using the condensed Q formula. Compare the results.

5. Determine the required amount of foam concentrate needed for 15 minutes on an ignited spill of gasoline 10 feet $\times$ 12 feet, using AFFF foam concentrate.

Chapter

4

INCIDENT MANAGEMENT SYSTEMS

Learning Objectives

Upon completion of this chapter, you should be able to:

- Describe the need for an incident management system.
- Identify the minimum requirements of an incident management system as set forth by the National Fire Protection Association (NFPA).
- Outline the major incident management components of the Fireground Command model, the NIIMS model, and the National Fire Service Incident Management System model.
- Describe how the National Incident Management System (NIMS) will improve coordination and integration of response agencies to an incident.
- Discuss the differences of terminology between FGC, NIIMS, and the National Fire Service Incident Management System models.
- Explain the need for and importance of standard operational guidelines.

CASE STUDY

On July 15, 1997, several southwest Florida fire departments responded to control a wildland fire that was 8 to 10 acres in size. An incident management system was in place, including an accountability system. Several wildland firefighting units along with a unit from the Florida Division of Forestry were operating to control and extinguish the fire, when it was noticed that a firefighter was missing. A few units were assigned to search for the missing firefighter, while attempts were made to try to contact the firefighter via the radio channel. When the radio attempts proved fruitless, a request for an air-recon unit was made to assist in the search. Shortly after the air-recon unit was in the sky over the area, the firefighter was spotted and ground units directed to him. The firefighter was found outside of the fire area with all protective clothing on with no sign of fire damage to it, yet he was unresponsive. Unfortunately, the firefighter was pronounced dead later that evening at an area hospital.

■ Note

An incident management system must be used on all incidents that involve more than several units so that safety, accountability, and operations can be controlled efficiently.

span of control
the ability of one individual to supervise a number of other people—usually three to seven, with five being ideal (the number depends on the complexity of the situation)—or units

■ Note

Span of control is variable depending on the type of incident.

INTRODUCTION

As described in the case study, even relatively small incidents involving just over a handful of fire department units can have disastrous results. It is imperative that an incident management system (IMS) be utilized on all incidents so safety, accountability, and operations can be controlled efficiently.

Organizations need to work within an incident management system for many reasons, covered in this chapter, but probably the most important reasons are that it provides flexibility and gives responsibility to select individuals for tasks and assignments, and it can create better accountability. Without an incident management system, firefighter safety and operations can become compromised. It is easy to become lost in the large scope of activities, and firefighters may not be found quickly enough when needing assistance; therefore, serious injuries and/or death can and have occurred.

There are many types of incident management systems. One item that they all have in common is that one person is in *command* of the entire incident. Command is the art of directing and controlling the resources and personnel assigned or requested to assist in the control of incident. Having such a system also helps the incident commander keep a manageable span of control. Typically, the normal span of control would be supervising anywhere from three to seven people (with five being rule of thumb). Of course this span of control is variable depending on the type of incident. For example, a hazardous materials entry team may require the span to be fewer people, while an overhaul of a room and contents fire with a lower risk could have a larger span.

Command procedures are designed to provide the framework to manage incidents with a strong command presence. With a strong command presence, integration of the efforts of other responding companies and agencies is more organized and practical while also providing for safer operation.

This chapter reviews and discusses the incident management system requirements of National Fire Protection Association (NFPA) 1561 and a few different types of incident management systems, along with their similarities and differences. This chapter will also cover how outside agencies can be integrated into the system and provide examples of standard operational guidelines that assist the incident commander prior to, during, and in the recovery phase of operations.

The three types of incident management systems to be described are Fireground Command (FGC), the National Fire Service Incident Management System Model Procedures Guide for Structural Firefighting, and the National Incident Management System (NIMS).

THE INCIDENT COMMANDER

The utilization of a single incident commander (IC) is a must for the operation to be effective. Without an incident commander and an incident management system, operations can, and will, break down. This may result in unsafe operations, increased life hazards to emergency personnel and occupants, increased fire losses, and a generally chaotic situation.

First and foremost, the incident commander must control the incident. An early, visible, and strong command presence starts the incident off in the proper direction. The IC should remain fairly stationary and increase visibility of the position by use of brightly colored vests or a specific color light to indicate the location of the command post. It is very important that the IC not become actively involved in the operation unless completely unavoidable. An IC who becomes actively involved loses the ability to see the overall incident picture and essentially becomes a company officer who is likely just seeing one piece of the puzzle. It is also imperative that the IC rely as much on what is being told or transmitted to him or her as on what it is that he or she is seeing, as conditions can vary widely depending on the size or scope of the incident.

Once command is established, the coordination of resources, corrective actions and/or action plans, safety of the overall incident, implementation of a supporting organizational (incident management) structure, and a standard communications flow, along with many other needs and responsibilities, need to be addressed, implemented, and utilized. It is very difficult to initiate a management system late in the incident and to change actions already taken.

There are many tools to assist the incident commander in the control of an incident, such as tactical worksheets (**Figure 4-1**), checklists, and elaborate incident management boards (**Figure 4-2**).

IMS WORKSHEET

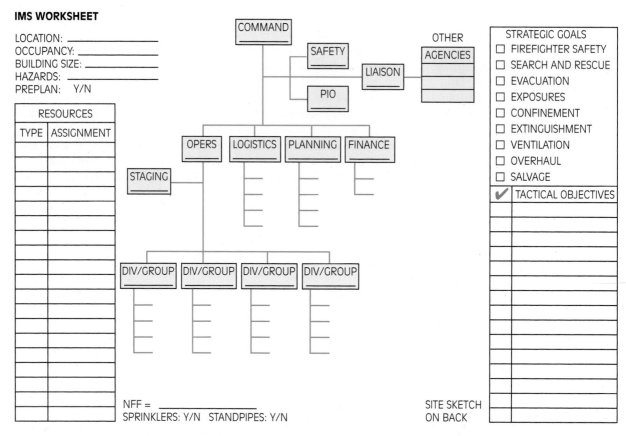

Figure 4-1 *One example of a tactical, or incident management, worksheet.*

■ **Note**

The purpose of any command system is to provide structure, coordination, and the integration of risk management, which in turn increases the level of a firefighter's safety and health, making the effort more efficient and effective.

NATIONAL FIRE PROTECTION ASSOCIATION (NFPA) STANDARD 1561

Within the NFPA 1500 series is NFPA 1561, *Standard on Fire Department Incident Management Systems.* This standard provides an outline of the minimum requirements of an incident management system to be used by all fire departments. The purpose of any command system is to provide structure, coordination, and the integration of risk management, which in turn increases the level of a firefighter's safety. It is also important to point out that this is the standard that the United States Department of Homeland Security (DHS) has adopted as its model.

    Fire departments must develop a command system with written policies and guidelines that define and describe the administration of the system, the system

Figure 4-2 *An incident command board.*

structure, the system components, and the roles and responsibilities of positions within the system.

The system structure shall describe implementation, communications, interagency coordination, command structure, training and qualifications, personnel accountability, and rehabilitation. It shall be utilized for all emergency incidents and, depending on the scope, shall be utilized for drills, exercises, and other situations that involve similar hazards as those encountered at actual emergencies. Standard operating guidelines that mandate the use of plain English for radio communications provide standard terminology for all types of incidents that will reduce the amount of, or eliminate, the confusion often created when radio codes are used. This system shall provide for the integration, coordination, and cooperation of outside agencies that may be involved in an emergency incident. The structure shall provide for a series of levels of supervision and an effective span of control to be utilized depending on the scope of the incident. All

Figure 4-3   *Vests used to identify positions in the IMS.*

personnel who may be involved in and/or expected to perform as the incident commander shall be trained in the system.

An important part of the structure is personnel accountability. Accountability helps maintain constant awareness of the location and function of personnel and helps keep track of all personnel who may enter or leave the hazard area. As part of accountability, the development of a standard way to alert and/or evacuate personnel of an imminent danger and account for them is required. The system structure shall also take into account the need for rest and rehabilitation.

The components of the system shall include and describe the duties and functions of the incident commander, command staff, planning, logistics, operations, staging, and finance/administration (see **Figure 4-3**).

■ **Note**

The Fireground Command system was developed in the Phoenix (Arizona) Fire Department.

FIREGROUND COMMAND

The Fireground Command system became popular in the late 1970s and has been widely used since. It was developed in the Phoenix (Arizona) Fire Department and gained national attention after the publication of the book *Fire Command* by Chief Alan Brunacini. Many other incident management systems are based on this system. This system works well for nearly all day-to-day incidents, although it is a little difficult to expand on for those large, unusual, and multi-agency responses.

Fireground Command (FGC) utilizes three basic levels:

1. *Strategic.* This is the role of the incident commander. At this level the incident goals and objectives are developed, priorities are set, resources are allocated, and control of the overall incident is maintained.

2. *Tactical.* In this system this is the position of a *sector officer* assigned to specific operational areas by the FGC. The sector officer directly supervises an area to ensure tactical objectives are carried out, evaluates the effectiveness of operations, monitors the situation area, and is the position that communicates with the FGC.

3. *Task.* This is the company level, where the physical functions take place to produce the desired outcome and meet the desired objective. Company officers report to the FGC or sector officer.

sector
a geographic area or function established and identified within the IMS for operational purposes

The Fireground Command system utilizes the term sector. Sectors should be named by using a standard system that clearly identifies and defines their locations or activities by using plain language. In this system, a sector may be assigned geographically, such as roof, front or back, or floor number within a structure. Sectors may also be assigned by function, such as safety, rehabilitation, and medical.

A typical incident management structure (IMS) for a small incident or fire is shown in the example in **Figure 4-4.** The typical IMS for a large incident is shown in **Figure 4-5.** As you can see, this is an easy system to work within because it does not use a lot of different terms, which may cause confusion.

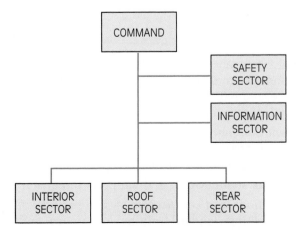

Figure 4-4   *A Fireground Command system organizational chart for a room and contents fire.*

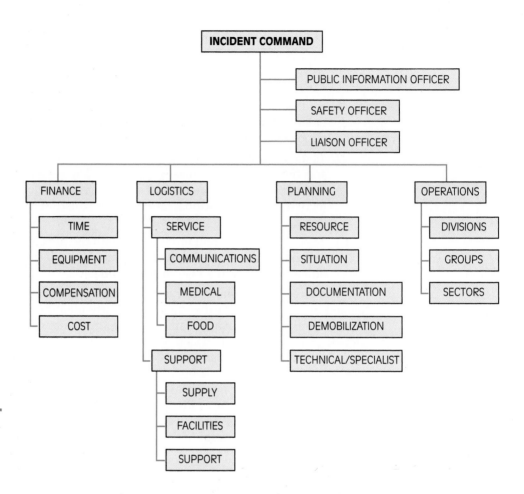

Figure 4-5 *An expanded IMS that would be used at a large incident.*

The National Interagency Incident Management (NIIMS) was developed to provide a common system that emergency service agencies can utilize at local, state, and federal levels. NIIMS consists of five major sub-systems that collectively provide a total systems approach to all risk incident management.

1. *Incident Command System (ICS)* includes operating requirements, interactive components, and procedures for organizing and operating an on-scene management structure.

2. *Training* is standardized and supports the effective operation of NIIMS.

3. *Qualifications and certification system* provides for personnel across the nation meeting standard training, experience, and physical requirements to fill specific positions in the Incident Command System.

4. *Publications management* includes development, publication, and distribution of NIIMS materials.

5. *Support technologies* include satellite remote imaging, sophisticated communications systems, geographic information systems, and so on, that support NIIMS operations.

The Incident Command System (ICS) was developed through a cooperative interagency (local, state, and federal) effort. The basic organizational structure of the ICS is based upon a large fire organization that has been developed over time by federal fire protection agencies; however, the ICS is designed to be used for all kinds of emergencies and is applicable to both small day-to-day situations as well as very large and complex incidents.

COMMON TERMINOLOGY

It is essential for any management system, and especially one that will be used in joint operations by many diverse users, that common terminology be established for the following elements:

Organizational functions: A standard set of major functions and functional units has been predesignated and named for the ICS. Terminology for the organizational elements is standard and consistent.

Resource elements: Resources refer to the combination of personnel and equipment used in tactical incident operations. Common names have been established for all resources used within ICS. Any resource that varies in capability because of size or power, for example, helicopters, engines, or rescue units, is clearly typed as to capability.

Facilities: Common identifiers are used for those facilities in and around the incident area that will be used during the course of the incident. These facilities include such things as the command post, incident base, and staging areas.

MODULAR ORGANIZATION

The ICS organizational structure develops in a modular fashion based upon the kind and size of an incident. The organization's staff will build from the top down, with responsibility and performance placed initially with the incident commander. As the need exists, four separate sections can be developed, each with several units that may be established. The specific organizational structure established for any given incident will be based upon the management needs of the incident.

If one individual can simultaneously manage all major functional areas, no further organization is required. If one or more of the areas requires independent management, an individual is named to be responsible for that area. For ease of reference and understanding, personnel assigned to manage at each level of the organization will carry a distinctive organizational title:

- Incident Command: "Incident Commander"
- Command Staff: "Officer"
- Section: "Section Chief"
- Branch: "Branch Director" (optional level)
- Division/Group/Sector: "Division/Group/Sector Supervisor"
- Unit: "Unit Leader"

In the ICS, the first management assignments by the initial attack incident commander will normally be one or more section chiefs to manage the major functional areas. Section chiefs will further delegate management authority for their areas only as required. If the section chief sees the need, functional units may be established within the section. Similarly, each functional unit leader will further assign individual tasks within the unit only as needed.

CONSOLIDATED ACTION PLANS

Every incident needs some form of an action plan. For small incidents of short duration, the plan need not be written. The following are examples of when written action plans should be used:

- When resources from multiple agencies are being used
- When several jurisdictions are involved
- When the incident will require changes in shifts of personnel and/or equipment

The incident commander will establish objectives and make strategy determinations for the incident based upon the requirements of the jurisdiction. In the case of a unified command, the incident objectives must adequately reflect the policy and need of all the jurisdictional agencies. The action plan for the incident should cover all tactical and support activities required for the operational period.

MANAGEABLE SPAN-OF-CONTROL

Safety factors as well as sound management planning will both influence and dictate span-of-control considerations. In general, within the ICS, the span-of-control of any individual with emergency management responsibility should range from three to seven with a span-of-control of five being established as a rule of thumb. Of course, there will always be exceptions (for example, an individual medical crew leader will normally have more than five personnel under supervision).

Just as in other incident management systems, the NIIMS was designed for and mandates a unified command. A unified command, as discussed in the section later in this chapter, is extremely important if an incident is to be handled effectively, because incidents do not come to an abrupt halt at jurisdictional boundaries. Within a unified command structure, keeping a manageable span of control is a must in order to have an effective, efficient, and safe operation.

This system encourages the predesignation of possible command post locations, incident bases for the performance of various support activities, possible staging locations, possible helibases and helispots, task forces (typically five unlike resources), strike teams (typically five like resources), and available single-unit resources.

The five major functional areas of this model are command, operations, planning, logistics, and finance. Depending on the size and scope of an incident, any or all of these components may be implemented. Also depending on the incident, each of these components may have a command staff assigned to and working for it (**Figure 4-6**).

The five functions begin with Command. The incident commander is responsible for all activities of the overall incident. The next functional area, Operations (**Figure 4-7**), is the position responsible for the management of all tactical objectives and activities. Therefore, the operations officer needs to have direct involvement in the development of action plans. The Planning section (**Figure 4-8**) is responsible for many areas beginning with the development of action plans. Along with plan development, planning must document and disseminate the plan information to all involved in the incident. Planning also collects

unified command
the structure used to manage an incident involving multiple jurisdictions or multiple response agencies that have responsibility for control of the incident

command
the highest level of responsibility and authority in the IMS at an incident

operations
part of the general staff of the IMS, responsible for all operational functions

planning
part of the general staff of the IMS, responsible for all incident planning functions

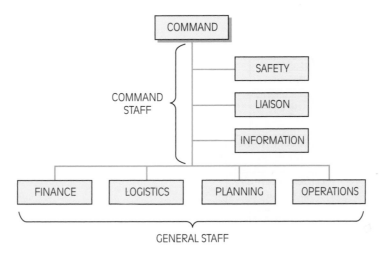

Figure 4-6 *IMS— Command and general staff positions.*

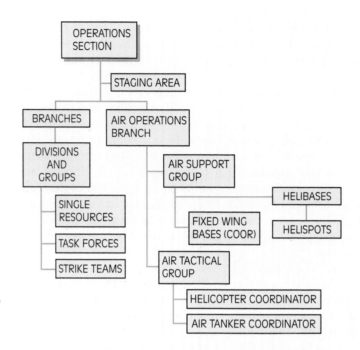

Figure 4-7   *The Operations section.*

and evaluates tactical information. It is extremely important for planning to maintain current information and records on the situation, publish possible and/or potential situation changes, and keep a current status of resources.

Logistics is responsible for all nonoperational support needs and activities (see **Figure 4-9**), including facilities (sleeping quarters, bathrooms, showers), transportation, supplies, equipment and apparatus maintenance, fuel for equipment

logistics
part of the general staff of the IMS, responsible for all logistical needs and supplies

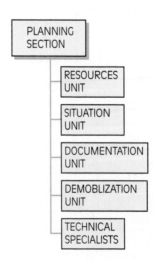

Figure 4-8   *The Planning section.*

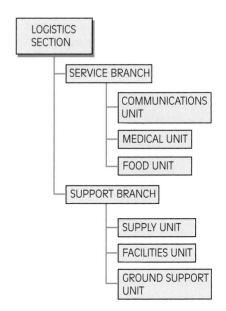

Figure 4-9 *The Logistics section.*

and apparatus, food for all involved in the incident, communications (radio, phone, computers, network), and medical services for the incident responders.

On very large or long duration incidents, it may be necessary to implement the last section of the system, which is Finance (see **Figure 4-10**). This position is responsible for incident cost analysis, tracking the time of equipment and personnel, procuring food and supplies by contract services, and processing the associated documentation.

Depending on the incident, each section may have its own command staff to assist and fill the roles of the various positions.

finance
part of the general staff of the IMS, responsible for all financial matters

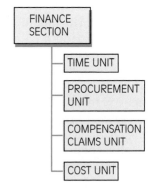

Figure 4-10 *The Finance section.*

■ **Note**

A division is established and named by geographical areas, such as the third floor of a structure being termed division three.

■ **Note**

Groups are established and named by a specific function that they are to accomplish.

The support positions for the main sections are then assisted as they are further broken down into branches, divisions, groups, and/or sectors.

A branch would generally be established when the number of divisions or groups has exceeded a manageable span of control. A branch could also be implemented when there are at least two specific and distinctively different operations that need to take place at the same time, such as fire suppression, hazmat, emergency medical services (EMS), and evacuation. When a branch is implemented, a branch manager is assigned and manages a number of divisions/groups, task forces, strike teams, or single-unit resources. A task force is typically a combination of five unlike resources, such as two engine companies, two ladder companies, and a heavy rescue. On the other hand, a strike team is a combination of five like resources, such as five engine companies or five ambulances.

To reiterate, those personnel assigned to operate within divisions and groups are the actual units that are physically trying to control the incident. A division is established and named by geographical areas, such as the third floor of a structure being termed division three. Assigned a division supervisor, a division may consist of a single unit, multiple single units, task forces, and strike teams. Groups are established and named by a specific function that they are to accomplish. For example, a need for ventilation would warrant the implementation of a ventilation group. Like divisions, groups have a supervisor that may manage single units, several single units, task forces, and strike teams.

NATIONAL INCIDENT MANAGEMENT SYSTEM (NIMS)

The National Incident Management Systems (NIMS) is a structure for management of large-scale or multijurisdictional incidents and is taking the place of NIIMS. It is being phased in at the federal, state, and local levels. Eventually, any jurisdiction seeking federal Homeland Security grant money will have to demonstrate that it is NIMS-compliant.

NIMS is the first standardized approach to incident management and response. In Homeland Security Presidential Directive-5 (HSPD-5), President George W. Bush called on the Secretary of Homeland Security to develop a national incident management system to provide a consistent nationwide approach for federal, state, tribal, and local governments to work together to prepare for, prevent, respond to, and recover from domestic incidents, regardless of cause, size, or complexity.

The Department of Homeland Security released NIMS in March 2004. NIMS integrates effective practices in emergency response into a comprehensive national framework for incident management. The NIMS will enable responders at all levels to work together more effectively and efficiently to manage domestic incidents no matter what the cause, size, or complexity, including catastrophic acts of terrorism and natural disasters.

Federal agencies also are required to use the NIMS framework in domestic incident management and in support of state and local incident response and recovery activities.

The benefits of the NIMS system will be significant:

- Standardized organizational structures, processes, and procedures
- Standards for planning, training, and exercising
- Personnel qualification standards
- Equipment acquisition and certification standards
- Interoperable communications processes, procedures, and systems
- Information management systems with a commonly accepted architecture
- Supporting technologies such as voice and data communications systems, information systems, data display systems, and specialized technologies
- Publication management processes and activities

The NIMS Integration Center was established by the Secretary of Homeland Security to provide strategic direction for and oversight of the National Incident Management System, supporting both routine maintenance and the continuous refinement of the system and its components over the long term. The NIMS Integration Center is a multijurisdictional, multidisciplinary entity made up of federal, state, local, and tribal incident management and first responder organizations. It is situated in the Department of Homeland Security (DHS).

The NIMS Integration Center will facilitate the development and dissemination of national standards, guidelines, and protocols for incident management training. They will facilitate the use of modeling and simulation in training and exercise programs and define general training requirements and approved training courses for all NIMS users.

In 1980, federal officials transitioned ICS into a national program called the National Interagency Incident Management System (NIIMS), which became the basis of a response management system for all federal agencies with wildfire management responsibilities. Since then, many federal agencies have endorsed the use of ICS, and several have mandated its use. An ICS enables integrated communication and planning by establishing a manageable span of control.

The ICS divides an emergency response into five manageable functions essential for emergency response operations: Command, Operations, Planning, Logistics, and Finance and Administration: The NIMS recognizes the National Wildfire Coordinating Group (NWCG) ICS training as a model for course curricula and materials applicable to the NIMS:

- ICS-100, Introduction to ICS
- ICS-200, Basic ICS
- ICS-300, Intermediate ICS
- ICS-400, Advanced ICS

The USFA's National Fire Academy and Emergency Management Institute both follow this model in their ICS training curricula.

UNIFIED COMMAND

Each of the system models described in this chapter mandate and depend on a unified command structure. When multiple agencies are involved in the mitigation of an incident, they must get together if the situation is to be handled effectively and efficiently. Unified command affords a strong, direct, and visible command along with a clear management framework in which incident functions and objectives are established. Through this unified command, the members must all agree on and develop the incident priorities, objectives, and overall strategy.

The actual physical command post can be within any of the jurisdictional boundaries as long as all jurisdictions are involved in the operation. There must be one command post from which all activities of the incident are managed. Without it many problems can and will occur. For example, a large wildland fire that covers or threatens multiple counties and the cities within them would quickly have many different agencies involved in the incident. If each of the different agencies had its own plan of attack, the situation would soon become chaotic, with potentially dangerous actions taking place, the possibility of duplication of effort, and the misuse of resources. With a unified command, competition between agencies and freelancing can be limited and corrected.

The use of a unified command structure makes the integration of outside agencies possible. An outside agency is not just another fire department, but could include representatives of local government, the military, the federal government, and many others.

STANDARD OPERATIONAL GUIDELINES

A key to the proper implementation and utilization of an incident management system is the development of standard operational guidelines (SOGs) **(Figure 4-11)**. Each organization must decide on an incident management philosophy and it must be supported throughout the organization.

SOGs, which spell out the standard course of action that should be taken and expected on every incident, are the foundation of an incident management system. SOGs help ensure an effective and efficient operation. In a perfect fire service world, all departments would utilize the same basic SOGs, which in the case of automatic and mutual aid is an absolute must.

FIRE DEPARTMENT STANDARD OPERATING GUIDELINE   No. 100

SUBJECT: RESPONSE, PLACEMENT, AND COMPANY FUNCTIONS

This S.O.G. provides guidance for units as they respond and position for operations at emergency scenes. It also establishes the primary assignments for units to begin work prior to Command becoming fully functional.

Response:

All units will respond in "emergency status" (warning devices activated) to all dispatches unless the unit determines otherwise based on pre-arrival information. Units responding from quarters shall clear the station within 60 seconds from 0700 to 2200 and within 90 seconds from 2200 to 0700. When in emergency status, apparatus shall follow the quickest route possible and maintain a safe speed considering the conditions and posted speed limit; personnel shall remain seated and wear provided restraints. Drivers shall use the utmost care and pay extra attention to safety when moving. All emergency units shall conform to State Law for response safety regulations. Responding units shall maintain radio contact on the appropriate channel with Dispatch and advise when on-scene. **Units shall not contact Command while en route to request assignment. Arriving first alarm units shall follow standard placement assignments (Level 1 staging) unless advised by Command to do otherwise.** Second and greater alarm units shall stage at a remote site from the scene (Level II staging) and notify the Staging Officer (if established) or Command of location and status. All Level II staged units should be at the same location. Personnel shall stay with their respective units in staged status until given an assignment through the Command system. Units not used in function positions (pumping, laddering, lighting, etc.) shall be parked out of the way.

Standard Placement (Level I Staging)

Upon arrival at the scene units shall position their vehicles based on standard practice, maximum effectiveness and safety. Placement of units at situations other than fire alarms, structure fires and smoke investigation will be covered in other incident-specific S.O.G.'s.

- **First Arriving Engine Company** - This unit shall be placed at the front of the fire structure, slightly to one side (to allow truck placement), or in best position based on size-up factors to begin suppression or investigation activities.

- **First Arriving Truck Company** - Unit to be placed at front of structure in position to reach roof surfaces for ventilation.

<u>Exception #1</u> Multi-story occupied structure with smoke or fire showing, placement shall be to enable rescue from upper floors.

Figure 4-11 *An example of an SOG for response, placement, and company functions.*

Exception #2 Large volume fire showing (already vented) and/or threatening exposures, placement shall be to enable establishment of aerial fire stream.

Exception #3 Mobile home parks or small structures (under 1,400 sq. ft.), placement shall be 250 ft. from fire structure out of way of engines laying lines but accessible for equipment or defensive operations, unless unit is used for other functions.

Exception #4 If a truck company is first due, it will follow first arriving engine functions, **if it has pumping capabilities**

- **Second Arriving Engine Company** - Placed in position to provide water supply upon direction of either Command or first arriving engine.

- **Third Arriving Engine Company** - Placed in a location to best utilize the company as a Rapid Intervention Team/Group (RIT/RIG).

- **First Arriving Squad** - Unit placement determined by outdoor light conditions; twilight and dark, unit shall be placed to allow lighting of primary suppression area. This can be from adjacent property or from behind first arriving engine. Elevated lighting enables unit to be outside of other units. Unit should also be placed to allow access to electrical cords and SCBA refill station while keeping a safe distance from suppression work areas. Must not block access of other function units (engines and trucks) to scene. In daylight, placement shall be primarily concerned with electrical cord access and SCBA refill safely removed from suppression efforts.

  First Arriving Command Unit.-.Placement shall be in position to afford view of structure and suppression area if possible and safe. Visibility of Command unit to other operating units is important.

  First Arriving Rescue - Placement will be 75 ft. from fire structure, out of the way of functioning units but accessible for equipment or EMS operations.

  First Arriving Ambulance - Placement shall be 300 ft. from fire structure out of way of functioning units but accessible for equipment or EMS operations.

  First Arriving Utility - Placement shall be in Level II staging.

  First Arriving Rehab Unit - Placement shall be with the ambulance, awaiting direction from Command.

  Company assigned to Rapid Intervention shall maintain readiness and visual contact with command or sector/division officer.

Figure 4-11
(Continued)

Subsequent arriving support, staff or command units shall be placed away from function areas and out of way of access. If possible, all units should be placed on the same side of the street or in a position that maintains an open access lane large enough for apparatus to get into or out of the function area. Size-up factors may indicate the need for the first unit on scene or Command to direct incoming units to other than standard placement.

Staging

Staging is established either by Command or by the first arriving unit (unassigned) en route as part of a second alarm. A large area (safe and secure, if possible) shall be identified at least 600 ft. from the function area for the assembling of incoming units, prior to assignment by Command. The first arriving unit is responsible for either assuming staging or passing staging to another unit. The next unit must acknowledge receipt of assignment. The staging officer shall maintain a list of units and personnel, control ingress and egress from staging, and communicate with command. Individual units in staging shall not communicate with command.

Standard Company Functions

In order to coordinate initial efforts on-scene, the following standard functions and priorities are assigned by Company. These standard functions are to be carried out automatically by the assigned Company until alternative assignments or stop orders are issued by Command.

- **First Arriving Engine**

    1.   Size-up or investigation
    2.   Forcible entry (if required)
    3.   Search and rescue
    4.   Placement of initial attack line
    5.   Establishment of own water supply (if necessary)

- **Second Arriving Engine**

    1.   Locate and establish initial water supply
    2.   Provide manpower to supplement secondary attack lines or rescue operations
    3.   Support sprinkler and standpipe operations if applicable (SOP #600-18)

- **Third Arriving Engine**

    1.   Establish RIT/RIG
    2.   Establish water supply if not already completed by second engine
    3.   React as directed by command

Figure 4-11
(Continued)

- **First Arriving Truck**

    1.   Size-up for ventilation method or aerial operations
    2.   Assist in search and rescue (if requested)
    3.   Position and set up tower (commercial or multi-story)
    4.   Prepare and position personnel to effect ventilation and communicate readiness to Command

- **First Arriving Squad**

    1.   Set up scene lighting (if needed)
    2.   Secure utilities (electric and gas) to fire structure
    3.   Assist truck company (if requested)
    4.   Secure scene with banner tape, if needed
    5.   Set up SCBA refill station

- **First Arriving Rescue**

    1.   Perform life safety and patient care activities as needed
    2.   Assist first-in engine with forcible entry, ventilation, etc.

- **First Arriving Ambulance**

    1.   Stand by for assignment from Command
    2.   Maintain visual contact with scene and anticipate possibility of patients

- **First Arriving Rehab**

    1.   Set up for rehabilitation activities as directed by Command

- **First Arriving Utility**

    1. Stage and await assignment from Command.

All companies shall be prepared to assume alternate duties, assignments or roles, depending on size-up factors and direction from Command. When possible, companies shall keep personnel together and function as a team. A team will consist of a minimum of two personnel. A company deviating from standard functions shall communicate the intent to do so to Command.

Figure 4-11
(Continued)

As stated earlier, every fire department must develop and utilize SOGs. SOGs should cover all aspects that may be encountered in an incident response including the following:

- An explanation as to by whom and how an incident management system is initiated and implemented
- A definition of the incident management structure and all of its components
- An explanation of the dispatching and communications system (including standard terminology) (see **Figure 4-12**)
- Safety
- Descriptions of company functions
- Outlines of the guidelines for tactical priorities
- Explanation of how personnel accountability is maintained

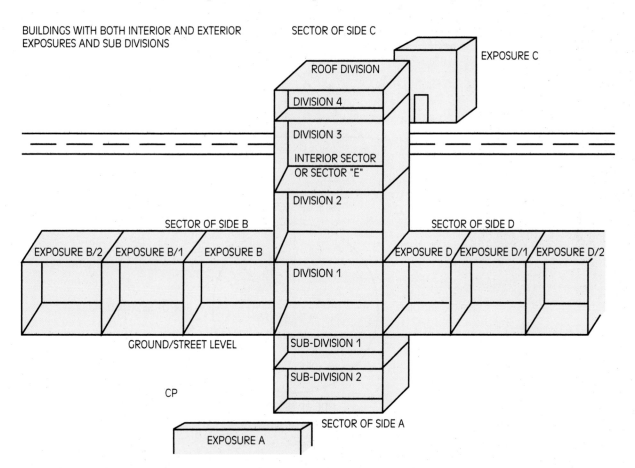

Figure 4-12   *IMS systems require the fire building and exposed areas to be divided into smaller units. Functional tasks may also be divided.*

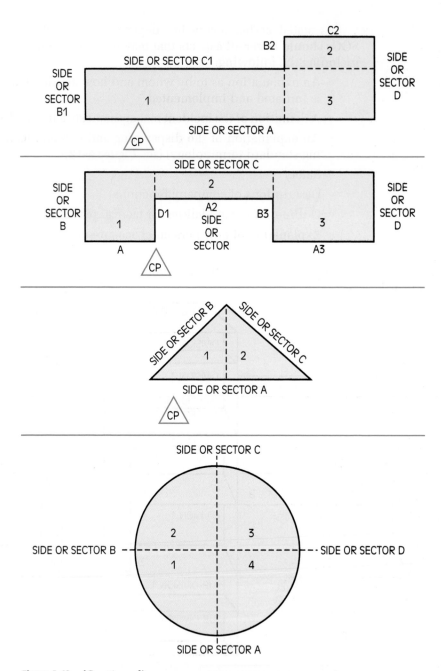

Figure 4-12   *(Continued)*

This list is just the beginning. Representatives of all the levels and areas of the organization should assist in the development of SOGs. After all, who is better to help develop a SOG on ventilation than a member of a ladder company? Also, if all areas are involved, any new guidelines will be more widely accepted and practiced. SOGs must be written, practiced, simple, attainable, official, applicable to all situations, and enforced.

When the foregoing areas have been addressed, an organization is on its way to more effectively controlling an incident. It is imperative that all members of the organization receive copies of, and training on, all of the SOGs. They cannot be expected to follow the SOGs if they do not know what they are and how to apply them.

■ **Note**
SOGs need continual review and revision, for as time passes and technology advances, they may become outdated and unneeded.

Another important consideration regarding SOGs is that they need continual review and revision, for as time passes and technology advances, they may become outdated and unneeded (**Figure 4-13**). Remember, SOGs must be applicable. It is also imperative that the incident management SOGs are utilized and practiced on a daily basis. Full-time IMS should be required, no matter how minor the situation. When the system is constantly utilized and practiced during minor incidents, the firefighter will be better prepared and have a better understanding of how to expand the system when the larger, more complex incidents occur. It will become second nature.

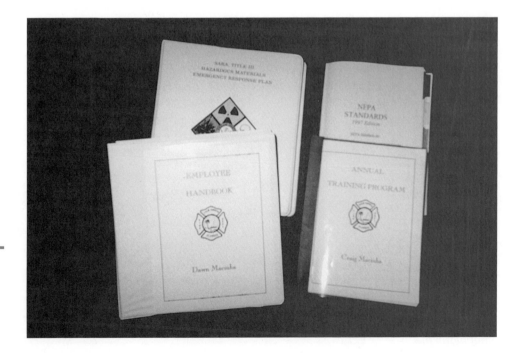

Figure 4-13 *SOGs must be kept up to date and current with trends, regulations, and standards.*

SUMMARY

Three popular incident management system models are described in this chapter. There are many different types of systems, along with many that have used one of the three models described herein as a core for their systems but have tweaked it a bit to fit local needs and desires.

NFPA Standard 1561 (see **Figure 4-14**) now mandates the use of an incident management system, but long before NFPA required the use of such a system, fire departments and other agencies saw the need for it. Large and long-term incidents involving multiple companies and multiple agencies pointed out some glaring faults that needed to be, and could be, addressed through the use and implementation of an IMS.

An IMS is more than just a diagram on a sheet of paper. SOGs need to be developed to mandate its use, describe its various parts, and support all of its sections and positions. All agencies and personnel that may be involved in an incident must have copies of the plan and related information; they must be trained on the use of the system; they need to know where they fit into the system; and they need to know who can implement it and how it is implemented.

It is imperative that the system be implemented very early in the incident (beginning with the first unit on scene). An incident commander must be visible and have a strong presence. When large incidents that involve multiple fire companies and/or multiple outside agencies occur, companies and agencies must come together to form a unified command structure. This unified command will help ensure the proper utilization of resources along with a safer, more efficient and effective effort. Remember, a unified command structure still mandates that there be only one incident commander.

Figure 4-14 *NFPA standards require the use of an incident management system. The IMS is necessary at all incidents, large and small. This figure shows a staging area for a multialarm fire.*

In conclusion, as stated earlier, it is imperative that an incident management system be utilized. The National Incident Management System (NIMS) is mandated by the federal government, and to help ensure this system is adopted, an agency must meet the requirements of NIMS to be eligible for federal funds. This IMS will work well for any organization and when utilized by other agencies allows for a seamless system to control any incident (more information can be obtained on the Web at http://www.fema.gov/emergency/nims/nims_compliance.shtm).

REVIEW QUESTIONS

1. What does the term *command* mean?

2. Why must only one person be in command of an incident?

3. What NFPA standard outlines the minimum requirements of an incident management system?

    A. 1561

    B. 1001

    C. 1403

    D. 1901

4. What are the three basic levels of an incident management system?

5. Define span of control and explain how it is applied.

6. Name the six main components of the National Incident Management System (NIMS).

7. Outline the responsibilities of the planning section.

8. Define the terms *branch, division, group, sector, strike team,* and *task force.*

9. Why are standard operating guidelines (SOGs) such an important component of any incident management system?

10. A team of firefighters sent to the roof to perform roof ventilation might be identified in the IMS as

    A. Roof sector

    B. Ventilation group

    C. Roof division

    D. Any of the above might be appropriate

ACTIVITIES

1. Draw a basic incident management system that would work with your organization and describe its components.

2. Review your department's SOG for incident command. Which system do you use? Do surrounding departments use the same system? What improvements may be necessary?

Chapter

5

COORDINATION AND CONTROL

Learning Objectives

Upon completion of this chapter, you should be able to:

- Describe why effective fireground communication is necessary.
- Discuss the process necessary to ensure effective fireground communication.
- Define size-up.
- List the components of an effective size-up.
- Define operational mode.
- List and describe the three incident priorities.
- Define strategic goals.
- Define tactical objectives.
- Define tactical methods.
- Describe action planning.
- Compare the relationship of incident priorities, strategic goals, tactical objectives, and tactical methods.
- Discuss the role of preincident planning to the overall strategic and tactical plan.
- Discuss the concept of recognition-primed decision making.

CASE STUDY

On October 24, 1997, at 0619 hours, the District of Columbia Fire (DCFD) and Emergency Medical Services (EMS) Department dispatched E-22 and T-11 to investigate an odor of smoke in the area of Fourth Street and Kennedy Street NW. While these units were responding, the communications center received additional phone calls reporting a building fire, and it subsequently dispatched a full box assignment of E-22, E-24, E-14, E-11, T-11, T-6, Battalion Chief 4, and Rescue Squad 2.

Upon arrival, E-22 reported smoke from the first floor of a grocery store, made a hydrant layout, and stretched an initial attack line. The crew of T-11 assisted the E-22 crew with forcible entry and positioned the aerial ladder for vertical ventilation. Engine 22 crew inside the first floor of the store moved about and located some fire that appeared to be gas fed. Engine 14 arrived and entered the first floor of the building to assist in fire confinement and extinguishment. Sergeant John M. Carter commanded the E-14 crew and entered the building with E-14's lineman and 200 feet of preconnected 1½-inch hose.

Crews from various other companies were attempting, with little success, to enter the basement, where most of the fire was located. As conditions on the first floor began to deteriorate, crews on that floor were ordered to evacuate. During this evacuation, some firefighters became separated from their crews and officer, and some became disoriented. However, all made it to the front exit and exited safely except E-14 officer Sergeant Carter. Once it was determined that Sergeant Carter was missing, aggressive search and rescue operations were undertaken. However, despite the heroic efforts of the firefighters, Sergeant Carter was not found until approximately 0830. He was transported to the hospital with cardiopulmonary resuscitation in progress by a DCFD medic unit; however, he could not be revived.

After this incident, a reconstruction committee was formed to investigate in order to identify and propose remedies to problems and deficiencies that occurred at this fire to prevent reoccurrence. After conducting thorough research and more than 98 hours of interviews with members of the DC Fire and EMS Department, the Reconstruction Committee identified five major areas of concern and developed recommendations in each of these areas: accountability, command procedures, communications, operations, and safety.

Source: This case study was adapted from the report of the Reconstruction Committee to the DCFD. The entire report is available on the Internet at www.dcfd.com. The fire service thanks the committee and the DCFD for sharing this information so that similar tragedies might be prevented.

INTRODUCTION

It is essential that fireground commanders (**Figure 5-1**) and firefighters have an understanding of fire scene coordination and control.

Chapter 4 discussed various incident management systems and standard operating guidelines. Other considerations that need to be studied to ensure safe, effective fireground operations and control of the incident include an effective communication process, effective size-up, incident priorities, strategic goals, tactical objectives, tactical methods, action planning, and preincident planning.

You probably have heard the phrase "no two fires are alike," but this statement is not entirely true. Because of their chemical and physical properties, fires usually behave in the same manner and therefore certain behavior—for example, fire spread—can be expected. Further, although buildings differ greatly, fires in similar buildings allow the incident commander to expect certain outcomes. For example, a fire in a row frame house that has gotten into the attic can be expected to spread throughout the cockloft. Another similarity is the application of the incident priorities and applicable strategic goals to bring the incident under control.

■ Note

Because of their chemical and physical properties, fires usually behave in the same manner and therefore certain behaviors;—for example, fire spread— can be expected.

Figure 5-1 *The incident commander is responsible for the overall incident management, regardless of the phase of the incident.*

COMMUNICATIONS

Having a good communication system is of primary importance in the development of strategy and tactics. This section is not intended to examine the many communication systems available, but it describes the essential steps in a good communication process.

The meaning of words in a communications process is based on prior understanding and knowledge. One can quickly see the ineffectiveness of fireground operation when common meanings of words are not known. Fireground terminology may depend on geographic areas or training received within an individual department. For example, what are fire hydrants called in your department? Plugs? Hydrants? What is a loom up? On the West Coast of the United States, fire service personnel might know that a loom up is the column of smoke produced by a fire, but this term is seldom used in East Coast departments. During the Florida wildfires in the summer of 1998, responders from Florida found out when responders from the West arrived that an engine may not be a class A pumper, but instead something the Florida firefighters would commonly call a brush truck (as shown in **Figure 5-2**). Clearly, common meanings and previous understandings can have a significant impact on the fireground operation, particularly if mutual aid is used in the area extensively.

Impact of previous knowledge and understanding is apparent in areas that rely on codes for communication. These codes or signals, often ten codes, should be replaced by a system of plain language. This argument, like the argument over fire apparatus color, has gone on for years. However, when responders from

communication systems
radios, computers, printers, and pagers, the numerous hardware and software that goes into a communications network

■ **Note**

The meaning of words in communications is based on prior understanding and knowledge.

■ **Note**

When responders from different departments, and sometimes different agencies, combine into the incident management system, there is no room for error due to lack of understanding of codes or signals.

Figure 5-2   *A wildfire firefighting unit. In most places this unit would not be thought of as an engine.*

Figure 5-3 *The six-step communications model as presented by the National Fire Academy.*

different departments and sometimes different agencies combine into the incident management system, there is no room for error due to lack of understanding of codes or signals.

Once the considerations of previous understanding and knowledge have been addressed, the firefighters, officers, and commanders must focus on the communication process. The National Fire Academy course "Managing Company Tactical Operations" introduces a six-step communications model (**Figure 5-3**) that can be applied to any emergency service communications and should be adapted and trained on for fireground communication.

In Step 1 in the communications model, the sender formulates an idea to be conveyed to another person or persons. The idea must be clear and concise, and in the case of fire scene communication, relatively brief. Because of the need to be brief, training of personnel in incident priorities, strategic objectives, and tactical methods can actually help improve the communication process. For example, a sender may formulate the thought of roof ventilation, specifically a trench cut between exposure B2 and B3. Although much more information may be in the sender's mind regarding this communication, all that might have to be said to the crew performing the operation is "Truck 1 trench cut the roof between exposures B2 and B3." There would be no need to discuss the specific methods of this operation, because that would have been previously trained on. In Step 1, an appropriate means of communication must also be considered.

Step 2 in the communication model is the sender actually sending the message. First, the sender must get the receiver's attention and then convey the information. In the fire service, this is commonly done via the radio. Getting attention is usually done by calling the unit number or by using an incident management system (IMS) division, group, or sector supervisor, for example, "Command to Truck 1" or "Command to Division 2." In some areas, the person being called is said first, for example "Truck 1 from Command." This choice is department or area specific, and particular advantages can be cited for both systems.

Step 3 is the message being transferred through a medium. The medium can be oral, visual, or written. Although most fireground communication is visual or oral, some functions, such as complex assignments or communications, for long-duration incidents involving complex action plans, and transferring of command in large incidents requires the communication to be in writing.

Step 4 is the receiver receiving the message, which immediately leads to Step 5, message interpretation. For the receiver to understand the message in the intended manner, the receiver must have background knowledge and experience to understand the message.

Step 6 is the receiver providing feedback to the sender that ensures that the message has been received in a manner that causes the receiver to understand the message. In the previous example, Truck 1 should respond by saying "Truck 1 copies, trench cut the roof between exposures B2 and B3." In essence, during Step 6 the receiver becomes a sender at Step 1.

Communication is necessary throughout the emergency incident. In the following sections and throughout the text, specific needs for communications are discussed.

■ **Note**
Communication is necessary throughout the emergency incident.

size-up
a decision-making process that starts before the incident and allows the firefighter or incident command to gather information and develop appropriate strategies

SIZE-UP

The next factor that should be considered is size-up, which can be defined as a step toward solving a problem in which information is gathered (**Figure 5-4**). It is an evaluation of critical factors that can begin before the incident and continue throughout. The incident priorities, as discussed in the next section, are not affected by specific incidents. However, through the size-up process, the incident priorities are applied to a specific situation.

So size-up is an ongoing information analysis that begins before the incident occurs and is specific to a particular incident. What factors should be examined during this information-gathering phase? Several fire service leaders have discussed their views of this concept in a number of texts. Acronyms are often used to help remember the areas of information that must be gathered.

All the information needed for size-up can fit neatly into three areas: the environment in which the incident occurs, the resources available, and the conditions, or the situation. A size-up triangle depicting these three areas is shown in **Figure 5-5**. These three areas have a number of subcomponents that should be examined at every incident. Some may be more applicable in some areas than others. Each area is discussed in the following paragraphs.

■ **Note**
Through the size-up process, the incident priorities are applied to a specific situation.

Figure 5-4 *Upon arrival, a fire officer continues the size-up process specific to the incident at hand.*

Environment

The environment in which the incident takes place includes the construction features of the fire building or, in the case of a wildfire, the terrain where the fire occurs. It also includes factors of time, weather, fuel load, height, area, building occupancy, and access to the fire area. **Table 5-1** lists further the areas that relate to the environment and the possible implications to the incident.

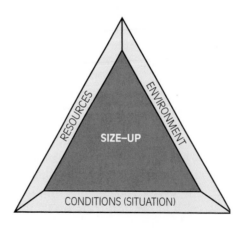

Figure 5-5 *The size-up triangle.*

Table 5-1 *Size-up considerations with examples.*

Environment	Resources	Conditions/Situation
Time	Personnel responding	Fire location
Building construction features	Personnel needed	Fire extension probability
Weather	Apparatus responding	Type of fuel
Height	Apparatus needed	Life hazard
Area	Water supply	Property conservation concerns
Occupancy	Built-in private fire protection	Possible incident duration
Access	Need for special extinguishing agents	
Terrain	Length of time for resources to be on scene	

Resources

■ Note

In evaluating the resources, the number of personnel, type of equipment, and need for specialized equipment are determined.

In evaluating the resources, the number of personnel, type of equipment, and need for specialized equipment are determined (**Figure 5-6**). As these needs are determined, an evaluation of what is available must also be undertaken. It would not make good tactical sense to plan for a three-tanker water shuttle for supply if only two tankers were available to the incident. Resources available

Figure 5-6 *The incident commander must consider which type and how many resources are needed to handle the emergency. At the first World Trade Center attack, many resources were needed from various agencies.*

for tactical objectives that require timeliness are also important. The incident commander must remember that the fewer personnel available to perform a trench cut, the longer it will take. Table 5-1 also lists further the resource considerations.

Conditions/Situation

As a part of size-up, the firefighter must consider the conditions and the current situation. Basically an analysis should be designed to answer four questions: What is on fire, where is it now, where is it going, and what harm has it caused or will it potentially cause? As with the environmental and resource considerations, Table 5-1 lists considerations for analysis of the conditions.

As can be seen, size-up is a conscious process that considers many variables. Not all of the considerations listed in Table 5-1 are applicable in every instance; however, each should be ruled out just to make sure all are considered.

INCIDENT PRIORITIES

Three separate functions come under the heading of incident priorities: life safety, incident stabilization, and property conservation. These functions should be the driving force for all incidents and the first items considered by the incident commander. Not only should they be considered first, they must be constantly evaluated until the incident terminates. The incident commander uses common goals to develop measurable tactical objectives that ultimately satisfy the incident priorities.

Life Safety

The first priority at any incident is maximizing life safety. Life safety involves the activities that ensure that the threat of injury or death to responders and civilians is reduced to the minimum possible. This is most easily done by limiting the exposure of danger to people to the absolute minimum. At a fire situation involving a room and contents, it may only mean that proper IMS is in place and safety and health considerations for the firefighters are in place, such as the use of proper protective clothing, an accountability system, or the placement of a rapid intervention crew. In a large wildland/urban interface fire, it may mean the evacuation of hundreds or thousands of people. Regardless of the extent of activity necessary, the incident commander must have life safety as the number-one priority at all times, at every incident.

Incident Stabilization

The second incident priority is the activities that are performed in order to solve the problem or bring the situation under control, stopping further damage. Some might expect this to be called fire extinguishment. However, because this system can and should be used for a multitude of different emergencies, *incident stabilization* is the preferred term.

Incident stabilization at a room and contents fire may require nothing more than a well-placed, coordinated interior attack, whereas at a larger wildfire, it may require the coordination of mutual aid fire departments and federal and state agencies in order to satisfy incident stabilization. Regardless of the resource commitment, incident stabilization is the incident commander's second priority. The appropriate mode of operation must be employed for incident stabilization. These modes are described in the following list:

- *Offensive attack mode.* When firefighters move close to the fire area to extinguish the fire (**Figure 5-7**), they are in an offensive attack mode. Generally, this would be considered an interior fire attack with handlines in the building and allows for several strategic goals to be accomplished. Firefighters are in close contact with the fire and exposed to all the inherent dangers, including burns, falling objects, being lost or overcome with toxic gases, and building collapse.

- *Defensive attack mode.* A defensive attack (**Figure 5-8**) is employed when the volume of fire does not permit an offensive attack, when insufficient resources are present to effect a safe offensive attack, or when the fire

Figure 5-7 *An offensive fire attack generally requires the firefighters to enter the fire building.*

Figure 5-8 *A defensive fire attack on a garden apartment building.*

building contains chemicals, explosives, or other known hazards, or the building being structurally unstable. Generally, a defensive attack indicates a writeoff of the fire building, with confinement and extinguishing efforts directed at the exposed building. However, in some cases, a defensive attack may precede an offensive attack. For example, a fire in a storefront of a five-story apartment building may be knocked down with master streams from a defensive position but would be followed by an offensive attack mode for checking the upper floors for extension. Defensive attacks are normally accomplished with large-caliber exterior streams and large handlines. Firefighters are outside the immediate area of fire danger and collapse zones are established. Personal protective equipment is still required. Dangers such as building collapse, falling objects, falling wires, and exposure to heat and fire products still exist.

- *Attack mode changes.* Can the operational mode change? It can and many times does. Chapter 6 discusses firefighter safety compared to risk versus gain. One reason to change operational modes might be a change from life-saving operations to property conservation. Another might be a switch from offensive to defensive because the offensive actions are not controlling the fire. Further, building construction materials can fail early after

■ Note
The third and last priority involves the activities performed in order to reduce the property loss caused by the incident.

■ Note
At a room and contents fire, property conservation may involve placing a few salvage covers or moving furniture and, with input from the occupants, removing valuables, whereas property conservation at the wildfire may involve land restoration after the fire by outside agencies.

■ Note
Life safety, incident stabilization, and property conservation: This order *never* changes.

exposure to heat. Truss roofs fail, floors and walls collapse. It is the responsibility of the incident commander to consider through size-up the situation and the risk versus the gain achieved through proper offensive attack. Whatever the reason for the change in operational mode, it is critical that *all* personnel at the incident know the current operational mode and are accounted for at the switch.

Property Conservation

The third priority involves the activities performed to reduce the property loss caused by the incident (**Figure 5-9**). In the firefighting arena these activities are often called salvage and overhaul or loss control. Whatever the name or activity, property conservation efforts should be directed at reducing the loss to property and the long-term health and welfare of the community. At a room and contents fire, property conservation may involve placing a few salvage covers or moving furniture and, with input from the occupants, removing valuables, whereas property conservation at the wildfire may involve land restoration after the fire by outside agencies.

Although the third priority, property conservation, is often the one that is overlooked, the incident commander must make this priority part of the plan and commit adequate resources to it as soon as resources are available.

Figure 5-9 *There are various ways to satisfy the property conservation priority.*

■ **Note**

Although priorities should always be considered in their order of importance, nothing rules out performing one or more simultaneously.

■ **Note**

Strategic goals are broad, general statements of what we are expected to accomplish.

REVAS
a fire incident management goal set: rescue, evacuation, ventilation, attack, salvage

RECEOVS
a fire incident management goal set: rescue, exposures, confine, extinguish, overhaul, ventilation, salvage

■ **Note**

Nine goals must be considered at every fire incident to ensure that the incident priorities are satisfied: firefighter safety, search and rescue, evacuation, exposure protection, confinement, extinguishment, ventilation, overhaul, and salvage.

Priority Order versus Order of Accomplishment

The incident priorities are arranged in order of importance: life safety, incident stabilization, and property conservation. This order *never* changes. However, sometimes the order of accomplishment may seem to differ.

For example, suppose an engine company arrives on the scene of a fire in a two-story apartment building with flames showing from one window on the first floor and numerous people at second-floor windows awaiting a rescue. The engine officer orders the crew to take a handline to the fire apartment and extinguish the fire. Poor action, right? Should have taken care of the first priority of life safety before incident stabilization, right? The answer, of course, is no on both questions and a good illustration of priority order versus order of accomplishment. Had the engine officer elected to rescue the people on the second floor, the fire would have extended further, reducing the chances for survival of others in the building. By attacking the fire, an incident stabilization action, the engine officer actually provided for the life safety priority. Of course, the best system would have been for multiple units to arrive at the same time and some could have performed rescue while the others performed incident stabilization. However, this best case is most likely not the actual case.

Remember that although these priorities are always considered in their order of importance, nothing rules out performing one or more simultaneously. After all, placing an exposure line between two buildings really is an effort at incident stabilization on the fire building and property conservation on the exposure building.

The incident commander, having a good solid understanding of the incident priorities and by using a good size-up, is able to apply the strategic goals presented in the next section.

STRATEGIC GOALS

Strategic goals are broad, general statements of what we are expected to accomplish. Strategic goals are related to incident priorities because the goals are developed in order to satisfy the incident priorities.

Over the years, a number of goal systems have been developed. These systems focus around five to seven goals from a firefighting perspective. The generally accepted firefighting goals include rescue, attack, exposure protection, evacuation, confinement, extinguishment, overhaul, salvage, and ventilation. Two common systems are REVAS (rescue, evacuation, ventilation, attack, salvage) and RECEOVS (rescue, exposure, confinement, extinguishment, overhaul, ventilation, salvage). The National Fire Academy's course Hazardous Materials Operating Site Practices lists eight strategic goals for hazardous materials operations. These eight goals are isolation, notification, identification, protection, spill control, leak control, fire control, and recovery and termination. Emergency medical service system responders are trained in six strategic goals for medical incidents including

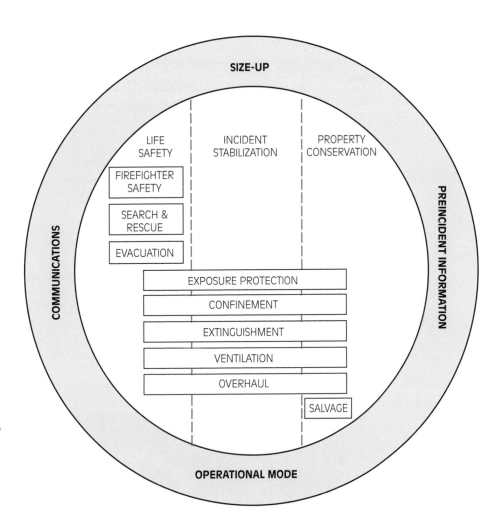

Figure 5-10 *The relationship of strategic goals to the incident priorities.*

gaining access, triage, stabilization, extrication, packaging, and transporting. In any of these goal systems, not all goals require an action on every incident, nor are the goals necessarily accomplished in the same order on every incident.

For the purpose of this text, nine goals must be considered at every fire incident in order to ensure that the incident priorities are satisfied: firefighter safety, search and rescue, evacuation, exposure protection, confinement, extinguishment, ventilation, overhaul, and salvage. These goals may be accomplished in different orders based on the situation; however, each is related specifically to one or more incident priorities. **Figure 5-10** shows the relationship between the incident priorities and the nine strategic goals. **Table 5-2** provides a brief description of these goals.

Not every incident requires an action to satisfy each goal; however, each goal must be considered at each incident. Further, size-up factors also help to

■ **Note**
Size-up factors also help to determine the strategic goals that need to be implemented or that can be implemented.

Table 5-2 *Nine strategic goals.*

Goal	Description
Firefighter safety	Design processes to minimize the risk to firefighters on the incident scene.
Search and rescue	Ensure that the occupants are located and removed from the incident area. Rescue implies that the firefighters must assist the occupants from the structure or area.
Evacuation	Remove persons who can help themselves from an area of danger. Evacuation implies that, although directed by firefighters, the persons being evacuated can do so on their own.
Exposure protection	Protect surrounding properties from fire extension.
Confinement	Limit the spread of a fire and keep it to the smallest area possible.
Extinguishment	Extinguish the fire.
Ventilation	Remove smoke, heat, and fire gases from the building.
Overhaul	Ensure complete fire extinguishment and search for hidden fire extension. See **Figure 5-11**.
Salvage	Limit property loss from the fire.

Figure 5-11 *Overhaul is a strategic goal that can satisfy one or more of the incident priorities.*

determine what strategic goals need to be implemented or can be implemented. For example, it may not be possible to implement the strategic goal of search and rescue at a fully involved house fire, at least not until some incident stabilization actions have occurred. In this case, the extinguishment goal would be satisfy both incident stabilization and life safety.

The strategic goals presented here and the tactical objectives that follow form the framework for the approach to the incidents described later in the text.

TACTICAL OBJECTIVES

■ **Note**

As strategic goals are designed to meet the incident priorities, tactical objectives are more specific functions designed to meet strategic goals.

As strategic goals are designed to meet the incident priorities, tactical objectives are more specific functions that are designed to meet strategic goals. Tactical methods, introduced in the next section, meet tactical objectives.

Whereas strategic goals are broad statements of actions needed to solve the problem, tactical objectives are much more specific and have measurable results. For example, to meet the broad goal of ventilation, one tactical objective might be to vertically ventilate the roof. Clearly, this goal is specific in that it describes the action and the location where the action should take place, and it is measurable in that either the smoke ventilates out the roof or it does not. If the latter is the case, then there is a problem with the tactical method employed, but the objective is still valid.

Another example might be an objective relating to the exposure goal. In this case tactical objectives might include the placement of large-volume hose streams on the exposed building to reduce the effects of radiant heat, or advancing a $2\frac{1}{2}$-inch attack line into the exposed building to prevent extension from brands.

One can quickly see the relationship in the goals and objectives. **Table 5-3** provides examples of tactical objectives related to strategic goals. This list is illustrative, not exhaustive. The strategic goals should all be given consideration at every fire; however, the tactical objectives could change based on size-up and the situation.

TACTICAL METHODS

■ **Note**

There is often more than one tactical objective for each goal; there is often more than one tactical method for each objective.

Just as strategic goals satisfy incident priorities and tactical objectives satisfy strategic goals, tactical methods satisfy tactical objectives. Tactical methods are the processes employed at the task level (**Figure 5-12**). For example for the strategic goal of ventilation, the tactical objective might be to provide vertical ventilation to the structure. The tactical method employed might be to cut a hole in the roof (see **Figure 5-13**). The objective is the measurable result, in this case vertical ventilation, whereas the tactical methods are the tasks employed to reach the objective. As there is often more than one tactical objective for each goal, there is often more than one tactical method for each objective.

Table 5-3 *The strategic goals and possible tactical objectives.*

Strategic Goals	Tactical Objectives
Firefighter safety/health	Implement accountability system.
	Assign safety officer.
	Provide for a rapid intervention team.
Search and rescue	Conduct primary search/location.
	Conduct secondary search/location.
	Conduct tertiary search/location.
	Rescue occupant.
	Provide for EMS needs.
Evacuation	Alert building occupants.
	Assist building occupants from the area as needed and as resources permit.
	Provide safe area of refuge.
	Provide EMS as needed.
Exposure protection	Place exposure lines to reduce radiant heat.
	Provide brand patrol/location.
Confinement	Place hose line of sufficient GPM flow on unburned side of the fire.
Extinguishment	Place a hoseline from the unburned side of the fire and apply a direct/indirect attack.
Ventilation	Provide natural vertical ventilation.
	Provide positive pressure ventilation (PPV) with the PPV fan from the front door and remove windows in the fire rooms.
Overhaul	Open the walls/ceiling/floor in the fire area and check for extension.
	Use a heat sensor to locate hot spots in the walls.
Salvage	Protect the interior contents on the first floor.
	Remove undamaged valuables from the fire area.

Figure 5-12 *An example of the relationship between strategic goals, tactical objectives, and tactical methods for firefighters' safety.*

Figure 5-13 *Vertical roof ventilation operation.*

The discussion of tactical methods is a course or text in itself. Many of these procedures are taught as part of basic firefighter training. The text *The Firefighter's Handbook* from Delmar is an excellent text for the basic firefighting tactical methods that a firefighter must perform. **Table 5-4** gives some examples of the tactical methods to be employed for the strategic goal of firefighter safety and related tactical objectives.

PLAN OF ACTION

■ Note
Developing the plan of action, or action planning, should occur at every incident. However, the incident determines the complexity of the action plan.

Developing the plan of action, or action planning, should occur at every incident. However, the incident determines the complexity of the action plan. For example, at a small room and contents house fire the action plan is probably nothing more than the incident commander's worksheet, whereas at a large multijurisdictional, long-duration wildfire, the action plan is written and formalized.

The action plan is developed after the size-up, or information gathering process, and gives consideration to the strategic goals and tactical objectives that have been identified. Once identified, the goals and objectives must be communicated to the operating firefighters. This can be accomplished with the action plan. The action plan organizes the course of action designed to bring the incident to conclusion.

Table 5-4 *Examples of tactical methods related to tactical objectives and strategic goals.*

Strategic Goal	Tactical Objective	Tactical Method
Firefighter safety/health	Implement accountability system	Have first-arriving drivers on each side of the building set up accountability boards.
		Each arriving unit gives passport tag to the accountability location prior to entering hazard zone.
		Both rehabilitation sector and staging sector will maintain accountability boards for crews in their respective sectors.
	Assign Safety Officer	Assign the on-call safety officer upon arrival at the scene.
		Safety officer should size-up the building and discuss operations or safety concerns with incident commander.
		The safety officer, using the fire scene worksheet, will ensure minimization of threats to crews in the hazard zone.
		Safety officer will ensure that rehabilitation needs are met by assigning a rehabilitation sector as necessary.
	Provide for rapid intervention	Assign the second-arriving ladder company as the rapid intervention crew (RIC).
		RIC to size-up the building and determine need for additional RICs.
		RIC to stage at a location with all necessary equipment available.

■ **Note**

In developing the action plan, the incident commander must consider not only the goals and objectives required but also department SOGs.

In developing the action plan, the incident commander must consider not only the goals and objectives required but also department standard operating guidelines (SOGs). In many cases, the SOGs might have standard actions that must be a part of the incident action plan. For example, on a two-engine, one-ladder, and one-battalion chief response to a single-family dwelling fire, the SOGs might state that the first engine will proceed directly to the scene and establish fire confinement and extinguishment using tank water, while the second engine establishes a water supply line, and the ladder performs search and rescue and ventilation. The battalion chief in this case has predesigned tactical objectives for each of the first-responding units. At anytime as the situation dictates, the officers of those units should be able to make exceptions to the SOGs and communicate the exception to the incident commander. The incident commander then may change the action plan as needed.

The action plan must be fully communicated to all personnel at the scene. This can be accomplished by several means, as follows:

- *Face-to-face communication.* This method is the most effective; however, often time does not allow for this type exchange. It requires the incident commander to meet face-to-face with on-scene officers and communicate the action plan (**Figure 5-14**).

Figure 5-14
*Operations at the
incident command
post. (Photo
courtesy of Deputy
Chief [Retired]
James Olson.)*

- *Use of aides or runners.* Aides or runners can assist with action plan communication via face-to-face communication with various officers in different sectors, divisions, or groups.
- *Radio communication.* The action plan and the assignment of objectives can be done via radio. It is important to practice the good communication steps described in the first section of this chapter.
- *Written plan.* Although often reserved for complex or long-duration incidents, the action plan can be written with specific assignments and copies provided to the companies to which the assignment or objectives are being made. Common forms should be used in written action plan communication.

■ **Note**
In the case of a fire, the preincident planning process allows the fire department to fight the fire before it happens.

PREINCIDENT PLANNING

The value of preincident planning can not be emphasized enough, both from an operational standpoint and a safety standpoint. Having preincident plans (see **Figure 5-15**) available is like a coach having the play book at a game. In the

Figure 5-15 *The preincident plan should be available to the incident commanders and the company officers.*

case of a fire, the preincident planning process allow the fire department to fight the fire before it happens. The preincident plan should provide the incident commander with the need-to-know information in a form that allows for quick recognition.

The preincident planning process will give you the tools and knowledge to become a more effective incident commander. If you do not have a preincident plan, you are going into an emergency situation "blind." The preincident plan provides information on the building or facility on paper or in a computer file to use when you respond to an incident.

By gathering information on a building, you can determine the construction methods, the strengths and weaknesses of the building, and the processes that occur inside the facility. Preincident planning helps your department to make better command decisions because important information is assembled before the emergency occurs. This can mean the difference between a successful incident or a failure because of unknowns.

One person should be assigned the responsibility of overseeing the preincident planning process. The process needs to be standardized so that all plans contain the same information and are laid out in the same manner for ease of reading and finding information. This consistency in standardized forms, inspection methods, and record keeping will result in a system that works.

The preplanning process begins with the identification of the buildings to preplan. The preplans are typically assigned to the first-due company in the response area. Upon receiving their assignments, the officer will schedule dates and times to conduct the preplan.

Using the standardized forms, the company officer and personnel will walk the exterior of the property to determine any access problems. Through Internet access, the officer can download satellite pictures of the buildings and grounds before they go to the site. They will identify hydrant locations and any exterior access to fire protection systems. They may photograph certain areas for use in training sessions at a later date. After exploring the exterior of the building, the officer will meet with the owner or manager and begin the interior exploration. The term *inspection* should be avoided, as company fire inspections and company preplanning are separate activities. The person to escort you through the building should be the building maintenance supervisor; he or she will typically know the building from top to bottom, inside and out.

Using the standardized form, the company officer will gather all the information necessary. Remember that too much information can be overwhelming. As you walk through the structure, imagine what it would be like to have to move through the structure in dark smoke, without lights, and so on. Identify and point out to your personnel the dangers of certain areas and what to look for when moving through the building.

Identify the building protection systems: What type of fire protection is there? Where are the exits? Are interior and exterior doors marked for easy identification? Do fire doors close automatically? How should the building be ventilated? Where are the utilities controlled? What are the required fire flows? Thoroughly analyze all features of the building and the processes that are operating inside.

Once the information has been gathered, it is sent to the preplanning department where the information is documented. The hazards are identified, potential building problems and other concerns can be predicted, and solutions can be detailed on the plans. From this information, standardized response plans can be set in SOGs for specific target hazards. Resources can be predetermined and alternate response plans can be formulated.

Once completed, the preincident plan becomes an excellent training tool. Use the preincident plan as a training tool to run companies through simulated exercises. Use overhead projectors to display pictures of the building and a layout of the plot plan and building. There are many computer simulator programs that allow you to place smoke, fire, or hazmat incidents in the picture. Train all companies that are listed on the first- and second-alarm assignments.

To be effective, the preincident planning process should have the following components:

- The preincident plan should be on a form used department-wide.
- The preincident planning should be done by the responders so that they can become familiar with the building during the planning.
- The process should provide for updating the plan at given time intervals, for example, annually. Remember occupancies change and renovations occur.

Figure 5-16   *The fire department should make every effort to preplan any target hazard within its jurisdiction.*

- Target hazards (**Figure 5-16**) should be identified and get priority in the planning process. Target hazards may include:
  - Health care facilities
  - Large industrial occupancies
  - Facilities using or storing hazardous materials
  - High-rise buildings
  - Malls or other high-occupancy locations, such as office complexes or stadiums
  - Schools
  - Hotels, motels, and apartment buildings
- The preplan should include text as a reference, site plan, and floor plan. Photographs are also useful.
- The preplan should show or provide the following information:
  - Address
  - Construction features
  - Building emergency contacts
  - Occupancy type and load
  - Hazardous materials or processes

- Location of utilities
- Location of shafts
- Location of entrances and exits
- Possible ventilation locations
- Nearest water supplies
- The need for specialized extinguishing agents
- Private fire protection equipment and systems on site, including fire department connections
- Fire flow requirements
- Fire behavior predictions and anticipated problems
- Possible placement of apparatus on a first-alarm assignment
- Mutual aid plans

One of the first requirements of preplanning is that a common form be used department wide. **Figure 5-17** shows a sample form that meets these requirements.

■ **Note**
One of the first requirements of the preplanning process is that a common form be used department wide.

RECOGNITION-PRIMED DECISION MAKING

Does the firefighter or incident commander actually go through a standard decision-making process each time he or she arrives on an emergency scene? Does the incident commander review, for each emergency, strategic goals and tactical objectives? The answer to both questions is both yes and no. A process has been researched by the Klein Association, termed recognition-primed decision making (RPD). This process was developed out of research that asked the question of how fireground commanders, military commanders, and other high-pressure, time-sensitive decision makers make decisions. The RPD model implies that a fireground commander in almost nine of ten cases (90%) does not make a decision based on a selection of choices, but rather based on previous experience. Further, the option or method selected is based on whether it has worked effectively in prior, similar situations.

■ **Note**
The RPD model implies that a fireground commander in almost nine of ten cases (90%) does not make a decision based on a selection of choices, but rather based on previous experience.

The implication of this research from a strategic goal and tactical objective standpoint is that incident commanders and firefighters must have a mix of experience and education to operate effectively. The application of the strategic goals and tactical objectives must be learned and practiced in real situations or through simulations so the incident commander can recognize similar situations and know immediately the proper action to take. But the incident commander must also be ready to adapt as additional information becomes available or the situation changes. The importance of postincident analysis is also underscored when examining this process. As the incident commander employs certain tactical objectives that do not work, the analysis points out those areas of weakness and allows for future correction.

ANY TOWN FIRE DEPARTMENT PREINCIDENT PLAN	
Building Address: Business Name:	Emergency Contact Name: Phone Numbers:
Building Description: Construction Type: Roof Construction: Floor Construction:	
Occupancy Type:	Hazards to Personnel:

Water Supply Location #1:	Available Flow:	GPM
Water Supply Location #2:	Available Flow:	GPM
Water Supply Location #3:	Available Flow:	GPM

Estimated Needed Fire Flow Based on a Single Story Involvement				
Level of Involvement	25%	50%	75%	100%
Est. Needed Fire Flow				

Fire Behavior Predictions:

Anticipated Problems:

Private Fire Protection:
Sprinklers: ☐   Standpipes: ☐   Fire Alarm/Detection: ☐   Other: ☐ (Note Below)

Completed by:	Date:	Last Update:

Figure 5-17 *A sample preincident planning form.*

NATURALISTIC DECISION MAKING

The RPD process is being replaced in the fire service curriculums by what the National Fire Academy is teaching in its Command and Control class as naturalistic decision making (NDM).This decision-making process is concerned with how real or ideal decision makers make their decisions. The NDM theory is concerned with identifying the best alternative to take, and assumes the decision maker is fully informed, rational, and able to compute the information at hand.

The NDM embodies a descriptive decision theory framework based upon the person's knowledge and experience, level of complexity, and the environment.

NDM came about after reviewing decision making as it happens to people in real life—from military commanders to fireground commanders. The theory states that people make decisions based on experience. The incident commander (IC), through his or her years of training, incident responses, and formal study, has built a knowledge base upon which he or she can make decisions. The information is embedded in the brain as if it were a computer hard drive. When incident commanders are confronted with an incident, they will associate the current incident with what has been learned through past experience, training, studies, and lessons learned by others.

ICs will associate former incidents involving similar situations and consider what has worked and what has not worked in the past. If it is an incident they have never confronted in the past, they can effectively break the situation down into bits of information they are familiar with, such as a hazmat incident for which they can only base their decisions on information received in training.

In NDM, the incident commander relies on his or her ability to recognize signs such as smoke movement and color, type of structure, visible location of the fire, and others to begin the decision-making process. He or she knows from past training how the command sequence needs to be followed and that a 360-degree walk around the fire building will provide the most information. This decision-making process ties past incident knowledge to what can be observed. Based upon this information, strategies can be determined and tactics decided upon to bring the incident under control.

The NDM theory is that the IC will use information based upon sight, sound, and odor to assist in identifying what is occurring. ICs will analyze this information and react in the most appropriate way based upon their past experience. The IC must use both her recognitional knowledge base (experience) as knowing about situations and her procedural knowledge base (SOG) as to knowing how to react in certain situations. The emphasis here is that people must be trained in the same manner as they will be required to function in the field. Realistic training exercises are essential.

CLASSICAL DECISION MAKING

The classical decision-making model is being taught by the National Fire Academy in its Command and Control curriculum. This model is used when the IC is confronted with a situation that has not been experienced in the past. He has no recognitional knowledge base upon which to draw information. The IC must then use his ability to identify recognizable cues and their procedural knowledge base to make strategical and tactical decisions. The classical decision-making model is based on a four-step process by which the IC can make his decision.

1. *Aim* is what the IC wants to accomplish, based upon the incident priorities and size-up.
2. *Factors* are anything that effect the decision-making process. Critical fireground factors include trapped people, rapidly spreading fire, frozen hydrants, and so on. The list of factors may be long, and the IC will have to determine which factor bears the most importance and should be addressed first.
3. *Courses* are the options an IC has to accomplish the aim. Based on the factors considered, an IC must determine the strength and weaknesses of each course of action.
4. *Plan* (action plan) is the course of action that best fits the situation. It also details how to implement the plan based upon the resources on scene and in staging.

The IC, confronted with an incident that she has not experienced before, must read the cues presented, compare these cues with others from the past, and remember the lessons learned from the past. She must complete a 360-degree size-up of the scene and, based upon the command sequence, look at any other factors that might be presented. Next, the IC should look at the actual problems presented and the strategies and tactics necessary to resolve these problems. Finally, the IC should consider her options in resolving the incident, and select and implement the most effective tactics.

While the tactics of Plan A are being implemented, begin the development of Plan B. If Plan A is not resolving the incident in an effective manner, the IC may have to make a quick shift to Plan B.

The IC then must make his decision based upon his own hypothesis as to what strategy and tactics will work best to resolve the incident. After the event, he should conduct a critique and determine what worked and didn't work. These are the lessons learned that will now become part of his recognitional knowledge base.

There are many decision-making models available for the IC to use to determine his or her plan of action and how to implement this plan. The most important factor for any incident command decision maker is to develop a logical thought process to evaluate the incident. Whether a form of the command sequence, tactical command sheets, REVAS, RECEOVS, or other mnemonics are used, the ability to determine the most effective and safe approach to the incident in a logical thought process will determine an effective or ineffective incident commander.

The only factors that never change among incident command decision making are the incident priorities:

- Life safety
- Incident stabilization
- Property conservation

SUMMARY

It is necessary to understand some common fire-ground coordination and control concepts in order to implement effective strategies and tactics. First the department must have an effective communication network and process. The need for effective communication at the fire scene is obvious. Messages that are missed or misunderstood can lead to injuries, deaths, and further property losses. Firefighters must understand that communicating is a six-step process involving both a sender and a receiver. To be effective, the process must be practiced and include the proper equipment.

Another consideration is size-up, the phase of the incident when information is gathered and analyzed. During size-up, the firefighter obtains information about the environment, resources, and the condition/situation. After analyzing this information, the firefighter can determine the course of action necessary to handle the incident. The information helps the incident commander determine the mode of operation. The operational mode is either offensive or defensive and may change throughout the incident.

The course of action is based on the three incident priorities, which are life safety, incident stabilization, and property conservation. In order to meet these three priorities, nine strategic goals have been developed. Each of these nine goals—firefighter safety, search and rescue, evacuation, exposure protection, confinement, extinguishment, ventilation, overhaul, and salvage—must be considered at every incident.

To meet the nine strategic goals, tactical objectives are developed. The number of tactical objectives needed to satisfy each goal varies with the type and magnitude of the incident. Once established and assigned, tactical objectives lead the task level groups or companies to tactical methods. Tactical methods include the actual task completed to meet a specific objective.

Size-up and consideration for strategic goals and tactical objectives leads the incident commander to develop an action plan. The action plan then must be communicated to all personnel working at the incident. There are various ways to communicate the action plan. Action planing to some degree must be done on every incident.

Preincident planning is also an essential element in coordination and control. It provides the firefighter with valuable information that can be obtained prior to the incident and then used if an incident occurs. Priority for preincident planning should be given to target hazards in a community. The preincident plan should include a text section, a site plan, and a floor plan.

Recognition-primed decision making is the process of decision making thought to be used by fireground commanders. Instead of making a choice from a group of options, which is done with classical decision making, the action taken is based on recognizing a similar incident and the action that worked then. Firefighters must understand this concept and develop skills to increase effectiveness. One area of skill development would be the application of the nine strategic goals to every incident.

REVIEW QUESTIONS

1. According to this text, which of the following is not a strategic goal?
   A. Firefighter safety
   B. Exposure protection
   C. Communication
   D. Salvage

2. The communication process introduced by the National Fire Academy has ___ steps.
   A. 3   B. 4   C. 5   D. 6

3. List the three broad areas that must be considered during size-up.

4. Using a saw to cut a 4-by-4-foot hole in a roof for vertical ventilation would be considered a _____.
   A. Tactical method
   B. Tactical objective
   C. Strategic goal
   D. Incident priority

5. True or False? Life safety is always the first incident priority.

6. True or False? Strategic goals are developed to meet tactical methods.

7. For the purpose of preincident planning, which of the following would not be considered a target hazard?
   A. A hospital
   B. A lumber yard
   C. A railyard
   D. A two-unit duplex

8. True or False? Attacking a room and contents fire from the inside of a structure would be considered an offensive fire attack.

9. List five requirements of an effective preplan process.

10. True or False? The recognition-primed decision-making concept implies that fireground incident commanders often make a choice of action from considering several alternatives.

ACTIVITIES

1. Review your department's communication policies and equipment. What changes might you recommend to make the communications more effective?

2. Using the strategic goals as a guide, review your department's SOGs. Do the SOGs provide for all of the strategic goals? Are all nine routinely considered at fires? What tactical objectives were used to meet the goals at the last fire incident that you responded to?

3. Examine your department's preincident planning process. How does it compare to that presented in the text? Do the forms provide adequate information? What changes or additions would you recommend?

Chapter

6

FIREFIGHTER SAFETY

Learning Objectives

Upon completion of this chapter, you should be able to:

- Discuss the relationship between firefighter health and safety, and strategy and tactics.
- Describe the reports available to study fireground injuries and deaths.
- List the most common causes of firefighter injuries and deaths.
- Describe the impact of regulations and standards on fireground operations.
- List general fireground safety concepts.
- Discuss the strategic considerations for safe fireground operations based on the general safety concepts, including incident safety officers, personal protective equipment, accountability, rapid intervention crews, and rehabilitation.

CASE STUDY

On April 11, 1994, at 0205 hours, a possible fire was reported on the ninth floor of a high-rise apartment building. This building had been the scene of numerous false alarms in the past. An engine company and a snorkel company were the first responders and arrived at the apartment building at 0208 hours. The engine company was the first on the scene and assumed command. Five firefighters from the two companies entered the building through the main lobby. They were aware that the annunciator board showed possible fires on the ninth and tenth floors. Command was set up in the lobby and radioed one firefighter that smoke was showing from a ninth-floor window. All five firefighters used the lobby elevator and proceeded to the ninth floor. When the doors of the elevator opened on the ninth floor, the hall was filled with thick black smoke. Four of the firefighters stepped off the elevator. The fifth firefighter, who was carrying the hotel pack, stayed on the elevator (which was not equipped with firefighter control) and held the door open with his foot as he struggled to don his self-contained breathing apparatus (SCBA). His foot slipped off the elevator door, allowing the door to close and the elevator to return with him to the ground floor. The remaining four firefighters entered the small ninth-floor lobby directly in front of the elevator. One firefighter stated that he was having difficulty with his SCBA and asked for the location of the stairwell. Another firefighter said, "I've got him," and proceeded with him into the hallway, turning right. Later, one of the four firefighters stated that he had heard air leaking from the SCBA of the firefighter having difficulty and had heard him cough. The remaining two firefighters entered the hallway and turned left, reporting zero visibility because of thick black smoke. Excessive heat forced them to retreat after they had gone 15 to 20 feet. They proceeded back down the hall past the elevator lobby. There they encountered a male resident, who attacked one of the firefighters, knocking him to the floor and forcibly removing his facepiece. The two firefighters moved with the resident through the doorway of an apartment, where they were able to subdue him. One firefighter broke a window to provide fresh air to calm the resident. At about the same time, the low-air alarm on his SCBA sounded. The other firefighter was unable to close the apartment door because of excessive heat from the hallway. Both firefighters and the resident had to be rescued from the ninth-floor apartment window by a ladder truck.

Firefighters from a second engine company arrived on the scene at 0209 hours. They observed a blown-out window on the ninth floor and proceeded up the west end stairwell to the ninth floor, carrying a hotel pack and extra SCBA cylinders. These firefighters entered the ninth floor with a charged fire hose and crawled down the smoke-filled hall for approximately 60 feet (the hallway was 104 feet long) before extreme heat forced them to retreat. As they retreated, they crawled over something they thought was a piece of furniture. They did not remember encountering any furniture when they entered the hallway. In the dense smoke, neither firefighter could see the exit door 6 feet away, and both became disoriented. After the firefighter from the first company rode the elevator to the ground floor lobby, he obtained a replacement SCBA and climbed the west end stairs to the ninth floor. When he opened the ninth-floor exit door, he saw the two firefighters from the second engine company in trouble. He pulled both into the stairwell.

When a rescue squad arrived at the scene at 0224 hours, lobby command could not tell them the location of the firefighters from the first company. They proceeded up the west end stairs to the ninth floor. The rescue squad opened the ninth-floor exit door and spotted a downed firefighter approximately 9 feet from the door. He was tangled in television cable wires that had fallen to the floor as a result of the extreme heat. The downed firefighter was from the first engine company; his body may have been what the firefighters from the second engine company encountered in the hallway. He was still wearing his SCBA, but he was unresponsive. The rescue squad carried him down the stairs to the eighth floor, where advanced life support was started immediately. The rescue squad then entered the first apartment to the left of the exit door and found a second firefighter from the first engine company kneeling into a corner and holding his mask to his face. He was unresponsive. The rescue squad carried that firefighter down the stairs to the eighth floor, where advanced life support was started. Both firefighters were removed within minutes and taken to a local hospital, where advanced life support was continued; but neither responded. Both victims died from smoke and carbon monoxide inhalation. Both victims wore personal alert safety system (PASS) devices, but because the devices were not activated, no alarm sounded when the firefighters became motionless.

Many factors contributed to the deaths and injuries that occurred in this incident. The key factors were as follows:

- The first five firefighters on the scene took an elevator to the floor of the fire—a violation of their department's written policy. Firefighter entrapment in automatic elevators is a recognized hazard, and the elevators in this incident had no firefighter control.
- At least one of the SCBAs leaked during this incident, and the respirator maintenance program appears to have been deficient. All four SCBAs tested by the National Institute for Occupational Safety and Health (NIOSH) failed at least two of five performance tests.
- When the rescue squad inquired about the location of the first firefighters at the scene, no one could account for them. Accountability for all firefighters at the scene is one of the fire command's most important duties.
- PASS devices were worn but not activated by the two firefighters who died. These devices should always be worn and activated when firefighters are working at the fire scene.

Although many factors contributed to the deaths and injuries reported here, they might have been prevented if the following essential precautions had been taken:

- Following established firefighting policies and procedures
- Implementing an adequate respirator maintenance program
- Establishing firefighter accountability at the fire scene
- Using PASS devices at the fire scene

These precautions are well known to fire departments and firefighters, but they require constant emphasis to ensure the safety of firefighters.

Source: This case study was taken in part from *Fire Fighter Alert,* DHHS (NIOSH) Publication No. 94-125. Further information regarding this incident can be found in this publication.

■ **Note**
Firefighter safety and health has always been an important concern for responders and incident commanders, but it has really surfaced as a priority in the past 15 years.

■ **Note**
Most firefighter injuries and deaths occur on the fireground.

INTRODUCTION

The study of fireground strategies and tactics would not be complete without a look at firefighter safety and health issues. Firefighter safety and health has always been an important concern for responders and incident commanders, but it has really surfaced as a priority in the last 15 years. Each year, thousands of firefighters are injured (**Figure 6-1**) and many are killed performing fire service functions. Not surprisingly, according to published injury reports, about half of firefighter injuries occur on the fireground and about one-fourth of the deaths occur on the fireground.

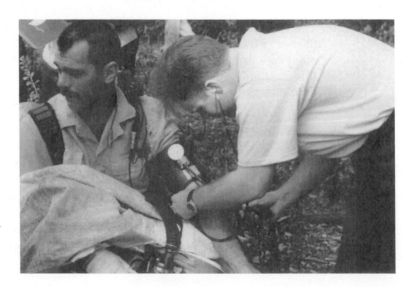

Figure 6-1
Paramedics treat a firefighter at a fire scene.

One of the major impacts on the fireground safety and health initiative was the publication of NFPA Standard 1500. This standard has provided a framework for occupational safety and health for firefighters. Other changes have also affected the way we respond and operate on the fireground. One is the ruling by the Occupational Safety and Health Administration (OSHA) that requires a team of at least two firefighters available for firefighter rescue to be assembled on the fireground prior, with few exceptions, to interior fire attack. This ruling has become known as the two-in, two-out rule and has had effects nationwide on fireground operations.

This chapter focuses on the relationship between fireground safety and health as it interrelates with fireground strategies and tactics; includes a review of current standards, regulations, and statistics; and provides general fireground safety concepts in order to further reduce common causes of injury and death.

Occupational Safety and Health Administration (OSHA)
the federal agency tasked with the responsibility for occupational safety of employees

two-in, two-out rule
the procedure of having a minimum of two firefighters standing by completely prepared to immediately enter a structure to rescue the interior crew should a problem develop; to be in place prior to the start of interior fire attack

■ Note
Life safety is always the number-one priority and all operations must be developed based on this premise.

FIREFIGHTER SAFETY, STRATEGY, AND TACTICS

The relationship between firefighter safety and strategy and tactics can be traced to the incident priorities. As introduced in Chapter 5, the three incident priorities are *life safety, incident stabilization,* and *property conservation.* Firefighter safety falls into the first incident priority, life safety.

Life safety is always the number-one priority and all operations must be developed based on this premise. Strategies and tactics developed and employed on a particular incident then must be based on the premise that life safety applies not only to citizens, but also to firefighters. Because life safety of the responders is a number-one priority, incident operations and strategies must be developed that provide for the highest level of safety to responders. In other words, for incident stabilization to occur, the incident commander should have considered life safety of the responders in formulating the strategy (see **Figure 6-2**).

Emergency scene risk management can be based on the following risk/benefit philosophy:

We will risk a lot to save a lot,

We will risk little to save a little,

We will risk nothing to save nothing.

In more useful terms, the incident commander would have little hesitation sending a search and rescue crew into a building fire if there were a possibility that people were inside. However, arriving at the same fire, meeting the occupants outside, and confirming they are all out may produce a different, safer strategy. Although some argue that a building is never unoccupied until a search proves it, more prudent and safety-minded incident commanders are changing strategies to make them more consummate with the risk.

Figure 6-2 *The incident commander must consider the safety of the firefighters when developing incident stabilization strategies.*

REVIEW OF FIREGROUND INJURIES AND DEATHS

As noted previously, almost half of firefighter injuries occur on the fireground. One might assume that this is because the fireground is basically an uncontrolled environment. However, with knowledge of fire behavior, good implementation of incident management systems, and an understanding of the historical causes of fireground injuries and deaths, the incident commander is better prepared to provide, in so far as possible, a safe fireground operation. In the effort to provide for strategies and tactics that consider firefighter safety, incident commanders should have a thorough understanding of what causes fireground injuries and deaths.

A start in this process is to review the annual injury and death reports from the fire service organizations that publish them and to evaluate where and how the injuries and deaths occur. The following list gives the organizations that publish reports and the information that they contain:

- *National Fire Protection Association (NFPA).* The NFPA publishes annual reports on both occupational injuries and deaths in the fire service. It has been compiling these reports since 1974. The NFPA death survey is a report on all firefighter deaths and includes analysis of the fatalities in terms of type of duty, cause of death, age group comparisons, and population-served comparisons.

Figure 6-3 *In order to define the national fire problem, it is important that departments participate in the National Fire Incident Reporting System.*

The NFPA injury survey is not an actual survey of all departments, but a sample used to project the national firefighter injury experience. Although a prediction, the survey has a high level of confidence and is representative of all sizes and types of departments. Both reports are published annually in the NFPA's *Fire Journal* magazine.

These reports are very comprehensive and useful in studying firefighter injuries and deaths.

- *United States Fire Administration (USFA).* The USFA oversees the National Fire Incident Reporting System (NFIRS). Part of the NFIRS report is data for firefighter casualties. This data can be obtained through the USFA and can be used to compare national data to local data; however, one problem with this system is that NFIRS is a voluntary system and departments may or may not participate (**Figure 6-3**). This creates the problem of not knowing what injuries or deaths may go unreported because a local jurisdiction did not participate in the program.

    These data are also formulated into reports in various formats and are available by mailing the USFA publication center. The USFA also has an intensive Web site providing a means to order publications, many of which are also downloadable and can be reviewed immediately. The Internet address for the USFA is http://www.usfa.fema.gov, or it can be contacted by telephone at 1-800-238-3358.

**FIREFIGHTER FATALITY INVESTIGATION AND PREVENTION PROGRAM:
PROGRAM DESCRIPTION**

The United States currently depends on approximately 1.2 million firefighters to protect its citizens and property from losses caused by fire. Of these firefighters, approximately 210,000 are career/paid and approximately 1 million are volunteers. The National Fire Protection Association (NFPA) and the U.S. Fire Administration estimate that on average, 105 firefighters die in the line of duty each year.

In fiscal year 1998, Congress recognized the need for further efforts to address the continuing national problem of occupational firefighter fatalities and funded NIOSH to conduct independent investigations of firefighter line-of-duty deaths.

FIREFIGHTER FATALITY INVESTIGATIONS

The NIOSH Firefighter Fatality Investigation and Prevention Program conducts investigations of firefighter line-of-duty deaths to formulate recommendations for preventing future deaths and injuries. The program does not seek to determine fault or place blame on fire departments or individual firefighters, but to learn from these tragic events and prevent future similar events.

The goals of the program are to:

- Better define the magnitude and characteristics of line-of-duty deaths among firefighters
- Develop recommendations for the prevention of deaths and injuries
- Disseminate prevention strategies to the fire service.

Figure 6-4 *NIOSH studies firefighter fatalities.*

- *International Association of Firefighters (IAFF).* Since 1960 the IAFF has produced an annual firefighter death and injury survey (**Figure 6-4**). This research reports on a number of issues relating to safety and health similar to those covered by the NFPA. However, the report is more in depth and provides information on lost-time injuries and exposure to infectious diseases. This is another resource for the health and safety program manager; however, the data is gathered from career fire departments that have IAFF affiliation, which limits the study to these departments.

- *Occupational Safety and Health Administration.* OSHA is an agency within the Department of Labor. Its regulations are applicable to many public fire departments and to all private fire departments. OSHA requires certain reporting requirements for occupation-related injuries and deaths. This requirement for record keeping and reporting allows OSHA to compile useful statistics and to study causes of occupational injuries in order to develop prevention strategies and countermeasures. Because not every state requires that public fire departments comply with OSHA regulations, this data also is not totally complete.

- *National Institute for Occupational Safety and Health.* NIOSH began a project in 1997, in which it will investigate firefighter line-of-duty deaths. The project, called Fire Fighter Fatality Investigation and Prevention Program, was funded in 1998. The overall goal of this program is to better define the magnitude and characteristics of work-related deaths and severe injuries among firefighters, to develop recommendations for the prevention of these injuries and deaths, and to implement and disseminate prevention efforts. A five-part integrated plan is used and the information disseminated to the nation's fire services (see **Figure 6-4**).

A review of the 2004 NFPA injury report and the 2005 USFA firefighter fatality report follows. These statistics from 2004 to 2005 are included to show the information that can be gathered from the reports. This review is also designed to provide a general overview of where the injuries and deaths are occurring, allowing the reader to consider these in developing strategies and tactics. The reader should study more recent reports.

NFPA 2004 Injury Report

According to the 2004 NFPA report on U.S. firefighter injuries, there were 75,840 firefighters injured in the line of duty, a 13.3% decrease from the 1998 total of 87,520. Of these 75,840 injuries, 36,880, or 48.6%, occurred on the fireground (**Figure 6-5**). The causes of fireground injuries are also reported. The causes are divided into nine categories; the nature of injury is divided into ten categories. **Figures 6-6** and **6-7,** respectively, show the causes of fireground injuries and the nature of the injury. **Figure 6-8** compares the number of fireground injuries to the

Figure 6-5
Firefighter injuries by type of duty, 2005. (Reprinted with permission from U.S. Firefighter Injuries, Copyright © 2005, National Fire Protection Association. All rights reserved.)

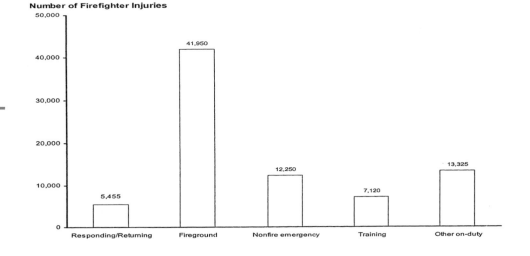

Number of Firefighter Injuries

Source: NFPA Annual Survey of Fire Departments or U.S. Fire Experience (2005)

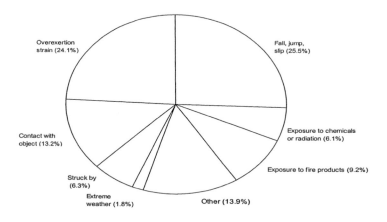

Figure 6-6
Fireground injuries by cause, 2005. (Reprinted with permission from U.S. Firefighter Injuries, Copyright © 2005, National Fire Protection Association. All rights reserved.)

Source: NFPA Annual Survey of Fire Departments for U.S. Fire Experience (2005)

Nature of Injury	Responding to or Returning from an Incident		Fireground		Nonfire Emergency		Training		Other on-duty		Total	
	Number	Percent	Number	Percent	Number	Percent	Number	Percent	Number	Percent	Number	Percent
Burns (Fire or Chemical)	155	2.8	2,930	7.0	80	0.7	270	3.8	215	1.6	3,650	4.6
Smoke or Gas Inhalation	75	1.4	2,485	5.9	110	0.9	135	1.9	70	0.5	2,875	3.6
Other Respiratory Distress	10	0.2	905	2.2	185	1.5	165	2.3	125	0.9	1,390	1.7
Burns and Smoke Inhalation	165	3.0	750	1.8	65	0.5	95	1.3	45	0.3	1,120	1.4
Wound, Cut, Bleeding Bruise	940	17.2	7,600	18.1	1,770	14.5	1,290	18.1	2,565	19.3	14,165	17.7
Dislocation, Fracture	210	3.9	970	2.3	515	4.2	170	2.4	450	3.4	2,315	2.9
Heart Attack or Stroke	55	1.0	315	0.8	85	0.7	125	1.8	185	1.4	765	1.0
Strain, Sprain Muscular Pain	3,075	56.4	18,620	44.4	7,150	58.4	3,970	55.8	6,925	52.0	39,740	49.6
Thermal Stress (frostbite, heat exhaustion)	255	4.7	2,480	5.9	285	2.3	390	5.5	155	1.2	3,565	4.4
Other	515	9.4	4,895	11.7	2,005	16.4	510	7.2	2,590	19.4	10,515	13.1
	5,455		41,950		12,250		7,120		13,325		80,100	

Source: NFPA Survey of Fire Departments for U.S. Fire Experience, 2005
Note: If a firefighter sustained multiple injuries for the same incident, only the nature of the single most serious injury was tabulated.

Figure 6-7   *Firefighter injuries by nature of injury and type of duty, 2005. (Reprinted with permission from U.S. Firefighter Injuries, Copyright © 2005, National Fire Protection Association. All rights reserved.)*

| Year | At the Fireground | | At Nonfire Emergencies | |
	Injuries	Injuries per 1000 Fires	Injuries	Injuries per 1,000 Incidents
1988	61,790	25.4	12,325	1.13
1989	58,250	27.5	12,580	1.11
1990	57,100	28.3	14,200	1.28
1991	55,830	27.3	15,065	1.20
1992	52,290	26.6	18,140	1.43
1993	52,885	27.1	16,675	1.25
1994	52,875	25.7	11,810	0.84
1995	50,640	25.8	13,500	0.94
1996	45,725	23.1	12,630	0.81
1997	40,920	22.8	14,880	0.92
1998	43,080	24.5	13,960	0.82
1999	45,500	25.0	13,565	0.76
2000	43,065	25.2	13,660	0.73
2001	41,395	23.9	14,140	0.73
2002	37,860	22.4	15,095	0.77
2003	38,045	24.0	14,550	0.70
2004	36,880	22.1	13,150	0.62
2005	41,950	26.2	12,250	0.56

Source: NFPA Survey of Fire Departments for U.S. Fire Experience (1988-2005)

number of nonfire emergency injuries per 1,000 incidents from 1994 to 2004. There has not been a significant change during the 10-year study period of fireground injuries per 1,000 fires.

Firefighter Fatalities in the United States 2004

The USFA's Firefighter Fatalities report lists 115 on-duty firefighter deaths in 2004. Firefighters included in the report are all members of organized fire departments, including career and volunteer firefighters; full-time public safety officers acting in a firefighter capacity; state and federal fire service personnel, including wildland firefighters; and privately employed firefighters, including employees of contract fire departments and trained members of industrial fire brigades, whether full- or part-time. Also included are contract personnel working as firefighters or assigned in direct support of fire service organizations. Of these 115 fatalities, 34 deaths were from career departments, and 81 were from volunteer organizations.

Twenty-seven deaths occurred while performing fireground activities (see **Figure 6-9**). Deaths that occurred on the fireground are the focus of this chapter.

The activities being performed answer the how and what of the deaths, but it is also important to note the cause of fatal injuries. The USFA study breaks cause down into eight areas. **Figure 6-10** shows the relationship of the causes.

Much can be learned from these reports about how and where these injuries and deaths are occurring. Before implementing fireground strategies, the incident commander should study these reports and employ methods to help to reduce the casualties associated with fireground activities.

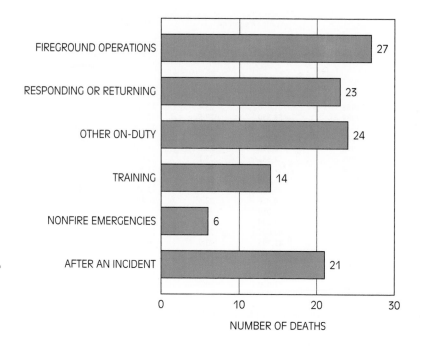

Figure 6-9 *Firefighter fatalities by type of duty, 2004. (Source USFA.)*

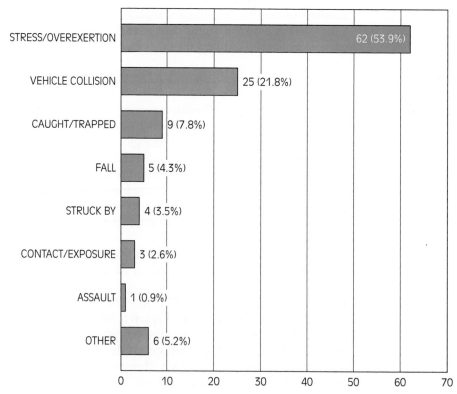

Figure 6-10 *Firefighter fatalities by cause of fatal injury, 2004. (Source USFA.)*

THE IMPACT OF REGULATIONS AND STANDARDS

As the fire service has progressed and the emphasis on occupational safety and health for firefighters has increased, so has the number of operationally relevant regulations and standards. There are differences between regulations and standards. Regulations are promulgated at some level of government by a governmental agency and have the force of law. Standards, sometimes called consensus standards, do not have the weight of law unless they are adopted as law by an authority having jurisdiction.

From the standpoint of strategy and tactics, an understanding of the applicable regulations and standards is essential for the incident commanders.

Regulations

Regulations carry the weight of law and are mandatory in their requirements based on federal, and in some cases, state and local legislation. The entire collection of federal regulations is contained within the fifty titles of the Code of Federal Regulations (CFR). Some of these mandatory requirements have an impact on fireground operations, primarily those regulations found in Title 29 CFR (see text box, "OSHA's Title 29"), which is the Occupational Safety and Health Administration's (OSHA) regulations. Remember, these regulations have the power of law and are enforced by the federal agency responsible for them.

Some states have adopted federal laws, which then become state mandatory requirements. When states adopt a regulation, then state officials provide the enforcement. Whether a public fire department is required to comply with OSHA standards depends on whether the state is one of twenty-five OSHA states and territories. If the state is an OSHA state, public employers must comply with the regulation.

Three OSHA regulations that have a direct impact on fireground operations follow. In fact, it was the formal interpretation of these three standards that resulted in the two-in, two-out ruling.

Standards

Other published documents do not mandate compliance. These are commonly known as consensus standards, because a group of professionals with a specific expertise came together and agreed on how a specific task should be performed. The NFPA is one such standards-making group. The NFPA has no enforcement authority or power; its standards are considered advisory. However, a jurisdiction can adopt an NFPA standard and then the adopting authority has legal rights to enforce the standard. The most common example of this process is the adoption of NFPA 101, *Life Safety Code*. This standard has been adopted by

---

regulations
requirements or laws promulgated at the federal, state, or local level with a requirement to comply

standards
often developed through the consensus process; standards are not mandatory unless adopted by a governmental authority

Code of Federal Regulations (CFR)
the document that contains all of the federally promulgated regulations for all federal agencies

■ **Note**
Regulations have the power of law and are enforced by the federal agency responsible for them.

consensus standards
standards developed by consensus of industry or subject area experts, which are then published and may or may not be adopted locally; even if not adopted as law, these can often be used as evidence for standard of care

OSHA'S Title 29

OSHA 1910.134 (29 CFR 1910.134) Respiratory Protection. This regulation requires that respirators be provided by the employer when such equipment is necessary to protect the health of the employee (including self-contained breathing apparatus). The employer is required to provide respirators that are applicable and suitable for the purpose intended. The employer is responsible for the establishment and maintenance of a respiratory protection program. The employee shall use the provided respiratory protection in accordance with instructions and training received. The regulation also sets forth standards for air quality and self-contained breathing apparatus maintenance programs (**Figure 6-11**).

OSHA 1910.120 (29 CFR 1910.120) Hazardous Waste Operations and Emergency Response. Although this regulation deals with hazardous material releases, it was referenced as part of the two-in, two-out ruling, specifically the requirement for a buddy system.

OSHA 1910.156 (29 CFR 1910.156) Fire Brigades. The OSHA fire brigade regulation applies to fire brigades, industrial fire departments, and private or contractual-type fire departments. The personal protective clothing requirements of this regulation apply only to those fire brigades that perform interior structural firefighting. The regulation specifically excludes airport crash fire rescue and forest firefighting operations. This regulation contains requirements for the organization, training, and personal protective equipment for fire brigades whenever they are established by employers.

As with the 1910.134 and 1910.120 regulations, this regulation was also referenced in the OSHA two-in, two-out ruling. Specifically the section was cited that requires that SCBA be provided for all firefighters engaged in interior structural firefighting.

fire brigades
the use of trained personnel within a business or industrial site for firefighting and emergency response

both local and state governments as part of the fire prevention program. Once a standard is adopted, the adopting governmental agencies have the power for enforcement.

The NFPA has developed a number of standards that impact fireground operations. The following list summarizes these standards:

NFPA 1500, *Fire Department Occupational Safety and Health Program*

NFPA 1521, *Fire Department Safety Officer*

NFPA 1561, *Fire Department Incident Management System*

NFPA 1021, *Fire Officer Professional Qualifications*

NFPA 1041, *Fire Service Instructor*

NFPA 1981, *Open-Circuit Self-Contained Breathing Apparatus (SCBA) for Fire Fighters*

NFPA 1982, *Personal Alert Safety Systems (PASS) for Fire Fighters*

■ **Note**
Once a standard is adopted, the adopting governmental agencies have the power for enforcement.

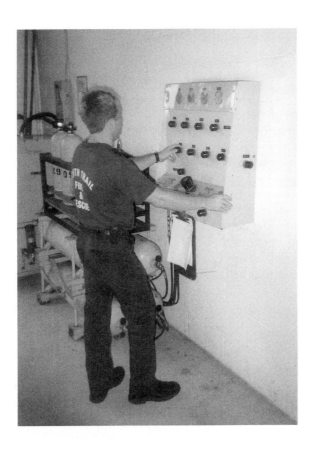

Figure 6-11 *Air cascade systems must meet air quality requirements.*

NFPA 1404, *Fire Department Self-Contained Breathing Apparatus Program*

NFPA 1403, *Standard on Live Fire Training Evolutions*

NFPA 1404, *Standard for Fire Service Respiratory Protection Training*

NFPA 1410, *Standard on Traaining for Initial Emergency Scene Operations*

NFPA 1710, *Standard for the Organization and Deployment of Fire Suppression Operations, Emergency Medical Operations, and Special Operations to the Public by Career Fire Departments*

NFPA 1720, *Standard for the Organization and Deployment of Fire Suppression Operations, Emergency Medical Operations, and Special Operations to the Public by Volunteer Fire Departments*

These are consensus standards that are not mandatory unless adopted into law by local or state legislation. But because a group of professionals with related interest and expertise have come together and agreed on some minimum level of

standard of care
the concept of what
a reasonable person
with similar training
and equipment would
do in a similar situation

■ Note
**The concept of
standard of care is
simple: Everyone has
certain expectations
when it comes to
performance.**

performance, these standards tend to become a standard of care in the particular subject area.

Standard of Care

The concept of standard of care is well-known in the emergency medical field. From the first day of training, this concept is introduced to make the student realize that to avoid liability he must perform in the same way as another reasonable person with the same training and equipment would perform. The concept is simple: Everyone has certain expectations when it comes to performance. When a deposit is made to your checking account at the bank, you have a reasonable expectation that the teller will put the money into the right account, although when mistakes or accidents do occur, the bank has to take responsibility for the teller's actions.

This same concept can be applied to fireground safety and health issues and closely relates to the existence of standards. The publication of a safety standard in and of itself has an impact on standard of care. For example, NFPA 1500 requires that all responders to hazardous situations be equipped with a personal alert safety system (PASS). Should an employee get lost in a building fire and die, someone could wind up in court answering the question of why a PASS device was not provided. But NFPA is not law and has not been legally adopted in my jurisdiction, this person might say. This defense will not be viable when the firefighter's family sues for not following reasonable industry standards, because the standard of care has been defined in the NFPA document.

Standard of care is not a static concept but instead is very dynamic. It changes with new technologies and the development of regulations, standards, and guidelines. It was an acceptable standard of care to allow firefighters to fight fires without PASS devices 25 years ago. They did not exist and if they did, no published documents required them; therefore, it was an acceptable practice. This is not the case today.

GENERAL FIREGROUND SAFETY CONCEPTS

Incident scene safety concepts are an important consideration for the incident commander. Clearly a number of concepts or needs can interplay here to provide for the highest level of safety at the scene. Some of the hazards have already been presented in earlier chapters, such as the hazards associated with fire behavior. Personal protective equipment, designed to protect the firefighter, may also be a cause of overexertion and heat stress. Other safety issues specific for situations are presented later in the text. However, a number of safety-related concepts are common to all fire emergencies. Some of these considerations can be undertaken prior to the incident, others when the incident occurs. Included in common

■ **Note**
Included in common incident scene concepts are assigning an incident safety officer, having the firefighter equipped with the proper personal protective equipment, having a fireground accountability system, having in place a rapid intervention crew, and providing for incident rehabilitation.

incident scene concepts are assigning an incident safety officer, having the firefighter equipped with the proper personal protective equipment, having a fireground accountability system, having in place a rapid intervention crew, and providing for incident rehabilitation.

Incident Safety Officer

The assignment and function of the incident safety officer (**Figure 6-12**) is a key component to minimizing risk to the firefighters and increasing scene safety. Depending on the size and type of the organization, the incident safety officer may be a dedicated position from within the safety division or a first-line supervisor assigned to the safety role on an incident-by-incident basis. Some departments assign extra units to large incidents and have these crews function in certain safety-related roles, including incident safety officer, accountability, entry control officer, or rapid intervention crews. In some cases, the incident safety officer is in name only; for example, the assignment is made to anyone who is free at the incident. This is not a good practice, as the incident safety officer must have additional knowledge, be well experienced with the type of fire incident at hand, and have no other duties. Different incidents may require different levels of safety officers to be assigned.

After arrival at the incident scene, the incident safety officer should evaluate the incident and what is happening. A prediction of what could or is going to happen should be part of this evaluation. This evaluation should involve a

■ **Note**
Depending on the size and type of the organization, the incident safety officer may be a dedicated position from within the safety division or a first-line supervisor assigned to the safety role on an incident-by-incident basis.

Figure 6-12 *A safety officer confers with the incident commander about scene safety issues.*

360° walk around the scene and examination of available information, such as preincident plans. The incident safety officer should talk to occupants or owners and question them about hazardous situations relating to the property or vehicle—for example, alternative fuel use in a car, gunpowder in a home, or the storage of explosives. In 1997, a Florida firefighter was killed by a .22 caliber rifle that was subjected to the intense heat in a house that was on fire.

The incident safety officer should assess the operation from a safety point of view and relay findings to the incident commander (IC). The safety officer must be given the authority to immediately stop unsafe acts that are dangerous to responders. The stopping of an assignment must be relayed to the IC immediately as well. During the incident evaluation, the incident safety officer should assess the operating personnel, including the use of proper protective clothing, accountability, and crew intactness. **Figure 6-13** shows a sample incident safety officer checklist.

The incident safety officer serves as a member of the IC's staff and must be a resource for them. The IC must use the safety officer's knowledge and expertise in helping with strategic and tactical decisions. The following list shows some of the varied responsibilities of the incident safety officer:

- Strategy to match situation—Do all operating members know the strategy?
- Crew intactness
- Incident Management System (IMS) in place
- Accountability system
- Proper level of protective equipment
- Building or vehicle structure status
- Establishment of collapse zone
- Rehabilitation setup
- Physical condition of personnel
- Proper scene lighting
- Communicating to operating teams of unsafe locations, for example, holes in floors, wires down, backyard swimming pools
- Means of egress for crews
- Risk assessment
- Rapid intervention crews ready
- Utilities secured
- Traffic controlled

Personal Protective Equipment

Personal protective equipment (PPE) designed for firefighting (**Figure 6-14**) is one strategy in minimizing exposure to hazards at the fire scene, meeting the

**Fire/EMS Department
Incident Safety Officer Worksheet
Structural Fire Incident**

Scene

❏ Traffic Controlled
❏ Utilities Secured
❏ Hazardous
    Materials

Structure

❏ Type
❏ Structural Stability
❏ Collapse Zone ID
❏ Fuel Load

Operations

❏ IMS
❏ Attack Mode Known
    to All Crews
❏ Rapid Intervention
    Crew (s)
❏
❏
❏

Personnel

❏ Adequate
❏ Proper Training
❏ PPE
❏ Teams Intact
❏ Within IMS
❏ Accountability
❏ Rehabilitation
❏

Site Sketch

Units Assigned

——————
——————
——————
——————
——————
——————

Safety Issues

Figure 6-13   *Incident safety officer checklist for a fire incident.*

Figure 6-14 *A firefighter trains in full firefighter PPE.*

priority of life safety as applied to responders. A discussion on PPE for response to fires can be broken down into three subject areas:

1. Design and purchasing
2. Use
3. Care and maintenance

Standards from the NFPA (1900 series) set forth requirements for the purchase of firefighting PPE. These standards give the minimum requirements for particular items of PPE. Once PPE is provided, guidelines or procedures must be adopted governing the use of the PPE. Procedures must be in place governing a care and maintenance program, including inspections.

The NFPA 1500 standard requires that new firefighting PPE meet the current editions of the respective standards. Older gear must have met the standard in effect when purchased. Therefore the purchasing agent for the department should reference the particular NFPA standard, in its entirety, in the specification for the PPE. The commonly expected components of PPE for firefighting is given in the following list:

- Approved fire helmet with eye protection
- Flame-resistant hood
- Turnout coat
- Turnout pants
- Firefighting gloves
- Firefighting boots
- Personal alert safety system (PASS) devices
- Self-contained breathing apparatus*

■ Note

Standard operating guidelines (SOGs) should be in place defining the use of the PPE.

Standard operating guidelines (SOGs) should also be in place defining the use of the PPE. Although this seems quite obvious, it is helpful to require the use of full PPE under specific conditions. Examples might include that full PPE be used for interior structure firefighting. Of particular importance is a procedure or requirement for the use of PASS devices and SCBA. Procedures should also provide incident commanders with guidelines as to when is it permissible to remove the SCBA or other PPE components.

The NFPA standards on PPE provide guidelines for the care and maintenance of the equipment. Generally, the standards require following the manufacture's recommendations. Although the care is somewhat delegated to the user, the PPE program SOGs must include a procedure for periodic inspection to ensure the PPE is in good condition (**Figure 6-15**). Monthly inspections are recommended. The monthly inspection can be performed by a station officer or, in some departments, the shift commander. PPE should also be inspected by the user at the beginning of each duty shift and after each use. SCBA and PASS devices should be checked and inspected at the beginning of each shift and after use.

Fireground Accountability

A personnel accountability system is a critical element of the incident management process. Accountability will occur by using several layers of supervision in various geographic areas or functional areas. Each sector, group, or division leader is responsible for his or her operating crews. An accountability officer should be assigned within the incident management system (IMS) or, on smaller incidents, should be handled by the IC or incident safety officer.

As with the IMS, there are several variations of accountability systems in use. Some use the passport system, where the crew's names are placed on small Velcro tags and placed on a small card used at the incident scene as a passport, as shown in **Figure 6-16.**

---

*Whereas the SCBA unit is commonly placed on the apparatus, many departments supply personal facepieces.

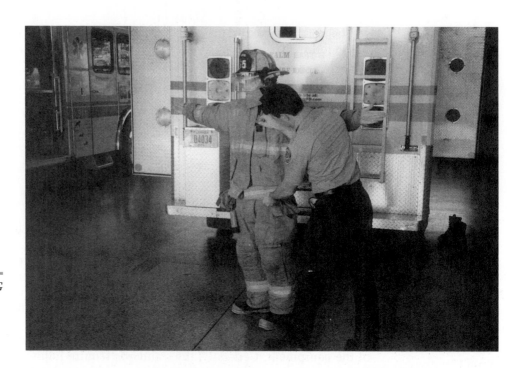

Figure 6-15   *The SOG should provide for routine inspection of the firefighting PPE.*

Figure 6-16   *Passport-style accountability tag.*

Others use two-dimensional bar codes and computers. Yet another system uses a card and clip attached to a large ring carried in the apparatus. Systems can be purchased commercially or designed locally, but the accountability system must meet the following objectives:

- Account for the exact location of all individuals at the fire scene at any given moment.
- Provide for expansion to meet the needs of the incident.
- Be adaptable to the IMS in use.
- Ensure that all individuals are checked into the system at the onset of the incident.
- Provide for visual recognition of participation.
- Provide for points of entry into the hazard zone.

Some systems may also include medical data and the level of an individual's training and some have information about apparatus for large, multijurisdictional incidents.

■ **Note**
Locally adopted accountability procedures are a strict requirement and must require participation by everyone at all levels and at all incidents.

Locally adopted accountability procedures are a strict requirement and must require participation by everyone at all levels and at all incidents. Each level has certain responsibilities to the system. Individual responders are responsible to ensure that their name tags are in the proper location on the apparatus and to stay in direct contact with their assigned crew at the scene. *Freelancing is not allowed!* Officers are responsible for knowing the location of each person assigned to them and to stay within their assigned work areas. Sector or group officers are responsible for accountability within the work area. Accountability officers are responsible for the accountability of any area of the incident from a specific point of entry. Incident command is responsible for putting the accountability function into the IMS and to provide a means for accounting for every individual at the scene. An accountability division/sector/group at a large incident may determine the points of entry, ensure communication with accountability officers, and provide the overall coordination of accountability at a specific incident.

personnel accountability reports (PARs)
verbal or visual reports to incident command or to the accountability officer regarding the status of operating crews; should occur at specific time intervals or after certain tasks have been completed

The accountability system requires personnel accountability reports (PARs). These PARs require the incident commander or accountability officer to check the status of crews and ensure that all personnel are accounted for during certain regular intervals or benchmarks during the incident. These benchmarks should be defined and may be related to time of activity. Examples of benchmarks include:

- A fixed time, as defined in local SOGs
- After primary search
- After fire is under control
- After a switch in strategic modes (offensive to defensive)
- A significant event such as a collapse, flashover, or backdraft
- After any report of a missing firefighter

Rapid Intervention Crews

Regardless of the effectiveness of the IMS in place or the greatest of accountability systems, nothing is gained if personnel are not available to respond when an unexpected event occurs on the fireground. The IMS and accountability system can provide the incident commander with the information that a crew is missing. But to intervene, the incident commander must have the resources close by to handle the situation. This is the concept of **rapid intervention crews (RICs)**. Although called by many other names, the RIC is a group of firefighters that are fully equipped with complete PPE who are on stand-by near the emergency scene and are prepared to handle any emergency that may occur to the operating firefighters (**Figure 6-17**). When the emergency occurs, this crew is ready to respond and assist, be it a lost firefighter, trapped firefighter, or injured firefighter requiring assistance. The RIC would meet the requirements of the OSHA two-in, two-out rule discussed earlier in this chapter.

The number of RIC necessary at an emergency incident can vary with the complexity of the incident (see text box, "Phoenix Rapid Intervention Study"). A room and contents house fire would probably only require a crew of two or three, whereas a large warehouse complex may require an RIC on each side of the building simply because of distance.

Once assigned to rapid intervention, the companies should generally not be assigned other tasks that would render them unavailable, and they should have

rapid intervention crews (RICs)
assignment of a group of rescuers with the sole purpose of rapid deployment to reports of operating personnel in trouble or missing

Figure 6-17 *The RIC should be in full PPE, have the equipment on hand specific to the incident, and be ready for immediate deployment.*

Phoenix Rapid Intervention Study

After the fire department experienced a line-of-duty death, Phoenix conducted a study on rapid intervention. Although the department had long ago developed procedures for the assignment of rapid intervention teams, it was not until after this tragic event that the effectiveness of these teams was evaluated. Phoenix performed 200 rapid intervention drills as part of the recovery process following this fireground death. The results of the study were widely published and in many cases surprised fire service leaders. The results showed that rapid intervention is not rapid; in fact, in the drills, it took rescue crews an average of 2.5 minutes to get to a ready state. From the firefighter distress signal to rapid intervention team entry, it was 3.03 minutes followed by 5.82 minutes for them to make contact with the downed firefighter. The average total time inside the building for each rapid intervention team was 12.33 minutes, with a average total rescue time of 21 minutes. They also found that it took twelve firefighters to rescue one downed firefighter, that one in five rapid intervention team members will also get into some type of trouble, and that a 3000 psi SCBA bottle has 18.7 minutes of air (plus or minus 30 percent).

■ **Note**
Once assigned to rapid intervention, the companies should generally not be assigned other tasks that would render them unavailable, and they should have direct communication to the incident commander or operations.

rehabilitation
the group of activities that ensures responders' health and safety at an incident scene; may include rest, medical surveillance, hydration, and nourishment

direct communication to the incident commander or operations. On arrival, the team should size-up the building and the operation and try to anticipate what could happen. It should determine possible entry and exit points. For example, in a two-story garden apartment building, the RIC may want to have a 24-foot ladder for rapid access to the second floor.

The incident commander must provide for an RIC early in an incident and for the duration. Some smaller departments are utilizing automatic and mutual aid in order to staff an on-scene RIC. A good idea is to automatically dispatch an RIC to all working fires.

Fireground Rehabilitation

The physical and mental demands placed on firefighters coupled with the environmental dangers of extreme heat and cold will have an adverse effect on responders.

Crews that have not been provided with adequate rest and rehabilitation (**Figure 6-18**) during fire operations are at increased risk for illness and injury.

Rehabilitation on the emergency scene is another essential element of the IMS. This need for rehabilitation is also cited in a number of the national standards relating to safety and fire scene operations. Although systems can vary, the basic plan should provide for the following:

- Establishment of a rehabilitation sector/group within the IMS
- Hydration

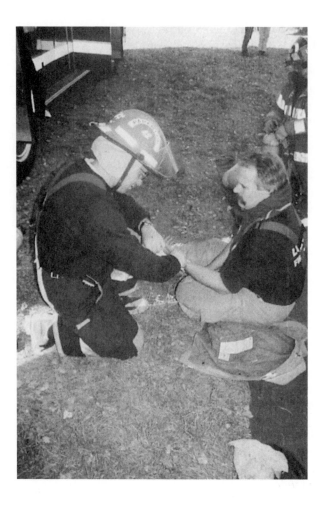

Figure 6-18
*Rehabilitation
efforts should also
include provisions
for quick medical
checks of the
firefighters.*

- Nourishment
- Rest and recovery
- Medical evaluation
- Accountability while in the rehabilitation sector

Also to be considered are supplies, shelter, and the number of people needed to operate the rehabilitation area (see text box, "Example of Fire Department Rehabilitation SOP").

Procedures should define the responsibilities within the IMS for the incident commanders, company officers, and personnel. In smaller agencies, having the staffing needed for setting up a rehabilitation sector/group may be a problem. In this case, examine options such as mutual aid, EMS personnel not involved in firefighting, or possibly the Red Cross if properly trained.

Example of Fire Department Rehabilitation SOP

FIRE DEPARTMENT STANDARD OPERATING PROCEDURE

SUBJECT: REHABILITATION

> **Definition:** The *Rehabilitation Sector/Group* is an incident function to provide rest, nourishment, comfort, and medical assessment to on-scene personnel.
>
> **Purpose:** To ensure that the physical and mental condition of personnel operating at an emergency incident or training exercise does not deteriorate to a point that affects the safety of the individual or crew, or the integrity of the operation.
>
> **Scope:** This procedure shall apply to emergency operations and training exercises of extended duration where strenuous physical activity or exposure to extreme heat or cold exists.

There obviously will be times when the incident commander (IC) determines that an informal rehab will suffice. One unit offering a short break and some fluids could staff an informal rehab.

Responsibilities

- Incident Commander

    Shall consider the circumstances of each incident and make adequate provisions early in the incident for the rest and rehabilitation of all personnel operating on scene. These provisions shall include: medical evaluation, monitoring, treatment, and transport; food and fluid replenishment; relief from extreme climatic conditions; and other environmental parameters of the incident. The provision of emergency medical services shall be at the basic life support level or higher. The IC shall establish a Rehab Sector/Group when conditions dictate that personnel operating on scene or at training evolutions need rest and rehabilitation. The Rehab Sector/Group shall report to and maintain contact with the incident safety officer.

- Sector/Division/Group Supervisors

    Shall maintain an awareness of the condition of each person operating within his or her span of control. Shall utilize the command structure to request relief and the assignment of fatigued crews to rehab.

- Personnel

    During any emergency incident or training evolution, all personnel shall advise their supervisor when they believe that their level of fatigue or exposure to environmental conditions could affect themselves, their crew, or the operation. Personnel shall be alert to the health and safety of other crew members. During periods of hot weather, personnel are encouraged to drink water throughout the day.

(continued)

Establishment of Rehab Sector/Group

- Location

   The rehab location may be designated by the IC or the rehab officer using the following criteria:

     Provides suitable protection from environmental conditions; cool and shaded in hot weather, warm and dry in cold weather

     Far enough away from incident to allow removal of protective clothing and allow mental rest from the stress and pressure of the emergency operation or training evolution

     Free of exhaust fumes from apparatus, vehicles, and equipment

     Large enough to accommodate multiple crews

     Easily accessible by EMS transport units

     Close or adjacent to SCBA air refill station, allowing prompt reentry back into the incident or training upon complete recuperation

- Resources

   The rehab officer shall secure all necessary resources required to adequately staff and supply the Rehab Sector/Group. The supplies should include the following:

     Fluids: water, activity beverage, and ice

     Food: high-energy nourishment (in other words, bars, crackers, granola)

     Medical: blood pressure cuffs, stethoscopes, thermometers, transport unit

     Tracking: status board, make-up (blank, white) passports, pencils, notepads or white boards, markers or grease pencils, tracking worksheets

     Other: Rehab units (HDD RH17, RH44, RH48, RH54), awnings, fans and barricade tape (to identify single ingress/egress point)

- Rehabilitation Procedures and Accountability: KEEP CREWS TOGETHER

   The IC or safety officer shall provide the rehab officer with a list of units and number of personnel assigned to the incident (see **Figure 6-19**). If needed, the rehab officer may contact dispatch on the admin subfleet to obtain crew information and arrival times.

   Crews are to report to rehab intact and present their accountability passport. Personnel should remove their SCBA and turnout coats and store them where directed by rehab personnel. All personnel shall wash their hands and begin hydration while having vital signs evaluated and rehab worksheets completed (see **Figure 6-20**).

   Personnel shall not leave the rehab area until authorized by the rehab officer. Crews leaving rehab shall obtain their passport from the rehab officer and report to staging.

(continued)

- Guidelines for the Rehab Officer

    Hydration: A critical factor in the prevention of heat injury is the maintenance of water and electrolytes. During heat stress, personnel should consume at least one quart of water per hour. The rehydration solution should be a 50/50 mixture of water and a commercially prepared activity beverage and should be administered at about 40°F. Caffeine beverages and carbonated beverages should be avoided, as both interfere with the body's water conservation mechanisms.

    Nourishment: Food (high-energy bars, crackers, or granola) should be provided to personnel on extended incidents, but is not to be used in lieu of hydration.

    Rest: The "two air bottle rule," or 45 minutes of work time, is recommended as an acceptable level of exertion prior to mandatory rehab. Personnel should try to rehydrate during SCBA bottle replacement. In all cases, the objective evaluation of an individual's fatigue level shall determine rehab time. Rehab time shall never be less than 10 minutes. Fresh crews and crews released from rehab shall be available in the staging area.

    Recovery: Personnel in rehab should maintain a high level of hydration. Personnel should not be moved from a hot environment into an air-conditioned environment because the body's cooling system may shut down in response to external cooling. An air-conditioned area is acceptable for personnel after a cool-down period at ambient temperature with air movement. Drugs such as antihistamines and diuretics may inhibit the body's ability to sweat and cool. Extreme caution should be taken with personnel who may have taken this type of drug. They may require extended time in rehab.

    Medical evaluation: An individual's vital signs shall be immediately evaluated upon entrance into rehab. Heart rate should be monitored for 30 seconds. If the rate is above 110 beats per minute (bpm), the person's temperature shall be taken. If the temperature is above 100.6°F, all protective clothing shall be removed. If the temperature is below 100°F and the heart rate remains above 110 bpm, rehab time should be increased. If the heart rate is below 110 bpm, the chance of heat stress is negligible. Out-of-range vital signs and vital signs that do not improve with rest shall be reported to the transport personnel and the incident safety officer. All medical evaluation shall be recorded on rehab worksheets. Rehab worksheets shall become part of the incident record and the rehab officer shall ensure that the IC receives copies.

    Tracking of personnel: Tracking and control of ingress and egress of personnel from the rehab area is facilitated by the use of accountability passports, which each company or group officer shall turn in to the rehab officer upon arrival at rehab. The rehab officer will use the passports to track units and personnel in rehab, returning them to the respective officers as companies/groups leave the rehab area.

REHAB SECTOR   **COMPANY CHECK-IN/OUT SHEET**

CREWS OPERATING ON THE SCENE _____

_____

_____

UNIT #	# OF PERSONS	TIME IN	TIME OUT	UNIT #	# OF PERSONS	TIME IN	TIME OUT

Figure 6-19   *Check-in/out sheet.*

Rehabilitation Worksheet   Incident _____   Date: _____

Name/Unit	Time(s)	Time/# Bottles	BP	Pulse	Resp	Temp	Skin	Taken By	Complaints/ Condition	Transport?

163

Figure 6-20 *Rehabilitation worksheet.*

■ **Note**

Medical information from a rehabilitation sector/group may fall under confidential medical records. Check with your local legal advisor.

Furthermore, the plan should provide guidelines for location, site characteristics, site designations, and resources required. Records should be kept on all individuals entering the rehabilitation area and medical evaluation documented. Remember that the medical information from a rehabilitation sector/group may fall under confidential medical records. Check with your local legal advisor.

SUMMARY

Along with all the experience and knowledge required of the IC associated with strategies and tactics, a thorough understanding of and an appreciation for fireground safety and health issues is also paramount. This begins with interrelationship of the three incident priorities and strategies and tactics. The first incident priority, life safety, should lead the incident commanders to have a strategic goal of firefighter safety. Examining and analyzing past injuries and death statistics will give the IC knowledge of common causes and types of injury that occur and how deaths are occurring.

General safety concepts can be applied to all fireground situations, including the assignment of an incident safety officer who possesses the ability to perform the duty required and has been given the proper authority at the scene. The firefighters must be equipped with properly designed and maintained PPE, and procedures should be in place governing their use. Within

the fireground IMS there must be provisions for a system to account for all firefighters working in the hazard zone. Should a firefighter need assistance, a rapid intervention crew must be available to respond immediately. Incident rehabilitation is also a necessary component for all but the smallest of fireground operations.

To think that all of the risk on the fireground could be eliminated would be akin to saying the fire department will quit going to fires. Obviously impossible. However having safety procedures as part of the overall strategic plan can minimize risk.

This chapter has examined safety from a fireground perspective. It does not consider many other safety and health issues related to fire departments, such as the physical and mental fitness of the firefighters, the safety within the station, safety in response, and safety related to the design and operation of the fire apparatus. Textbooks on safety and health programs are available that discuss these areas in detail.

REVIEW QUESTIONS

1. Describe the three incident priorities.
2. How do the incident priorities interface with fireground safety.
3. True or False? Regulations, sometimes called consensus regulations, do not have the

weight of law unless they are adopted as law by an authority having jurisdiction.

4. What three authorities must an incident safety officer be given to be effective on the fireground?

5. List five characteristics of an effective accountability system.

6. How often should fire fighting personnel inspect their PPE?

   A. Annually

   B. Weekly

   C. Monthly

   D. At the beginning of a duty shift and after each use

7. Discuss the meaning of standard of care.

8. True or False? According to this text, once assigned to rapid intervention the companies should not be assigned other tasks.

9. List three fireground situations when a personnel accountability report should be performed.

10. What three OSHA regulations were cited in the two-in, two-out ruling?

ACTIVITIES

1. Discuss the case study presented at the beginning of the chapter. How does your department's procedures compare to the recommendations from NIOSH?

2. Review you department's SOGs for fireground operations. Do the SOGs meet or exceed applicable regulations and standards?

3. Does your department routinely assign an incident safety officer on the fireground? Is the assignment effective and consistent with the recommendations presented in this chapter?

Chapter 7

COMPANY OPERATIONS

Upon completion of this chapter, you should be able to:

- Describe the responsibilities of the engine company.
- List three ways water can be delivered to the incident scene.
- List the basic ladder company functions.
- Derive the formula for tender delivery rate.
- Describe the support functions of the ladder company.
- List the considerations for apparatus placement.

CASE STUDY

It is 2:00 A.M. on a warm summer morning as the teleprinter in the house watch booth of the firehouse spits out an alarm reporting a fire in a residential dwelling. The house watch acknowledges receipt of the alarm, then turns out the engine, ladder, and chief. As the units are responding, the dispatcher reports that numerous phone calls are being received at the dispatch center reporting a fire on the top floor of a three-story row frame dwelling. Units responding realize that they are going to a working fire. As the units respond, various thoughts are going through the firefighters' minds. The first-due engine officer is concerned with apparatus placement, getting an adequate water supply from a nearby fire hydrant, determining what size hose line should be stretched and to what location, and finding the location of the fire. The driver is concerned with arriving safely at the scene, securing a water source, and operating the pump, while the firefighters are concerned about properly stretching the hose line. If too little hose is pulled, the seat of the fire will not be reached, whereas too much hose will lead to kinking and increased engine pressures. As the engine officer determines a stategy and needed tactics, the officer of the first ladder company is determining what tactics the ladder company will implement. What makes the ladder officer's job a little easier is that standard operating guidelines (SOGs) are in place for this type of fire. As the ladder company arrives, the members see that fire is showing out of three windows on the top floor of the three-story row frame building. As the ladder officer dismounts the apparatus, the engine company is hooking up to a serviceable fire hydrant and the engine firefighters are already stretching a 1¾-inch hose line to the top floor of the fire building. The reason for this size hose line is the mobility it affords in getting it into operation. The ladder company in this scenario is staffed with an officer and five firefighters. The ladder officer and two firefighters proceed to the top floor. They force entry into the fire apartment, search it, and vent it. At the same time the ladder company driver places the roof firefighter to the roof via the aerial ladder to perform vertical ventilation. This firefighter removes skylights and scuttle covers at roof level and watches as smoke, heat, and flame quickly escape the building through these vertical openings. The outside ventilation firefighter proceeds to the rear of the fire building to determine if civilians are trapped and to look for rear fire escapes. The driver remains at the turntable, ready to move the aerial ladder to other locations. As the ladder company makes forcible entry, the engine officer determines that the hose line has been properly stretched and flaked out and orders the engine operator to start water. As the ladder company searches for victims, the engine company begins an aggressive interior attack on the fire. The chief, who arrived with the first-due units, ordered the second engine to assist in stretching the first hose line and then a second hose line to back up the first.

The third engine was ordered to stretch a hose line into the top floor of an adjoining exposure while the second ladder company was ordered to split up and examine the top floors of exposures for fire extension. After placing small examination holes in the ceilings of both exposures, the second ladder company reports no extension. The first-due engine has extinguished the fire and all searches for victims have turned up negative. There were no injuries at this fire and 1 hour after the initial receipt of the alarm, the incident commander declared it under control and all units returned to service.

INTRODUCTION

The case study in this chapter depicts a building fire with predetermined SOGs and common company assignments and operations used to control and extinguish the blaze. The concepts were simple and standardized by most fire departments. The strategic goals and tactical objectives at the fire were those common to every fire. The engine companies followed common practices. They took a position at an adequate or reliable water supply to provide a water source and stretched hose lines to proper locations to facilitate the fire's confinement and extinguishment and to prevent extension to adjoining exposures. At the same time, ladder companies rapidly deployed their personnel to force entry, ventilate, and search. They followed up-to-date SOGs for this type of building fire. Engines and ladders worked together to achieve a common goal. In the case study there were adequate resources to cover all sides of the building, stretch multiple hose lines, and conduct rapid searches, achieving the goals of firefighter safety, search and rescue, exposure protection, confining, extinguishing, ventilating, overhaul, and salvage. Some departments have adequate resources to respond to and extinguish fires in a timely fashion, whereas other departments may require assistance from surrounding departments. In this chapter we discuss basic engine and ladder company operations. These are the functions that must be carried out in order to successfully fight fires. It makes no difference whether the department is paid or volunteer or if the department runs engines or ladders—fire growth, flame travel, and smoke travel are all the same. Understanding the engine and ladder concepts and their roles at fires will better prepare you to apply effective strategies and tactics.

When we speak of engine and ladder functions, we speak of a **company**. This can be any type company and does not necessarily refer to personnel only assigned to a pumper or ladder truck. The terminology is used to describe the required functions, the fire apparatus, the tools and equipment carried on it, and most important, the number of trained personnel.

■ **Note**
Understanding the engine and ladder concepts and their roles at fires will better prepare you to apply effective strategies and tactics.

company
a team of firefighters with apparatus assigned to perform a specific function in a designated response area

■ **Note**

**Because the engine
company's primary
function is to supply
the water to extinguish
the fire, a fire
department would not
be a fire department
without this basic
operating unit.**

■ **Note**

**For the purpose of
this book, the term
engine company refers
to the commonly
carried fire
equipment, tools,
hose, and water tank
on an NFPA-compliant
class A pumper and
the assigned
personnel.**

ENGINE COMPANY OPERATIONS

General

The engine company (**Figure 7-1**) is the basic building block of every fire department. Because the engine company's primary function is to supply the water to extinguish the fire, a fire department would not be a fire department without this basic operating unit. In various regions of the country, engine companies may be referred to as something other than an engine, such as a pumper or hose company. Therefore, for the purpose of this book, the term *engine company* refers to the commonly carried fire equipment, tools, hose, and water tank on an NFPA compliant class A pumper and the assigned personnel. The minimum requirements for a class A pumper from NFPA standard 1901 are given in the following list:

- Minimum water tank of 500 gallons (**Figure 7-2**)
- Must have a hose compartment of at least 55 cubic feet for $2\frac{1}{2}$-inch or larger supply hose and two compartments of at least 3.5 cubic feet for $1\frac{1}{2}$-inch or larger attack hose (**Figure 7-3**)
- Minimum pump size of 750 gpm (**Figure 7-4**)

Figure 7-1 *The
engine company
apparatus,
equipment, and
personnel are the
basic building
blocks of the fire
department.*

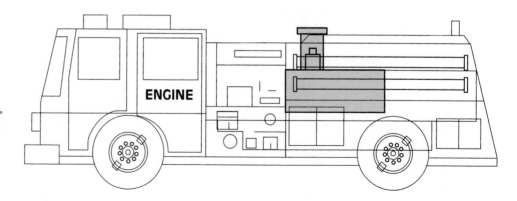

Figure 7-2   The class A pumper must have a water tank with a 500-gallon capacity.

■ **Note**

Firefighters should have a very good knowledge of the hose loads, types and sizes of hose, and the various nozzle types and sizes available.

The Apparatus

All firefighters should be familiar with the particular apparatus that they are assigned, including a basic understanding of pump principles and, most important, location and operation of equipment. The firefighters should also have a very good knowledge of the hoseloads, types and sizes of hose, and the various nozzle types and sizes available. This sounds like common sense, but often in fire departments, firefighters may be assigned to an engine company on the other side of town where they may never have worked. This is another good case for apparatus standardization.

Figure 7-3   A compliant hose compartment.

Figure 7-4 *A typical fire pump arrangement.*

In many fire departments, driver/operators are usually designated and provided with additional driver training and pump operation training. This designation may come by seniority or by examination administered for a driver. Regardless of how your department selects its driver/operators, all firefighters must be well versed in the operation of the apparatus and equipment.

Engine Company Responsibilities

The responsibilities of the engine company focus around the premise that the engine company is the basic building block of the fire service. As with any first-arriving unit, size-up is of utmost importance for an engine company. Size-up starts with the receipt of the alarm and, in many instances, before the alarm

Figure 7-5 *Mobile data computers can provide various types of information to responding firefighters and a way to communicate with dispatch centers.*

■ **Note**
The engine company must secure an uninterrupted water supply and pump water to supply a water stream.

■ **Note**
Although all pumpers carry some water with them, the amount is generally not enough for a complete fire suppression operation.

comes in. Computer programs such as Critical Information Dispatch System (CIDS) can send pre-entered building data over the mobile data computer (MDC) while enroute to an alarm. This information may come from written resources in the cab, such as preincident plans or maps (**Figure 7-5**), or it may come from the dispatch center or from the crew's personal knowledge of the response area. As the unit arrives on the scene, the officer must open his or her eyes and ears to everything. Look up, down, and all around. Listen for the radio and for information gained from bystanders. What odors are in the air? Is it wood, structure fire, vehicle fire? The officer should look for immediate hazards, locate a water supply, determine what and how much is burning, and determine the need to respond more units or other agencies such as emergency medical service (EMS), police, or utility companies. Knowledge of the response zone is critical in locating a water supply. The engine company must secure an uninterrupted water supply and pump water to supply a water stream.

Water Supply In order to ensure that water is available for the entire fire suppression operation, an uninterrupted water supply must be established. Although all pumpers carry some water with them, the amount is generally not enough for a complete fire suppression operation. The **booster tank** water is useful for some

booster tank
the onboard water tank for an engine

vehicle fires, some small brush fires, and other small, outside fires. Fires in structures require an additional water supply, which generally comes from fire hydrants or static water sources, such as lakes, pools, or ponds. In areas where hydrants are plentiful and have adequate water and pressure, engine companies have three options for moving the water from a fire hydrant to the fire scene.

The following are options for moving water from a water source to the fire scene:

1. *Forward lay.* In a forward lay (**Figure 7-6**), the engine company stops at a fire hydrant or other water source while en route to the fire scene. The engine company prepares the hose connection(s) for the hydrant and lays

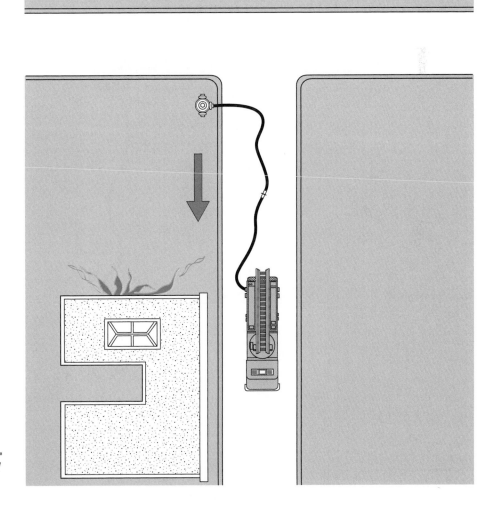

Figure 7-6 *A forward lay.*

the hose toward the fire scene. Once the engine stops at the fire scene, the supply line(s) are then connected to the operating engine, after which the supply line(s) are connected to the hydrant and charged (the water is supplied by hydrant pressure only). A second engine may connect to the hydrant using a **capacity hookup** to boost the pressure.

capacity hookup
a hookup to a fire hydrant designed to supply the full volume of the pump

2. *Reverse lay.* In the reverse lay, the engine company goes to the fire scene, drops the supply hose or hoses and any other needed equipment, and then lays the hose toward the water source (**Figure 7-7**). If the water source is a

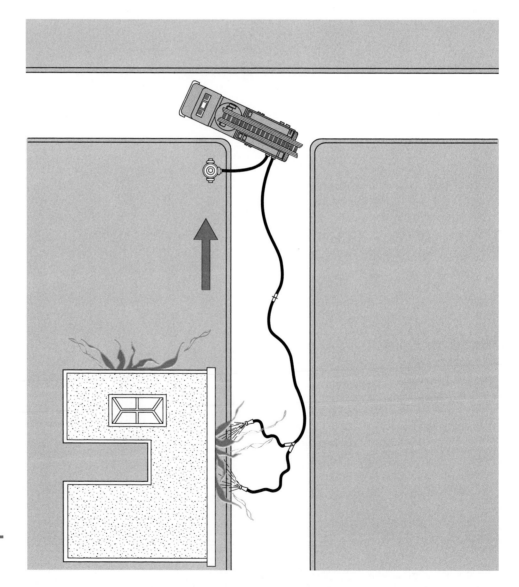

Figure 7-7   *A reverse lay.*

fire hydrant, the engine laying the supply hose may or may not connect to the fire hydrant, depending on water flow needs and the pressure of the hydrant. Again, if the supply engine connects to the hydrant to augment the water supply, the connection should be a capacity hookup.

3. *Split lay.* The split lay (also referred to as a driveway lay) (**Figure 7-8**) is a combination of two methods. One option for a split lay is when the initial engine drops its supply line at a corner or entrance of a drive and a second engine lays from another location (a water source or other preplanned spot) toward the supply line of the initial engine. In this scenario the second

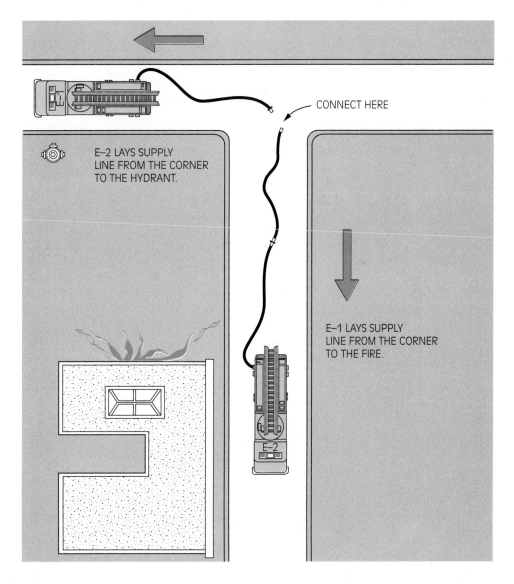

CONNECT HERE

E–2 LAYS SUPPLY LINE FROM THE CORNER TO THE HYDRANT.

E–1 LAYS SUPPLY LINE FROM THE CORNER TO THE FIRE.

E-2

Figure 7-8   *A split, or driveway, lay.*

engine would supply water by relay pumping or may just connect to the first supply line if the hydrant pressure is sufficient. A second option is when the second engine drops the supply hose at the end of the first and connects to it, then proceeds to the water source. Again, whether the engine connects to the hydrant depends on the needs of the incident and the water system.

The second type of water source can be a static supply. This source may be the norm in areas where hydrants are not readily available. The options available in these areas include drafting and water tender shuttles described in the following list:

- *Drafting.* Drafting water sources can come from any static water location, such as lakes, rivers, pools, or portable water tanks (**Figure 7-9**). Further,

static supply
a water supply that is always in the same location, such as a lake or pool

water tender
fire apparatus that is a mobile water supply; may be termed a *tanker* in some areas of the United States

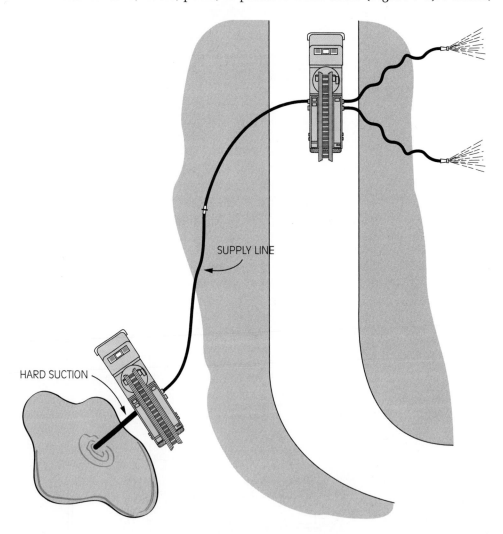

SUPPLY LINE

HARD SUCTION

Figure 7-9
Apparatus positioned for drafting operation.

many municipalities are utilizing dry hydrants. Dry hydrants are installed by developers in order to meet various flow requirements. Dry hydrants are essentially hard suction tubes that are piped into a close static water source. When the engine company arrives on the scene, it must attach the hard suction hose to the dry hydrant and pull a vacuum in order to establish a water supply. Care should be given in dryer seasons to be sure the water source remains a viable option.

- *Tender shuttles.* Mobile water tenders are another option in areas where a fixed water supply is scarce. Water tenders may carry water in amounts from 1,000 to 8,000 gallons or greater in some cases. Depending on the need and the resources available, tender shuttles may be established wherein several mobile water tenders respond to the incident and shuttle water back and forth from a fixed water supply site (**Figure 7-10**). Tenders may connect and directly supply the operating engine or they may fill portable water tanks that the operating engine can draft from. It is important when choosing a tender shuttle as a water supply option that the resources available match the water supply need. A helpful tool in this analysis is the tender delivery rate (TDR) formula. The flow requirements for the incident are calculated and the time it takes for a tender to make an entire cycle is calculated. This cycle is the time it takes to dump the water at the scene, return to the static source, fill with water, and return to the scene. The TDR allows the incident commander to determine how many similar-size tenders are needed for the shuttle.

    As an example, suppose we have a tender with a capacity of 2,500 gallons and it takes 20 minutes for the refill cycle; the TDR for this tender is 125 gpm (2,500 ÷ 20). If the fire required 750 gpm as a needed flow rate, then six tenders of the same size would be needed to have a continuous water supply (750 ÷ 125). As a note, tenders with differing capacities affect the rotation in terms of refill and offload times.

Many departments are very efficient at establishing a water source where none is available. A combination of drafting and tender shuttles wherein the tenders continually dump water into portable tanks and engines draft from the tanks can be used to supply an endless amount of water to the fire scene. Obviously, training and preplanning are crucial in making such an operation a success.

■ **Note**

Besides stretching a hose line to the right location, choosing the right size and length of hose line is essential for a successful firefighting operation.

Fire Attack Lines The keystone of a fire department is the engine company and its ability to supply the water to the hose lines needed to control a fire. These lines may be handlines, aerial streams, deluge guns, or any combination thereof.

Besides stretching a hose line to the right location, choosing the right size and length of hose line is essential for a successful firefighting operation. Many factors must be considered when arriving on the fire scene in deciding what type, length, and size of hose to stretch. For example, should a 1½-inch, 1¾-inch, or a

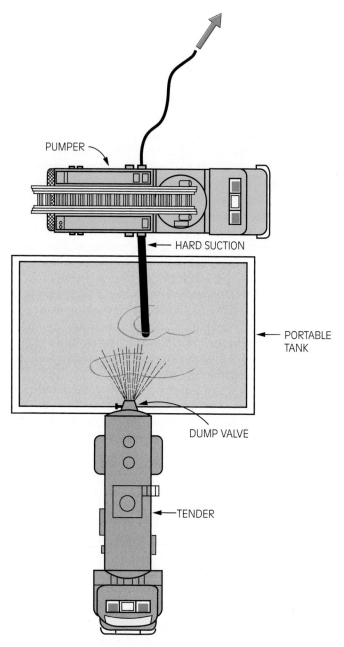

PUMPER

HARD SUCTION

PORTABLE TANK

DUMP VALVE

TENDER

Figure 7-10   *A portable tank put in place as part of a tender shuttle.*

larger 2½-inch hose line be used? The factors that must be considered in selecting the proper hose line are given in the following list:

- What is the location of the fire?
- What materials are burning—plastics or natural?
- How much is burning and in what storage configuration?
- Where is the fire extending to?
- Where is the line going to be placed or stretched to?
- What areas need immediate attention (such as a structure's means of egress)?
- How much staffing is available?
- What is the building's construction?
- What is the building's height?
- What are the building's dimensions?

■ **Note**

It is of no use to stretch a smaller 1¾-inch hose line on a fire that requires a 2½-inch hose line for extinguishment.

It is of no use to stretch a smaller 1¾-inch hose line on a fire that requires a 2½-inch hose line for extinguishment. For a room and contents fire, be it in a private dwelling or a multiple dwelling, the smaller 1½-inch or 1¾-inch will probably produce an adequate flow. The shorter period of time needed to get a 1½-inch or 1¾-inch into operation and the fact that it can be handled easily by two firefighters makes it the hose line of choice, but it is important to understand the difference in the possible flows between the various line sizes and nozzles. A 1¾-inch hose line with a solid bore nozzle is capable of flowing approximately 180 gpm at a nozzle pressure of 50 psi, whereas a 2½-inch hose line operated at a similar nozzle pressure is capable of flowing approximately 320 gpm, which is almost double. Different nozzles have varying flows. Obviously, the larger the line, the greater the flow; however the larger line, as shown in **Figure 7-11,** is more difficult to maneuver and requires more personnel to operate safely. The volume of fire must be assessed when determining the size attack line and nozzle to be used.

■ **Note**

The right line for the right fire is necessary for an effective and safe fire attack.

The right line for the right fire is necessary for an effective and safe fire attack. The decision of what type and size of hose line to stretch must be made instantaneously by the arriving engine officer with consideration to all the foregoing factors.

Locating, Confining, and Extinguishing In order to meet the strategic goals of confinement and extinguishment, the fire must be located, which, as shown in **Figure 7-12,** is sometimes apparent, but most times may not be easy.

■ **Note**

Locating the fire starts with information received on receipt of the alarm.

Locating the fire starts with information received on receipt of the alarm. A specific address or general location is given by dispatch. Upon arrival at the scene, the building's exterior should be surveyed for signs of smoke or fire. Further, building occupants may have vital information such as the apartment number, information regarding other people in the building, or where the smoke or

Figure 7-11   *A 2¹⁄₂-inch hose, though it has higher flows, is more difficult to move.*

Figure 7-12
Sometimes the location of the fire is apparent, but many times it is not. (Photo courtesy of South Trail Fire Department.)

fire was seen. If a specific apartment is given, then it should be examined. If there is no answer at the apartment or structure, then an examination must be conducted from all sides, if possible. Look in the windows, check the windows for heat, soot, and condensation. The doors should be shaken, which may cause smoke to show from around the door. A particular odor may indicate what is burning. It is common for people to leave their homes while cooking. An irritating odor of an aluminum pot burning may indicate food burning on the stove. A fuel oil odor may indicate an oil burner malfunction, and an electrical odor may be from a defective or failing lighting ballast, whereas an odor of burning wood or plastic may indicate a structure fire. Listening for the crackling of the fire and the buildup of heat can also help to determine the fire's location. When the exact location of the fire has been verified, all units on the scene should be notified, given an assignment, or placed in staging. It is important to consider what may be occurring and whether a hose line should be stretched if the exact location of the fire has not yet been determined. This does not preclude the engine from stretching its hose line to the front door of the building, but remember that sometimes the address of the person reporting the fire is inadvertently given as the fire location, although the actual fire may be at another location (for example, across the street or on another floor). The use of some equipment that is relatively new to the fire service, such as handheld and thermal imaging cameras, can make the job of locating the fire much easier and quicker.

■ **Note**
Confining the fire is the second step in fire control.

Confining the fire is the second step in fire control. This action is taken to prevent fire spread from one room to another, one apartment to another, or one building to another. Simply stated, confinement is keeping the fire in the area, room, or building of origin (**Figure 7-13**). Closing a door to a room or apartment may temporarily be sufficient to confine the fire until extinguishment efforts can take place. In some situations, an action as simple as operating a water extinguisher may slow the fire's growth enough while a line is placed in operation. Likewise a single hose line may be used to confine a fire while a second hose line extinguishes the fire, for example, a fire in a cellar of an apartment building, as described later in the text. Vertical ventilation is also a tool that can be used to draw fire, heat, and smoke up and out of the fire building, thus slowing and often preventing possible horizontal extension.

■ **Note**
Confinement means keeping the fire in the area, room, or building of origin.

■ **Note**
Extinguishment is the final step in the suppression effort.

Fire extinguishment is the final step in the suppression effort. Extinguishment includes putting out all visible flames and any hidden pockets of fire. The search for concealed fire areas and eventual total extinguishment is called *overhauling*. However, before the fire can be extinguished, a hose line or hose lines need to be properly stretched and charged with water. Getting a hose line into operation and obtaining an adequate water supply are the two important basic functions of an engine company. It is crucial that the engine company personnel understand that more lives are actually saved by properly positioning and operating hose lines than any other rescue technique. Engine companies must stop the fire spread and put the fire out. Getting the hose line into operation typically falls under the responsibility of the engine company officer and firefighters.

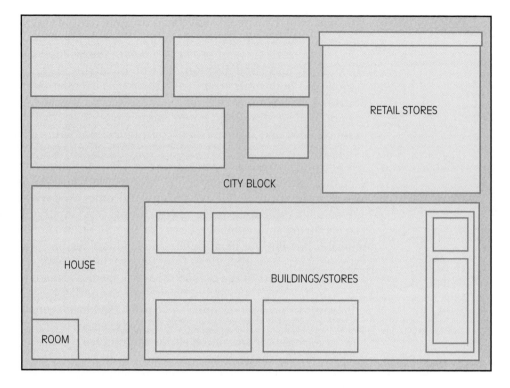

Figure 7-13 *The concept of confining the fire can be thought of as a series of boxes. Once the fire has extended from one box, efforts must be focused on confining the fire to the next box.*

Proper operation of the hose line is the officer's responsibility. In some cases, once a hose line is called for, due to staffing levels, the officer may assist in the stretch, whereas in other cases the officer calls for a hose line and proceeds into the building while the firefighters stretch the hose line. In the best case, the officer should be responsible for communicating with command, for crew safety, and for planning an escape route. Regardless of the department's SOGs, all engine firefighters should be able to perform the following procedures:

- Lay lines from hydrant to pumper.
- Position or reposition the apparatus.
- Determine the amount of hose needed to reach the seat of the fire.
- Stretch a hose line.
- Supply tank water to hose lines while awaiting hydrant supply.
- Make a hydrant connection.
- Make all hose connections.
- Maintain adequate pressure in hose lines using the pump.
- Utilize deck guns.
- Supply foam.
- Supply auxiliary protection systems.

Figure 7-14   *The engine company officer must ensure that the crew is ready to enter for fire attack.*

Once the hose line is properly stretched and charged from the pumper location, the engine officer must ensure, through size-up, that all of the crew are properly geared to begin the fire attack (**Figure 7-14**).

The nozzle person should bleed all air out of the hose line before advancement is made into the fire area. Before entering a door to the fire area, all firefighters should remain down low and should be on the same side of the hose line and on the same side of the doorway opening. After the initial blast somewhat subsides, the room can be entered. Once the room is entered, the officer must monitor conditions and ensure the safety of the firefighters. This may include ensuring that the floor is swept with water to prevent leg and knee burns to firefighters. If fire does not meet the engine company at the door, firefighters will want to stay low. The officer may be able to see where the fire is coming from before the smoke banks down to the floor. When fighting a fire in an apartment

■ **Note**

When fighting a fire in an apartment building, a quick look at the layout of the apartment directly below the one involved may assist the engine company's movements toward fire attack.

building, a quick look at the layout of the apartment directly below the one involved may assist the engine company's movements toward fire attack. Apartments that are vertically aligned usually have similar layouts. Never advance into a fire apartment without a charged hose line. As conditions become increasingly hotter, it may indicate that you are getting closer to the fire. Most of the time the hose line should not be opened and used on smoke; however, the line may be used to cool down a superheated ceiling to prevent flashover. Once the fire is located, the nozzle should be opened and operated in a clockwise direction. Another way to operate a nozzle is in a Z pattern, which sweeps the ceiling area, room area, and floor area. Examples of the O and Z patterns are shown in **Figure 7-15**.

Unless directing the stream on a localized fire, the hose line should also be operated overhead and in front of the nozzle team. The stream should also be deflected off the ceiling. The reach of the hose stream should always be used to the firefighters' advantage. Hose lines have great reach and cover large areas—for

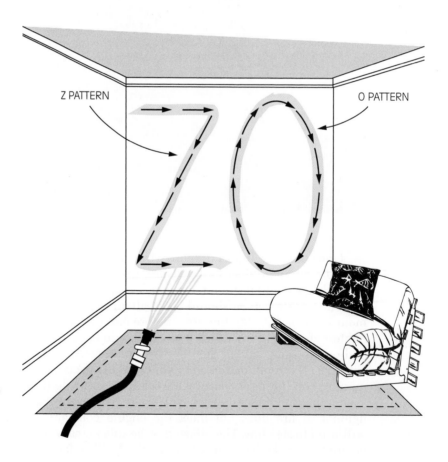

Figure 7-15 *Two acceptable patterns for fire attack.*

■ **Note**

If progress is not being made toward extinguishment, then advancing several more feet may be the solution or the flow may not be enough for the volume of fire.

■ **Note**

In the incipient stage, the smoke, heat, and fire conditions are minimal. A direct attack on the fire can be made immediately.

ladder
the fire apparatus with aerial and ground ladders, referred to sometimes as a *truck;* may or may not have pumping capabilities

■ **Note**

Many departments rely on engine companies, squad companies, or rescue companies to fill the void.

squad
a unit that may carry firefighters, firefighters with specialized tools, or a medical rescue (EMS) unit; may be referred to as a *rescue truck*

rescue
see *squad*

instance, when fighting a fire that involves one or two rooms of an apartment, taking a position at the doorway of a room is usually sufficient for a quick knock-down of the fire without entering the room itself. If progress is not being made toward extinguishment, then advancing several more feet may be the solution or the flow may not be enough for the volume of fire. Fires burning in walls and the ceiling overhead generate tremendous heat, and extinguishment can only be made after these areas are opened up. As the fire begins to darken down, the nozzle can be lowered and directed at the burning contents. Once the fire is knocked down, the nozzle should be shut down and the area should be allowed to vent. Once this occurs, the officer and firefighters should inspect the area for any remaining pockets of fire that flare up. When it is determined that the fire has been fully extinguished, the final steps in the overhaul and salvage operations can take place.

Stages of Fire and Fire Attack As discussed in Chapter 2, it is crucial to understand what stage the fire is in. In the incipient stage, the smoke, heat, and fire conditions are minimal. A direct attack on the fire can be made immediately. In the free-burning stage, the fire produces tremendous amounts of heat and heavy fire conditions. In the free-burning stage, rooms adjacent to the fire area may be at the point of flashover. Videos have been produced showing that opening the nozzle and operating it into the superheated gases at the ceiling level will drive back the fire and gases from uninvolved areas and prevent flashover. In this stage it is suggested that the ceiling, then the floor be swept with a hose line, while continually operating the hose line on the fire. In the final stage, the smoldering stage, firefighters must be ever vigilant and alert. In the smoldering stage, fire has passed through both the incipient and free-burning stages and, due to a lack of oxygen, the fire has burned itself out, yet dangerously high levels of heat and flammable gases still remain. All that is needed for this fire to violently begin to free burn is the introduction of oxygen. In this stage vertical ventilation must precede forcible entry. Once vertical ventilation has been performed, the superheated gases should ignite and escape through vertical openings, eliminating the backdraft potential. As soon as this is accomplished, forcible entry can begin and fire suppression efforts commence.

LADDER COMPANY OPERATIONS

Ladder company operations are just as critical as engine company operations, and all ladder company personnel must be acutely aware of their responsibilities and roles on the fireground. Although many fire departments operate ladder companies, such as the one shown in **Figure 7-16**, others do not. Many departments rely on engine companies, squad companies, or rescue companies to fill the void. As shown in **Figure 7-17**, these units must have the proper equipment for the ladder company functions. No matter what the department utilizes, the

Figure 7-16 *Ladder company operations are essential to the fireground even if a department does not operate apparatus with a ladder on it. (Photo courtesy of South Trail Fire Department.)*

Figure 7-17 *Some departments operate squads or rescues to perform ladder company duties.*

basic ladder company functions must still be in place and still be accomplished. The term *ladder company function* is used in this text to describe a certain group of fireground activities. Again, whether your department operates a ladder company or not, these functions and activities must still be undertaken on each and every fire to meet the strategic goals.

In order for ladder company functions to be met on the fire scene, various ladder company specific tools and specialized equipment may be required, such as hand tools for forcible entry, various saws that cut through wood, metal, and concrete, vehicle rescue equipment, air bags, cutting torches, and portable lighting and generators, just to name a few. Portable ladders of assorted sizes (**Figure 7-18**) are also an essential part of the company's arsenal of equipment. These tools may be located on an engine company, squad, rescue truck, or any apparatus, but they must be provided so the ladder company functions can be completed efficiently and effectively.

Ladder companies play varied and dynamic roles in the fire service however, this text focuses on ladder company operations as it pertains to the fireground. The major role of the ladder company at the fire scene involves forcible entry, search, and ventilation. The ladder company is also responsible for other support functions such as laddering the fire building, overhauling, and salvage operations, as well as shutting off gas, electric, and water service to the fire building, when necessary.

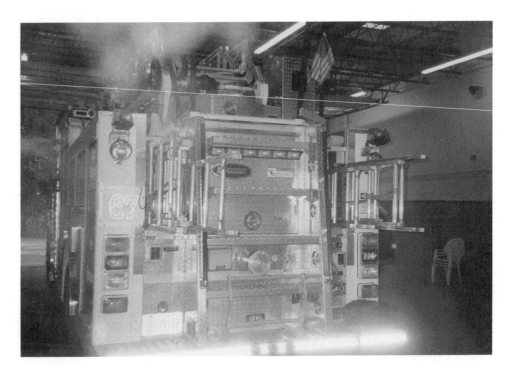

Figure 7-18   *An assortment of ground ladders.*

Ventilation

Ventilation is a technique used by firefighters to remove smoke and heat from a structure. When performed properly, it greatly assists search and rescue operations while allowing the engine companies to advance their hose lines to perform extinguishment duties. In many cases, ventilation determines the outcome of a fire (**Figure 7-19**). In fact, when the tactical method of extinguishment includes fog nozzles, ventilation is critical and proper timing is needed. The extinguishment must be done in conjunction with the ventilation so the steam that is created has an avenue out of the building, or the fire attack team will be caught in the steam.

Two basic types of ventilation are performed by firefighters: vertical ventilation and horizontal ventilation. Each is performed at different stages in the fire and for different reasons. There are also different methods of performing ventilation—for example, natural ventilation and mechanical ventilation.

Vertical Ventilation The purpose of vertical ventilation is to draw heat, smoke, and fire up and out of the fire building through vertical openings. Depending on the type of structure and occupancy, these openings can include bulkheads, scuttles, and skylights, as shown in **Figure 7-20**. Venting all these openings is considered

Figure 7-19 *Proper ventilation is crucial in the outcome of a fire.*

Figure 7-20   *Possible vertical roof openings include bulkheads, scuttles, and skylights.*

part of the roof division's primary roof duties. Benefits of vertical ventilation include:

- Prevents fire gases and heat from banking down from upper floors
- Increases survival time of victims by channeling gases and heat away from them
- Clears hallways and stairwells of smoke and heat, assisting escape of occupants
- Assists other firefighters in attaining a position on the floors above the fire for search and rescue
- Assists the engine company in advancing its hose lines, allowing a rapid interior attack

Determining how the roof division gets to the roof is another consideration. The roof division must get to the roof as soon as possible without compromising the firefighters' safety. There are three basic ways to get to the roof, and the type of building usually dictates which method is best. In large multiple dwellings or apartment houses the preferred method of getting to the roof would be via an adjoining building, an aerial or tower ladder, or a rear fire escape, if one is provided. As shown in **Figure 7-21**, in most cases, for security reasons front fire escapes or fire escapes facing a street stop at the top floor and do not go all the way to the roof. The preferred choice in this situation would be using an adjoining building, as shown in **Figure 7-22**. It is often an easier and quicker way to get

Figure 7-21 *Note the fire escape does not go the whole way to the roof, therefore would not be an option for roof access or egress.*

to the roof, while allowing aerial or tower ladders to be used for other positioning at windows. If the fire building is isolated, as in **Figure 7-23**, then an aerial or tower ladder would be the preferred method.

Getting to the rear of these large buildings to use the fire escape is often both difficult and time-consuming, making it the least preferred method. If the fire is in a smaller building, the preferred way to the roof would be via an aerial or tower ladder. Using adjoining buildings of smaller structures is often problematic because they may have scuttles that are sealed up, which makes gaining entry into these smaller, basically private dwellings difficult and very time-consuming. Using the aerial ladder ensures the roof division of getting to the roof and a safe means of escape. The interior stairs of a fire building are *to be avoided* in getting to the roof. Of course, in some buildings, such as high-rises or an isolated building above the reach of ladders, the interior stairs would have to be used.

The first thing the roof division must consider when getting on the roof is a way of getting off the roof. The firefighters must always have an alternate way off the roof (**Figure 7-24**), if fire should cut off the primary means of egress. The firefighters on the roof are the eyes of the incident commander and must inform the incident commander of conditions visible at the roof level such as:

- Visible fire and smoke, color, and volume
- Location of shafts
- Paths of fire extension
- Persons in immediate distress

■ **Note**

Avoid the interior stairs of a fire building in getting to the roof because of the potential of trapped gases, smoke, and heat in the stairwell.

Figure 7-22 *Using an adjoining similar building is a preferred way to get to the roof.*

The roof division must immediately force open any existing roof openings, such as bulkheads, scuttles, and skylights, to immediately allow ventilation to take place. After forcing open a bulkhead or other interior entrance to the roof, the bulkhead landing and stairs should be probed with the blunt end of a 6-foot pike pole to check for the presence of overcome occupants, particularly if the bulkhead was found to be locked from the interior. Additional ventilation can be provided by breaking the glass on the skylight over the bulkhead, if present. Scuttle covers can be removed with conventional hand tools. However, if they have been tarred over, a saw with a carbide tip blade may be necessary. When breaking skylights for ventilation, the roof division should notify the firefighters operating below by radio as to the operation going on above them. Again, after removing the glass, a 6-foot pike pole should be used to check for the presence

Figure 7-23   *In some cases, the aerial ladder is used to get to the roof.*

Figure 7-24   *An alternate way off the roof must be provided to ensure firefighter safety.*

of a draft stop, especially if little or no smoke is showing at this opening. All of these procedures are part of the roof division's primary tactical objectives.

Secondary roof procedures involve roof-cutting operations with power tools. This operation would take place at top-floor fires, fires involving the cockloft or attic, or fires in single-story structures. The operations that involve power tools should take place after primary roof procedures using existing openings have been completed. When cutting the roof, the hole should be cut in a way that makes it easy to pull and should be at least 4 feet by 4 feet. The opening should not be so large that it cannot be easily opened. The opening should be made as directly over the fire as is possible and safe without cutting any parts of the roof support assembly. When trying to determine where the roof should be opened, the roof firefighters should use the following reliable signs to locate the fire:

- Melted snow
- Steam on a wet roof
- Bubbling tar
- Soft areas of the roof
- Sense of touch
- Looking over the roof's edge
- Visible location of fire

> **■ Note**
>
> When cutting the roof, the hole should be cut in a way that makes it easy to pull and should be at least 4 feet by 4 feet. The opening should not be so large that it cannot be easily opened.

After the roof is cut and sections are pulled out, it is imperative that top-floor ceilings are pushed down with a pike pole or other tool or pulled open from below. If this action is not performed, then the roof operation will not be effective. As critical as vertical ventilation is at fire operations, there are times when firefighters do not want their members on the roof. Naturally when the roof condition is in doubt, such as in vacant buildings or heavy fire conditions, firefighters should not be committed to roof operations. It is equally important to understand how roofs are constructed in the area. Membrane roofs and gypsum-poured or gypsum-slab roofs could be disastrous to firefighters. Membrane roofs that ignite can spread fire rapidly across the entire roof, whereas gypsum roofs can fail rapidly under fire conditions and easily absorb moisture. If a member cuts a roof and a white paste is observed on the saw blade (indicating wet gypsum), the roof firefighter should immediately notify the incident commander and evacuate the roof. In many areas of the country, large commercial stores have lightweight trusses supporting their roofs. There is no justifiable reason to cut a roof supported by these trusses.

Horizontal Ventilation Horizontal ventilation involves removing windows and outside doors of the fire area or building. The purpose of horizontal ventilation is the same as vertical ventilation—to remove smoke and heat from the fire building in order for the engine company to make an aggressive attack on the fire and to decrease the chances of fire rolling over the attack crews by allowing heat,

> **■ Note**
>
> Horizontal ventilation is usually performed after the engine company's hose line is stretched, charged, and ready to begin the fire attack.

smoke, and fire to escape through these openings. Removing products of combustion and heat makes searches of the fire floor and floors above easier and increases the survival time of overcome occupants. Horizontal ventilation is usually performed after the engine company's hose line is stretched, charged, and ready to begin the fire attack. This ventilation should be coordinated between the inside and outside fire teams. If horizontal ventilation is performed prematurely, it could lead to rapid extension of the fire and possible autoextension. Another problem facing the fire service today is the use of building materials such as polystyrene foam and thermal pane windows, which effectively insulate a structure. Firefighters are more frequently finding fires in their smoldering stage, with tremendous heat and no visible fire present. These conditions are prime for a backdraft, which can definitely be caused by improper horizontal ventilation preceding vertical ventilation. Communication between the outside and inside teams is crucial under these conditions in order to reduce the chances of a backdraft. Although it is not the practice of fire attack crews to operate a handline on smoke, it would be advisable under these conditions to reduce the temperature of the gases below their flammable range by completely cooling the area.

Venting for Life versus Venting for Fire Two terms used in the fire service are *venting for life* and *venting for fire.* When firefighters vent for fire as described in horizontal ventilation, they are facilitating the engine company's advance into the fire area. The heat and products of combustion are removed from the structure so that the fire can be rapidly extinguished. This operation is a coordinated process between the inside and outside teams. When firefighters vent for life, they are doing this in order to enter an immediately dangerous to life or health (IDLH) area where there is a known or suspected victim. This task can prove to be most difficult. There is a tremendous risk venting a window that may be pulling the fire in that direction. Windows must be completely removed before entry is made. The incident commander and the inside team must be aware that a firefighter is venting and entering a window in search of trapped or unconscious victims. Venting windows, when necessary, makes life easier for firefighters operating inside along with the possible overcome occupants. Be sure that crews working outside below the window being opened are aware of what is happening above them. Compared to the damage smoke causes to a building and its contents, windows are easier and less expensive to replace. Windows should not be taken indiscriminately, causing unnecessary damage; but when necessary, one should not hesitate to remove them.

Natural Ventilation versus Mechanical Ventilation Natural ventilation involves the opening of doors, windows, skylights, bulkheads, and any other building openings to permit smoke, heat, and flames to escape through these openings with no assistance from fire personnel other than by creating the opening. The gases and smoke are looking for a means to escape from the structure and the firefighters are

autoextension
when a fire goes out a window or door and extends up the exterior of a building to the floors above or the cockloft

■ **Note**
When firefighters vent for fire, they are facilitating the engine company's advance into the fire area.

■ **Note**
When firefighters vent for life, they are doing this in order to enter an immediately dangerous to life or health (IDLH) area where there is a known or suspected victim.

immediately dangerous to health or life (IDLH)
used by a number of OSHA regulations to describe a process or an event that could produce loss of life or serious injury if a responder is exposed or operates in the environment

providing that means by performing ventilation. Not so with mechanical ventilation. As the term implies, mechanical ventilation is a means to remove smoke from a structure with the assistance of mechanical equipment such as smoke ejectors, positive pressure blowers/fans, exhaust fans, building ventilation systems (HVAC), and fog streams from hose lines. Generally a fog line is used to assist with smoke removal from a structure where the fire has been knocked down and the smoke will not vent. To accomplish this, the nozzle team stands 6 to 8 feet from a window and directs a fog stream roughly the size of the window opening, out the window. This is a proven, effective way to quickly ventilate a room or area. With the use of an exhaust fan, smoke can be pulled from a structure to the outside atmosphere. The fans can be set up at windows or doorways and assist in smoke removal; however, fans are not the most efficient method. It is time-consuming to get fans into operation, and if they are used near a door, they become a hindrance for firefighters trying to quickly enter and exit the fire area. Fans used in the exhaust mode do not operate at peak efficiency and foreign objects can be easily pulled into them.

In many fire departments, when the construction is favorable and the locations of occupants are known, positive pressure ventilation is used during initial fire operations. The concept of positive pressure ventilation is shown in **Figure 7-25**. This method of mechanical ventilation entails setting up a fan or fans that circulate clean air into a room or structure in conjunction with horizontal ventilation of doors and windows opposite the fans. This is an effective means of ventilation and an asset to firefighting forces because heat and smoke are being pushed away from the attack team, making it easier for them to move in and extinguish the fire. A potential drawback to positive pressure ventilation is that there is a good possibility of pushing smoke and fire into uninvolved areas of the building, possibly extending the fire or endangering occupants. Therefore, proper size-up and creation of an escape exit to vent the fire area from the exterior of the smoke and heat is necessary. All horizontal ventilation should be well coordinated between the inside and outside teams to reduce injuries and maximize the effectiveness of smoke, heat, and fire removal from the involved structure.

Search

The search for life, fire, and fire extension are several functions of ladder companies. Along with the proper placement of hose lines there is no greater tool than a well-developed search plan. A search can be summed up as a plan of action to cover all areas of a structure in a probe for victims or fire. While looking for possible fire victims, searches should be broken down into two stages: primary search and secondary search. The primary search is an immediate search for trapped occupants. The search should be performed quickly, but thoroughly. The secondary search can be defined as a meticulously thorough search by a different group of firefighters to ensure that no victims were missed

■ **Note**
A potential drawback to positive pressure ventilation is that there is a good possibility of pushing smoke and fire into uninvolved areas of the building, possibly extending the fire or endangering occupants.

■ **Note**
A search can be summed up as a plan of action to cover all areas of a structure in a probe for victims or fire.

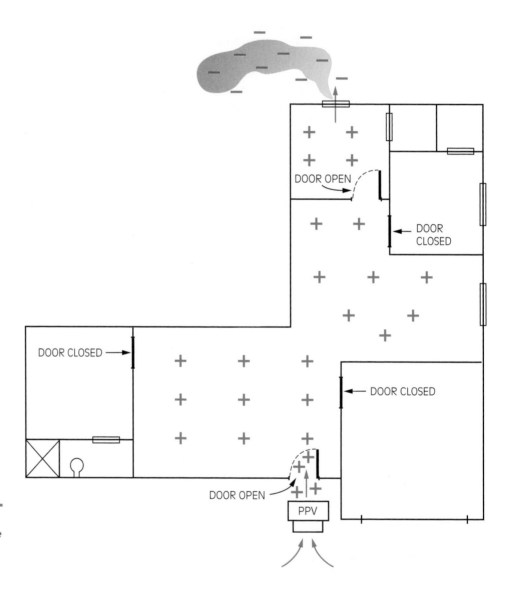

Figure 7-25 *The principle of positive pressure ventilation.*

during the primary search. It is recommended that the firefighters who perform the primary search should not perform the secondary search. This search should include the structure's exterior, including shafts, basements, and cellars. Search protocols for different departments depend on available staffing and equipment. In many large urban departments, at least two ladder companies respond on the initial alarm for a structure fire. The first ladder company is responsible for the fire floor, while the second ladder company is responsible for the floors above the fire. Squad and rescue companies can also be utilized to augment search teams on the fire scene. It is imperative, though, that no company freelance: Each should be given specific assignments by the incident commander, although in some cases this may not be possible. The incident commander should also ensure that the firefighters who performed the primary search are out of the structure before having different firefighters perform the secondary search. It is impractical to conduct a secondary search while the primary search is still underway. All searches must be performed within the limits of safety. A risk analysis must be performed by the incident commander prior to committing firefighters into a structure in search of victims. This is all a part of size-up. The time of day, building occupancy, and information available should be evaluated. If the fire building is occupied, great risks are sometimes taken during a search; however, if the building is unoccupied risks should be minimal. In the case of known abandoned and vacant buildings, no risk should be taken at all. Regardless of the time of day, firefighters must realize that there could always be a life hazard in a structure. An example of this would be cleaning and maintenance personnel working in office buildings at night. It is sometimes the practice of commercial building owners to lock cleaning crews or security people in the structure at night.

How to Search When possible, searchers should determine and work toward an alternate means of escape, such as a window or fire escape. When searching the fire area, the firefighters should proceed to the seat of the fire and start their search at this point, working their way back toward the door they entered. When searching the floors above the fire, the search should be initiated immediately upon entering the door. In both cases, the area behind the front door should be searched immediately for overcome victims. In many cases, unconscious victims are found behind entrance doors or below windows, overcome while attempting to flee the area. Before opening a door it should be felt for heat. If the door is hot to the touch, there is a good chance fire will be found on the other side. When forcing a door to begin a search, a piece of rope should be tied to the door knob, so in the event fire extends out of the door, a firefighter can use the rope at floor level and close the door, protecting firefighters in the hallway. When entering a window to search a room, be careful when breaking windows and entering. Probe the floor below the window carefully to make sure there is a floor present and to check for overcome victims. Once you have entered a window, work around the walls and use a tool to probe toward the center of the room. Use the walls as your reference point. If you find a door, close it so the room can vent, thereby improving vision as the smoke lifts. Remember to take the window out completely in the event you are forced out of the room due to heat and fire. If entering an apartment from the front door, follow the walls as you search, venting as you move, unless this action will extend the fire. Do not leave the wall you are working from or cross over to another wall. Doing so can cause you to become disoriented.

Here are some searching tips:

- Plan your search.
- Check behind doors and under windows immediately.
- If the door opens easily, then stops, there is a good chance there is a victim behind it.
- Use caution when searching with tools to avoid injuring victims.
- Do not let a door lock behind you. Chock it or leave a firefighter at the door.
- Confine a fire by closing a door, then continue with the search.
- Work around walls, probing toward the center of the room.
- Use a tool to your advantage; it increases arm length.
- Vent during the search, as long as it will not extend the fire.
- Listen for crying, moaning, or coughing.
- Check all closets and cabinets.
- Do not assume locked rooms are empty.
- Look under beds.
- Treat furniture as an extension of the wall. Do not move it.
- Be aware of bunk beds: The top bunk must be checked.
- Narrow furniture legs are found on bunk beds, cribs, and baby highchairs.
- Search through piles of clothes thoroughly.
- Check refrigerators, toy boxes, and dressers for children.
- If one victim is found on a bed, check for additional victims.
- Outward opening doors could indicate a utility closet or elevator shaft.
- Look out a window, if disoriented.
- The primary way out is the way entry was made.
- Plan your own escape route.

Search procedures and protocols should be a subject of much practice and hands-on training. Firehouse bunk rooms and offices make great areas to conduct mock search drills.

Forcible Entry

■ **Note**

The type of doors and locks encountered dictates the tools and techniques used to gain entry.

Prior to searching a fire building, firefighters must be able to successfully gain access into the building or occupancy. Gaining entry is also a very important function of the ladder company. Forcible entry, when required, must be initiated immediately. The type of doors and locks encountered dictates the tools and techniques used to gain entry. The strategy should be to do the least amount of damage possible while preserving the integrity of the door. It must be emphasized that a door is a firefighter's best friend when fire and heat are on the other side. Losing the control or integrity of a door can have disastrous effects for firefighters on the fire floor and especially on the floor above the fire. For some fires and emergencies, an alternate means of access, such as a portable ladder or fire escape, may prove more beneficial than forcing a door. This section is not intended to describe all aspects of forcible entry, but to give the reader some insight and hints for the forcible entry team.

Doors When dealing with doors, firefighters may find that they open either inward or outward. A quick look at the hinge area will show which way the door opens. Hinges on the outside indicate outward opening doors. Remember that the first thing that must be done prior to forcing entry is to try the doorknob. If the door is open, then forcible entry is not required. A quick way to force an inward opening door is to place a halligan tool 6 inches above or below the lock, with the bevel side of the fork against the door slightly angled up or down. The halligan tool is then struck with a flathead axe and driven past the door jamb. The tool is now pushed toward the door, popping it open. Remember to tie a piece of rope on the doorknob and have a firefighter hold the rope so the door can be quickly pulled closed if necessary. If the door is difficult to force, place the bevel side of the halligan tool against the door jamb and apply pressure toward the door. If multiple locks are found, place the tool between the locks. Another method of forcing these type doors would be to attack the side of the door opposite the lock with the back of an axe or maul, knocking the door's hinges off the wall. Always attack the top hinge first. Many ladder companies are equipped with hydraulic rams (**Figure 7-26**) or rabbit tools that are easily operated by one person. Four to six pumps of the ram will force most inward opening doors.

Figure 7-26 *A hydraulic ram used for forcing entry on doors.*

Outward opening doors are usually indicative of doors to commercial occupancies, elevators, and closets. They are usually not found as entrance doors to houses and apartments. If they are encountered, the adz end of a halligan tool can be driven in above or below the lock. The firefighter can then force the tool down and out to pop the door open. Padlocks are often found on doors and gates of commercial occupancies. They can be attacked at either the staple, shackle, or point of attachment. They can also be attacked with an arsenal of equipment such as saws, torches, and hand tools. The weakest link should be attacked. Many padlocks now have toe and heel locking, which requires both shackles to be cut. The use of a duck bill in conjunction with an axe is more than enough to pop the shackles of the lock. A pipe wrench placed across both shackles and then twisted will also snap the shackles. A piece of pipe should be used with the wrench for leverage.

Windows Depending on the amount of ventilation required, windows can be either completely removed or just opened up. Double-hung windows should be opened two-thirds from the top and one-third from the bottom for adequate ventilation at smaller fires, such as stove or mattress fires. If there is moderate to heavy smoke, complete removal of the window may be required (see **Figure 7-29**). Windows can

Figure 7-27 *In most cases by removing the entire window and frames, you are in essence creating another door.*

be forced by placing the fork end of a halligan tool under the bottom of the window and exerting downward pressure on the tool, popping the window lock open. In the case of newer windows, there are clips that can be depressed, making removal of the entire window possible without breaking them. When forced to break double pane or thermal-insulated windows, both panes of glass must be removed. It is important to understand the types of windows that are installed in your response area. From thermal pane windows to hurricane windows or high-impact glass, firefighters must be prepared to attack the glass or window frames to perform ventilation. In many instances, members may need to use these windows to immediately escape an IDLH atmosphere. In most cases, by removing the entire window and frames you are in essence creating another door.

LADDER COMPANY SUPPORT FUNCTIONS

Ladder Placement

■ **Note**
Portable ladders should be placed on as many sides of the fire building and at as many windows as possible. At a minimum, one ladder must be to each floor on which firefighters are operating.

The proper placement of ladders at fire operations lies with the ladder company. From the moment it arrives on the scene, portable, aerial, and tower ladders should be placed into operation. There is nothing worse than a firefighter or civilian appearing at a window with no ladders in sight. Portable ladders should be placed on as many sides of the fire building and at as many windows as possible. At a minimum, one ladder must be to each floor on which firefighters are operating. Firefighters operating on the inside should be informed of the locations of these portable ladders. Portable ladders come in many sizes and are either straight or extension ladders. The right size ladder should be used to safely accomplish your objective. The preferred portable ladder would be an extension ladder, because it can be adjusted to the exact height needed, a luxury not found with straight portable ladders. Aerial and tower ladders should be positioned for maximum coverage. There should be little reason to have an aerial ladder in the nested position at a working structural fire. A tower ladder provides an excellent platform to perform ventilation and search and rescue and to apply a large-caliber elevated stream. All firefighters of both engine and ladder companies must be well versed in getting ladder apparatus into position and operation.

Overhaul

Overhaul at fire operations is performed to expose hidden pockets of fire and prevent reignition. It is performed during the entire fire operation and can be broken down into two stages: precontrol and postcontrol. The precontrol phase entails pulling ceilings, examining baseboards, shafts, and so forth to determine the fire's path of travel and location. Knowledge of different construction features and building designs assists the firefighter in overhauling operations. In most cases, the precontrol stage is performed under adverse conditions, such as high heat, heavy smoke, and limited visibility. If fire involves just one room with contents, overhaul

Figure 7-28
Overhaul must be performed by opening ceilings and walls until areas that have not been burned are found. (Photo courtesy of Palm Harbor Fire Rescue.)

■ **Note**
Indiscriminate overhauling can cause unnecessary damage and hardship to the building's occupants. However, when and where it is necessary, one should not hesitate to overhaul.

is usually limited to exposing baseboards, pulling ceilings where necessary, and examining the walls. The intent is to work from charred areas back toward clean areas to determine the fire's extent, as shown in **Figure 7-28.** If fire involves a large area, extensive overhauling may be necessary to ensure no hidden pockets of fire remain.

The postcontrol phase of overhauling is performed under more favorable conditions than the precontrol phase and after the fire is under control. This is the time to be extremely meticulous in determining whether all fire has been extinguished. The fire area must be thoroughly examined as well as contents such as furniture, mattresses, or piles of clothes. When possible, postcontrol overhauling should be performed by a fresh company that is not fatigued from the fire attack operation. Indiscriminate overhauling can cause unnecessary damage and hardship to the building's occupants. However, when and where it is necessary, one should not hesitate to overhaul.

In either phase, the use of thermal imaging will help the firefighters performing the tasks.

Salvage Operations/Loss Control

Salvage duties involve saving possessions in danger of being damaged by fire, smoke, and water. It begins with the firefighters' arrival on the scene and continues until fire department operations have concluded. Salvage is usually a function of the ladder company. It is inevitable that damage will occur at fires, but fire personnel should treat each individual's home and property as though it were their own. Many items, although having little monetary value, have priceless sentimental value, and firefighters have a responsibility to protect and salvage as much as possible. Attempts should be made to protect and cover furniture and belongings where possible, including placing valuables, videos, and pictures in dressers and closets where water and smoke will not affect them. Doors, windows, and roof openings should be covered to protect the building and its contents from further damage from the outside elements. The small acts of salvage that can be performed at fire operations may be remembered by the building's occupants for a lifetime.

■ **Note**

Doors, windows, and roof openings should be covered to protect the building and its contents from further damage from the outside elements.

Monetary reasons for salvage:

- Keeps insurance rates down
- Building remains occupied
- Occupants do not need to relocate
- Keeps repair costs down

Professional reasons for salvage:

- Saves occupants from unnecessary hardship
- Fire service viewed as professional
- Obligation to the community
- Feeling of satisfaction

ENGINE AND LADDER COMPANY APPARATUS PLACEMENT

■ **Note**

The function that the company is to perform predicates where the apparatus should be placed.

After forming an understanding of the company functions, a question to be answered is where to park the apparatus. The function that the company is to perform predicates where the apparatus should be placed. The following list gives some considerations for apparatus placement:

- Apparatus capabilities (100-foot aerial versus a 75-foot aerial).
- Standard operating guidelines (SOGs; first engine to the front, second to the rear).

■ **Note**

Engine companies should be placed in a manner that allows room for aerial trucks to access the front of the structure.

scrub area
the area or the building that can be reached with an aerial ladder once the apparatus is set up; a consideration for ladder truck placement

■ **Note**

You can stretch a hose, but you cannot stretch a ladder.

- Prearranged SOGs for staging (only first-due engine to scene with nothing showing, all others to stage in direction of travel uncommitted).
- Order from incident commander (Engine 1 hook to the fire department connection on side 3).
- Prearranged placement based on preincident planning (according to the preplan, ladder 1 will position on the northeast corner of the building and place the aerial to the roof).
- Fire location and extent (defensive operations may require the use of truck-mounted deck guns, however, collapse zones must be considered).
- Apparatus staged with personnel assigned.
- Overhead hazards must also be considered in all apparatus placement decisions.

Engine companies should be placed in a manner that allows room for aerial trucks to access the front of the structure. Ladder companies should be placed in locations that allow the ladder to have the most versatility. Aerial trucks should be placed in order to obtain the best **scrub area** (see **Figure 7-29**). A good rule of

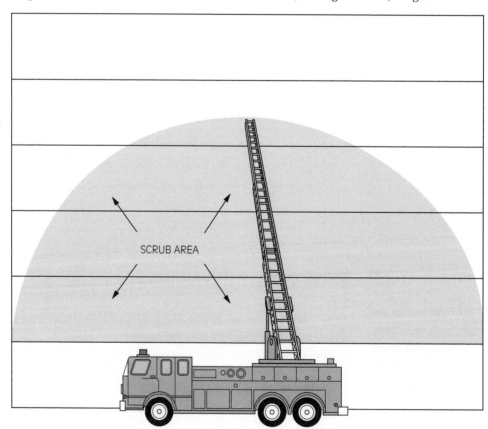

Figure 7-29 *An example of scrub area.*

SCRUB AREA

collapse zone
the safety zone set up around a fire building where the potential for a collapse exists; should be the full height of the highest wall

thumb is that you can stretch a hose, but you cannot stretch a ladder. Again, placement greatly depends on what actions are going to take place. Keep aerial apparatus capable of flowing elevated streams ahead of the fire. In well-involved buildings that have the potential of collapse, keep apparatus out of the **collapse zones** and as close to the corners of the structure as possible.

SUMMARY

Engine and ladder company operations are the foundation of the fire rescue service. Nothing is more important than an engine company getting a hose line in operation and water on the fire. Knowing the proper hose line to stretch, nozzles to use, and method of attacking the fire rests with the engine company. While the engine company is focusing on fire suppression, the ladder company is focused on forcing entry, ventilation, and search. The ladder company also performs additional duties such as laddering the building, overhaul, and salvage. Although engine and ladder companies concentrate on their individual responsibilities and duties at fire scenes, well-trained companies understand that these units must work hand in hand in the fire operation to save lives, suppress fire, and prevent firefighter injuries. More and more fire departments are placing a heavy emphasis on handling medical emergencies, and in many departments medical emergencies far exceed the number of fire responses. For this reason, vigorous training programs must be carried out by all departments as often as possible. Good company operations result in safe and effective operations.

REVIEW QUESTIONS

1. True or False? Only apparatus drivers should be concerned with how to operate pumps.

2. List three considerations when choosing which size hose line to use.

3. True or False? Operating a hose line in a counterclockwise direction pushes heat and smoke away from the nozzle team.

4. Which of the following is *not* a ladder company support function?

   A. Search and rescue

   B. Overhaul

   C. Salvage

   D. Utility control

5. Which one of the following describes mechanical ventilation?

   A. Opening a door

   B. Breaking a skylight

   C. Operating a fog nozzle out a window

   D. Cutting a roof with a saw

6. True or False? When attacking the hinge of a door, always attack the top hinge first.

7. List two monetary and two professional reasons salvage duties are performed.

8. List five considerations in apparatus placement decisions.

9. Which of the following is not an option for getting a water supply to the fire scene?

A. Split lay   B. Forward lay

C. Reverse lay   D. Hydrant lay

10. List ten search tips as presented in this text.

ACTIVITIES

1. Select a target hazard in your response zone and discuss the various functions required for the engine company and the ladder company.

2. From a postincident analysis of a recent fire (from your department or from another department), determine how the engine and ladder company operations worked together to extinguish the fire.

3. Compare your department's engine and ladder company operations to the operations presented in this chapter. What changes do you feel would enhance your department's performance?

Chapter

8

BUILT-IN FIRE PROTECTION

Learning Objectives

Upon completion of this chapter, you should be able to:

- Explain the need for built-in fire protection systems and why they are beneficial to building occupants and firefighters.
- Describe what the main water control valve is and what it does.
- Identify the three main types of main water control valves and how to determine whether they are in the open or closed position.
- Given a diagram, locate and identify the fire department connection (FDC).
- Describe the various means by which sprinkler and standpipe systems may be supplied.
- Outline the major components/valves of a sprinkler or standpipe system.
- Identify and explain the operation of pressure-reducing valves found on standpipe systems.
- Explain the differences between the types of sprinkler systems and how they activate and operate.
- Explain the differences between a residential sprinkler system and those installed in commercial facilities.
- Explain the differences and similarities, along with the minimum requirements, of the three classes of standpipe systems.

- Describe the types of special extinguishing agents and the hazards associated with each.
- Explain the need for fire department support of built-in fire protection systems.
- Outline the minimum items that should be addressed within a standard operating guideline for the support of built-in fire protection systems.

CASE STUDY

There is a small city in west central Florida that has the potential for large fire losses. This city realizes that the local fire protection resources can quickly be overwhelmed and because of this, they strongly pursue, enforce, and rely on built-in fire protection systems.

In this city, in December 1998, just such an incident occurred. At 12:15 A.M., the local fire department was dispatched to a report of a fire alarm. The location of the alarm was a large multibuilding complex within the heavy industrial portion of the city. This complex is located within 1 mile of the closest fire station.

The first-due engine company arrived at the scene at 12:18 A.M. and observed a moderate amount of smoke pushing from a building that housed a box manufacturing plant. The box plant is about 400 feet wide, 1,200 feet long, and about 50 feet high and has a very heavy fire load. This building is protected by an automatic sprinkler system, which the engine began to immediately support.

The officer of the first-due engine company quickly requested the balance of a first-alarm structural assignment and requested a second alarm. Because of the building's size and fire loading, the officer set up a command post and began to orchestrate the assignments of the other arriving companies. This incident eventually grew to four alarms.

As companies began arriving and going to work, entry was made from multiple locations around the building to perform a search for occupants and to try to locate and extinguish the fire. Once the building was accessed, it was realized that the smoke conditions were much worse than what was visible from the outside. The building was completely filled with dense, dark smoke, which made the search for any trapped occupants and fire very time-consuming. The cool, humid air kept the smoke down, making ventilation efforts (forced, horizontal, and vertical) more difficult.

As the incident went on, no victims were located, and the fire was eventually located and extinguished by two 1¾-inch handlines flowing a combined 250 gpm. Until the fire was located and extinguished by firefighters, the fire was held in check by five activated sprinkler heads.

With the presence and utilization of the built-in fire protection equipment, the incident ended up as a relatively small fire with a relatively small dollar loss and minimal interruptions to the company's service and production. Without this presence and utilization this incident could have had a very different outcome and would have likely destroyed the building not to mention entailed the loss of the company's production line and the loss of employee jobs (temporary or permanent). It also would have kept fire and rescue companies tied up for a much longer time.

INTRODUCTION

Built-in, or private, fire protection systems are and will continue to be an important ally of the fire service. They will become ever more prevalent due to owners' desires to protect their property and the ever increasing local mandates on their installation. With today's structures being built of lighter weight and more combustible materials, built higher, built larger and then packed with a higher fire loading than ever, control and confinement of fires in these structures has become more labor-intensive and time-consuming.

Fire officers and firefighters alike must have a good working knowledge and understanding of fire protection systems and how to use them to their advantage.

This chapter discusses various common types of fire protection systems and how the fire department can and should support them.

SPRINKLER SYSTEMS

■ **Note**
According to statistics, about 96% of fires in sprinklered buildings are either extinguished by the sprinklers or held in check until they can be completely extinguished by the fire department.

main water control valve
the main water supply valve in a sprinkler or standpipe system

There are four main types of sprinkler systems:

1. Wet pipe systems
2. Dry pipe systems
3. Deluge systems
4. Preaction systems

These systems have many common components that the firefighter must be able to recognize and understand the operation of.

According to statistics, about 96% of fires in sprinklered buildings are either extinguished by the sprinklers or held in check until they can be completely extinguished by the fire department (**Figure 8-1**).

All sprinkler systems have a main water control valve along with other test and drain valves and piping. The main control valve is the valve that controls the flow of the water from the domestic water supply system and/or on-site fire pump(s). The main control valve is an indicating valve: At a glance a firefighter can tell if it is open or closed. This valve is manually operated and, along with

Figure 8-1 *When properly designed and maintained, sprinkler systems control 96% of the fires when they are activated.*

Figure 8-2 *The most common main water control valve is the outside screw and yoke.*

outside screw and yoke (OS&Y)
one type of main water control valve characterized by the visible screw and yoke

post indicator valve (PIV)
one type of main water control valve characterized by the visible window, which indicates the position of the valve

other valves, should always be chained or locked in the open position. This valve is typically located just under the sprinkler alarm valve. There are three common types of this valve.

The most common valve is the outside screw and yoke (OS&Y) (**Figure 8-2**). This valve has a threaded stem that controls its opening and closing. When the stem is visibly out of the valve, it is open; when the stem is not visible, the valve is closed. Another common type of valve is the post indicator valve (PIV). This valve has no visible stem for it is located inside the post (**Figure 8-3**). This type of valve has a window through which can be viewed a moveable target that has the words *open* or *shut* printed on it. These words indicate the position of the valve. The operating handle for this valve is attached and secured on the side of the post. The third type of main control valve is the wall post indicator valve (WPIV). This valve is generally the same

Figure 8-3 *An example of a post indicator valve.*

wall post indicator valve (WPIV)
one type of main water control valve mounted on a wall and characterized by the visible window, which indicates the position of the valve

as the PIV except that it protrudes out of a wall of a structure horizontally, as shown in **Figure 8-4**.

Other common components of sprinkler systems include valving such as stopcock valves, globe valves, check or clapper valves, automatic drain valves, and alarm test/inspector test valves, illustrated in **Figure 8-5**. These various valves each have a different functional purpose ranging from sprinkler system testing to system draining. Stopcock valves are for both system drainage and alarm silencing; likewise, globe valves are used for system drainage and as test valves. These valves are manually operated but unlike the main control valve, they are of the nonindicating type. Check or clapper valves ensure that water can only flow in one direction. A simple example of this is on a ground monitor/deluge gun that is fed by two or more lines. Each input connection of the monitor has one of these valves so that if only one line feeds the monitor, the water will not back-pressure the other line(s) or flow backward out of it. Automatic drain valves are used to automatically drain the sprinkler system once the pressure on the system has been relieved. The alarm test/inspector test valve is used to simulate actuation of the sprinkler system to ensure it is working properly.

Figure 8-4 *An example of a wall post indicator valve. The chain and lock are to ensure the valve cannot be inadvertently closed.*

Figure 8-5 *Various other sprinkler system valves and components attached to the sprinkler riser.*

Figure 8-6 *Outside waterflow alarm.*

main drain
the drain for a sprinkler system that drains the entire system

waterflow alarm
a mechanical or electrical device attached to a sprinkler system to alert for water flow.

■ **Note**
Every sprinkler system must have a water supply that is reliable, automatic, and of adequate volume and pressure.

fire pump
a stationary pump designed to increase water flow or pressure in a sprinkler or standpipe system

The **main drain** is also a common component of all sprinkler systems. This drain piping and valve is used to drain the system to replace heads (activated or otherwise) or to conduct system repairs. The main drain can also be used to test the system much like the alarm test/inspector test valve.

Another common component of sprinkler systems is the **waterflow alarm** (**Figure 8-6**). This alarm is activated in either one of two ways, hydraulically or electronically, each of which is to indicate water flow. The hydraulic alarm uses water movement within the system to drive the alarm gong (water gong). This alarm is a local alarm used to alert building personnel or passersby of water flow in the building system. The second way, electronically, the alarm is activated when water movement presses against a diaphragm that activates a switch to operate the alarm. This alarm, like the hydraulic alarm, can be a local alarm or it can be connected to a monitoring company that can notify the fire department when the alarm is activated.

One of the most important commonalities, at least from the fire service perspective, is water supply. Every sprinkler system must have a water supply that is reliable, automatic, and of adequate volume and pressure. The flows are dependent on the size and height of a structure as well as any specific hazards of an occupancy. In many cases a secondary means of water supply may be required. The flow must minimally be able to provide a residual water pressure of 15 psi to the highest and furthest sprinkler head. The water supply can be provided from the domestic water supply, gravity tanks, or pressure tanks. These supplies (specifically the domestic water supply) may also be supported by an on-site **fire pump** or pumps.

test header
is a group of outlets used to test the capacity of a building fire pump system

fire department connection (FDC)
a siamese connected to a sprinkler or standpipe system to allow the fire department to augment water volume or pressure

■ **Note**
When the fire department connects to the FDC, it must make sure to pull its water supply from a different source from the one the sprinkler system is supplied from so as not to take water away from the activated system.

riser
the vertical piping in a sprinkler or standpipe system

An indication that a fire pump is present is a test header, which looks like a wall hydrant with multiple 2-1/2" outlets that are used for testing the fire pump. The number of outlets is determined by the required flow for the occupancy.

Along with the presence of an on-site water source, a fire department connection (FDC) or connections will be provided for fire department use. The fire department should connect an engine to the FDC to supplement the system, serve as an auxiliary water source, or increase the water pressure or volume of the sprinkler system. **Figure 8-7** depicts a sprinkler system FDC.

When the fire department connects to the FDC, it must make sure to pull its water supply from a different source than the one the sprinkler system is supplied from so as not to take water away from the activated system. For this reason the firefighter must be aware of how and from where a building system is supplied. On sprinkler systems with a single riser, the FDC is connected to the sprinkler side of the system, which means that if the main water control valve were closed, the only way to stop the flow of the sprinkler(s) would be to shut down the lines from the engine supplying the system. On systems with multiple risers, the FDC is connected to the supply side of the system; thus, the flow can be controlled by shutting down the main water control valve. In either case, there are check valves in the piping to ensure that through the FDC the domestic supply cannot be contaminated or back-pressured and no harm could be caused to the gravity and/or pressure tanks. Much like the clapper valves in the ground monitor/deluge gun, the FDC has clapper valves that allow the system to be supplied with only one line from an engine. With these clapper valves, water backflowing out of the second connection FDC is prevented and the second line can be added later in the operation.

Figure 8-7 *One example of a fire department connection.*

The FDC for sprinkler systems should be marked accordingly. There must be some type of sign, letters stamped into the connection, or other visual indicator that the connection is for the sprinkler system and, if the system is broken into more than one zone, which zone it supplies.

The utilization and support of a sprinkler system should be of utmost importance to the fire department. Every fire department should have standard operating guidelines (SOGs) for sprinklered buildings. A sample SOG is presented in Figure 8-18.

Wet Pipe Sprinkler Systems

A wet pipe sprinkler system constantly has water throughout the system. The water in the system is also kept under pressure so that when a sprinkler head activates, water immediately begins to flow from the head, thus activating the waterflow alarm and beginning control of the fire.

Dry Pipe Sprinkler Systems

A dry pipe sprinkler system has no water in the system piping beyond the check valve. This system replaces the water with air that is under pressure to keep water from entering the system until a sprinkler head is activated. Only a minimal amount of air pressure is needed to hold the check valve closed. Dry pipe systems are typically used in buildings that have the potential for the water in the piping to freeze due to insufficient heating capabilities or other situations where some piping is exposed to the outside elements. In this system, once a head is activated, the air begins to flow out of the open head, which in turn reduces the air pressure holding the water back, thus allowing the water to enter the system and begin flowing out of the open sprinkler head. As it is plain to see, this system takes longer to provide water onto the fire. For this reason, many dry pipe systems (especially large ones) utilize accelerators or exhausters to speed up the process. Although accelerators and exhausters are somewhat complicated devices, they provide for quicker water flow to the fire area.

accelerator
device designed to speed the operation of a dry pipe valve

exhauster
device designed to speed the operation of a dry pipe valve by bleeding off pressure

Preaction Sprinkler Systems

A preaction sprinkler system utilizes separate additional alarm equipment as part of the system. These systems are typically used in locations where water damage must be prevented. A preaction system is basically set up the same as a dry pipe system. The piping is dry and only fills once the separate additional alarm detection equipment is activated. After the detector activates, the system fills with water, but a sprinkler head still needs heat to activate it in order to allow water to flow in the affected area. What makes this a preaction system is that once water begins to flow in the system piping, an alarm is activated and sounds to provide warning prior to the activation of a sprinkler head.

Deluge Sprinkler Systems

A deluge sprinkler system, like the preaction system, utilizes separate additional detection equipment as part of the system. A deluge system is typically used in locations where the occupancy is extra hazardous. This system is also basically a dry pipe system. In this system the heads are all open and once the separate fire/smoke detection equipment is activated, the system begins to flow and water is discharged from all of the open heads at once. Once the water begins to flow, an alarm is activated as well as any alarm associated with the separate detection equipment.

Residential Sprinkler Systems

Residential sprinkler systems are becoming more and more popular and in some cases mandated. These systems comply with NFPA 13D, *Residential Sprinkler Systems*. For the most part, these systems are designed in the same manner as the other systems that have been described. There are a few exceptions to this statement. First, residential systems are not required to have an FDC for the fire department to connect to supplement, however some do. Second, the system piping is made of copper or polyvinyl chloride (PVC) rather than steel like the other ones are. Although the fire department does not have much to do with these systems, as far as fire department utilization, it is important that firefighters know that they are out there and becoming more popular. Firefighters should have a basic understanding of how these systems operate.

STANDPIPE SYSTEMS

Standpipe systems are another important form of built-in fire protection. A standpipe system provides connections for the fire department and in some cases, the occupants of the building, at various locations throughout a building. The connection can be located on each floor or several places on a floor, the roof, and even on the exterior. Standpipe connections are basically an extension of the discharge outlet of a fire engine. Like sprinkler systems, a standpipe system must be able to provide a water supply that is reliable, of adequate volume, and of adequate pressure. Standpipe systems are divided into three classes, each depending on the intended use of the system.

- *Class I* is for use by fire departments and those trained in the handling of heavy fire streams, as shown in **Figure 8-8**.
- *Class II* is for use primarily by building occupants. An example is shown in **Figure 8-9**.
- *Class III* is for use by fire departments and those trained in the handling of heavy streams or for the use of the building occupants (see **Figure 8-10**).

Figure 8-8 *Class I standpipes are designed for use by fire departments and those trained in heavy streams.*

Figure 8-9 *Class II standpipe systems are for use primarily by building occupants.*

Class Characteristics

Class I Class I standpipe systems must be capable of providing effective fire streams that would be required of fire departments during the advanced stages of a fire. A class I system must have a water supply adequate enough to flow 500 gpm for at least 30 minutes. The supply must provide a minimum residual pressure of 65 psi at the highest furthest outlet while flowing the required 500 gpm. In systems that require more than one standpipe riser, the system is still required to flow the 500 gpm for at least 30 minutes for the first standpipe in the system, and each additional standpipe must flow a minimum of 250 gpm for at least 30 minutes.

Class II This standpipe system is provided for building occupants so they may try to quickly control a fire while it is still in the incipient stage. Class II systems must

Figure 8-10   *Class III standpipe systems are for use by both fire departments and building occupants.*

have a water supply adequate enough to flow 100 gpm for at least 30 minutes. This standpipe class must also be able to provide a minimum residual pressure of 65 psi at the highest furthest outlet while at the same time flowing the required 100 gpm.

Class III Because it is also for fire department use, the class III standpipe system must be capable of providing fire streams that would be required of fire departments during the advanced stages of fire, as well as providing building occupants the means of quickly controlling a fire while it is in its early stages of growth. A hose cabinet similar to the one shown in **Figure 8-11** is provided. The class III system must be capable of flowing the same required minimums as the class I system.

It is important to point out that in many cases, occupancies are removing the hose from the class II and class III systems and simply replacing it with the appropriate standard fire extinguisher.

General Characteristics

All firefighters must be aware of the possibility of pressure-reducing devices being present on standpipe systems, such as those at the One Meridian Plaza fire described in the Chapter 3 case study. Specifically in class III systems, which are designed for occupant and fire department use, pressure-reducing devices can

■ **Note**

All firefighters must be aware of the possibility of pressure-reducing devices being present on standpipe systems.

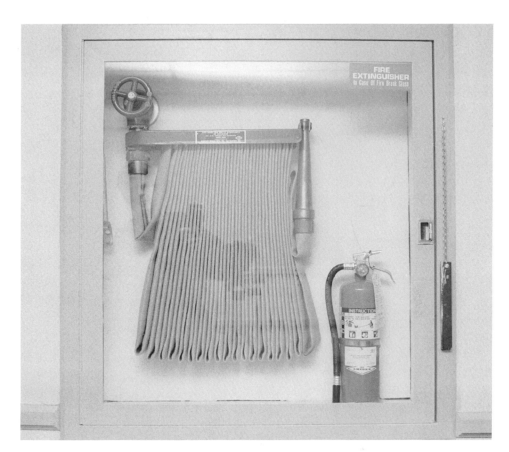

Figure 8-11 *A typical hose cabinet configuration for a class III standpipe system. Firefighters should always bring their own hose and not use hose lines that are supplied in the cabinet.*

be found. These devices can range from the type that resembles a washer that reduces the outlet flow diameter so occupants may be able to control the line, to manual control valves that can be used to increase or decrease the pressure. **Figure 8-12** is a example of a pressure-reducing device that works at the valve stem. Any system that has flow pressures of 150 psi or more should be tagged or marked to notify of the higher pressures.

Standpipe systems can be either wet or dry systems and are broken down into four different types as follows:

- A wet standpipe system has water throughout the system at all times. When an outlet is opened, water immediately begins to flow.
- A dry standpipe system that is under no air pressure, much like a deluge-type sprinkler system, and must be activated by manual operation of a valve at the beginning of the system or by a remote control located at each standpipe outlet.

Figure 8-12 *One example of a pressure-reducing device on a class III standpipe system.*

- A dry standpipe system that is filled with air under pressure, much like a dry sprinkler system, that allows water in the system when a standpipe connection is opened and allows the air to escape the system while allowing water in.

- A dry standpipe system that has no water supply and must be supplied by the fire department.

■ **Note**
The installation, maintenance, and minimum requirements of standpipe systems are covered within NFPA 14, *Standard for the Installation of Standpipe and Hose Systems.*

The installation, maintenance, and minimum requirements of standpipe systems are covered within NFPA 14, *Standard for the Installation of Standpipe and Hose Systems.* Within this standard, it mandates that standpipe systems shall be limited to a maximum height of 275 feet. When a structure is higher than the 275-foot limit (**Figure 8-13**), an additional zone or zones must be installed; however, no two zones are allowed to exceed a height of 550 feet.

Again, like sprinkler systems, the fire department must be very interested in how standpipe systems are supplied with water. Each building along with the specific hazards of the building must be studied to develop the best means of supplying the system with water. The supply may come from the domestic water system or on-site water tanks and either or both need some type of fire pump to support them. Class I and class III standpipe systems also have an FDC so the fire department can support and/or supply the system. The systems that have more than one zone due to building height or size have an FDC-designated

Figure 8-13
*Standpipe systems
are limited to a
height of 275 feet.
Multistory buildings
such as this one
have additional
zones. The fire
department must
connect to the
proper FDC for the
zone to be used.*

for each zone. As in sprinkler system FDCs, the standpipe FDC may be wall-mounted or located where it is in an easily accessible location for fire apparatus. The FDC connections must be of the female type, and the FDC must be designated as a standpipe FDC by the use of stamped letters on the connection itself or by some type of metal plate or disc that says the word *standpipe,* as shown in **Figure 8-14**.

In addition to the indication that the FDC is for standpipe use, it must also indicate which zone it supplies if the system has more than one zone. Some fire departments mandate that the FDC be colorcoded for fire department identification. **Table 8-1** is one example of color-coded FDCs. The standard mandates that there should be no type or way of shutting off the supply between the FDC

Figure 8-14 *Fire department connections should be clearly marked* Standpipe *or* Sprinkler.

Table 8-1 *Fire department connection color coding.*

Cap Color	Code
Red	Standpipe System
Green	Automatic Sprinkler System
Silver	Non-Automatic Sprinklers System
Yellow	Combination Sprinkler/Standpipe System

and where it enters into the system. SOGs need to be developed and used in buildings that are equipped with standpipe systems.

SPECIAL EXTINGUISHING SYSTEMS

There are a multitude of specialized extinguishing systems. This section discusses some of the most common ones that firefighters will encounter, including the following:

- Carbon dioxide extinguishing systems
- Halogenated agent extinguishing systems
- Dry chemical extinguishing systems
- Wet chemical extinguishing systems

These systems typically activate and exhaust their respective agent prior to the arrival of the fire department.

The carbon dioxide and halogenated agent types of extinguishing systems are typically found in computer rooms, spray booths, areas of intricate electrical equipment, or areas of flammable liquid storage, to name a few. These extinguishing agents are very useful in areas where it is important to control and/or extinguish fires without causing water damage to the facility components or product and where leaving a harmful residue on potentially expensive equipment that was not affected by the fire could be more harmful than the fire itself. Many of the more advanced types of these systems have become quite elaborate in that they have detectors within the room or area to detect human presence and give some type of prealert signal to notify occupants to exit just prior to the system activating. This helps to prevent any injuries to building occupants. Depending on the size of the system, the agent maybe stored outside the building (see **Figure 8-15**).

A carbon dioxide system works by diluting the air in the room to displace oxygen; thus, the fire will go out. With the carbon dioxide system there still exists the potential for the fire to flash back to life once the CO_2 has been exhausted and begins to escape the fire room or area. CO_2 is not very effective at extinguishing a smoldering fire, so it is imperative that the fire department quickly locate and extinguish any remaining pockets of fire with water. The main hazard to firefighters associated with this system is that it dilutes oxygen. Proper protective equipment including self-contained breathing apparatus (SCBA) must be worn when entering these atmospheres.

Halogenated extinguishing agents, although used in many of the same areas as CO_2, work by interfering with and interrupting the chemical chain reaction of fire. Halogenated agents, or **halon** as it is commonly called, are halogenated hydrocarbons in a chemical compound that contains carbon and one or more of the elements from the halogen series (bromine, chlorine, fluorine, and iodine).

■ **Note**
The carbon dioxide and halogenated agent types of extinguishing systems are typically found in such locations as computer rooms, spray booths, areas of intricate electrical equipment, or areas of flammable liquid storage.

■ **Note**
With the carbon dioxide system there still exists the potential for the fire to flash back to life once the CO_2 has been exhausted and begins to escape the fire room or area.

halon
an extinguishing agent that works by interrupting the chemical change reaction

Figure 8-15 *The exterior components of a carbon dioxide extinguishing system installation.*

Halogenated agents are typically used in very low concentrations and are very effective at extinguishing fire, even class A fires. The main hazard associated with halon to the building occupants and firefighters is the potential for toxicity from the halogenated agent. As the concentration percentage increases, so does the potential for toxicity, not to mention the possibility of oxygen displacement as is common with CO_2 systems. For the reasons already stated, it is very important for firefighters to wear all protective equipment and SCBA when entering these atmospheres.

Dry and wet chemical extinguishing systems can be used in many common areas of application. Dry chemical systems are used to control fires involving flammable liquids, flammable gases, grease, and electrical equipment. This system can be found in areas such as commercial kitchens and the like, above deep fat fryers and cook tops, as shown in **Figure 8-16**. Dry chemical systems work by interrupting the chemical chain reaction of fire. This system leaves a powdery residue that can be cleaned up by brushing or vacuuming. The main hazards associated with the activation of this system are limited visibility and respiratory irritation. When expelled, the system discharges in a cloud that can severely limit visibility and can cause breathing irritation and coughing associated with inhaling high concentrations of the powder (although the agent is considered to be nontoxic). Firefighters need to wear all protective equipment including SCBA so as to reduce the potential side effects.

Wet chemical, or class K, extinguishing systems are used in many of the same locations as the dry chemical systems but have an added benefit. The wet chemical system has the extra ability of cooling and removing the fuel by coating it. Because of its cooling and coating properties, wet chemical extinguishers

Figure 8-16 *A dry or wet chemical hood system protects a cooking area.*

saponification
process that occurs when wet chemicals come into contact with grease and the like, forming a soap-like product

are also capable and very effective in extinguishing class A fires, even those that are deep-seated. When this system is discharged on a fire and comes in contact with grease, fat, and the like, it turns into soap. This process is called **saponification,** and the resulting product is easily cleaned up. The main hazards of this extinguishing agent are possible inhalation hazard, as with dry chemical agents, and slippage. Class K extinguishing systems have become the new NFPA standard for fire protection of cooking surfaces. One other point of information is that class K is a subclass of class B, as was explained earlier in the text.

On a side note, there are some newer extinguishing agents that are evolving, called *clean agents,* as they are just that—they leave no residue nor cause any harm to computer or electrical equipment. The actual extinguishing agent may be FE-36 or Centrimax ABC 40. These agents are effective for class A, B, and C fires and are considered to be nontoxic, nonconductive, and ecologically safe.

FIRE DEPARTMENT SUPPORT OF BUILT-IN FIRE PROTECTION

The built-in fire protection systems of a building are typically very effective in the control and extinguishment of fire, provided that the built-in protection system is properly designed, installed, maintained, and supported by the fire department in one way or another. Each type of built-in protection system, when activated, requires some sort of support from the fire department. This support may range from providing and supplementing the water supply, as shown in **Figure 8-17**, to a sprinkler or standpipe system or controlling the flow of water from the activated head or heads, to the typical fire department operations, such as search for victims and occupants, ventilation, and total extinguishment of the fire.

It is not recommended that it be the responsibility of the fire department to place these systems back in service after an activation. However, the department should be as helpful as possible and ensure that the system is properly restored by a qualified person.

It is essential that every fire department develop standard operating guidelines (SOGs) for buildings that have these protection systems. NFPA 13E, *Fire Department Operations in Properties Protected by Sprinkler and Standpipe Systems,* is an excellent reference as to what a SOG should contain along with adding any local operational requirements or needs. A SOG should include the following:

- Mandate that the first- or second-arriving engine company report directly to the FDC and prepare to support the system.

- Secure a water supply. This supply should not rob the system of its water supply.

Figure 8-17 *Fire department support of a sprinkler system with two hose lines.*

- Maintain contact with interior crews and command to monitor when the system is to be charged.
- Initially develop and maintain a pump pressure of 150 psi. Depending on the system, such as a deluge-type sprinkler system or a multizone standpipe system in a high-rise, the pump pressure may need to be increased anywhere from 175 psi to 200 psi.
- Mandate that a minimum of two hose lines supply the sprinkler or standpipe system. When more than two hose lines are utilized, the pump pressure may also need to be increased from 175 psi to 200 psi.
- Mandate that a firefighter be assigned to locate and ensure that the main water control valve is open. This firefighter must stay assigned to the main control valve to ensure nobody will turn it off, and to turn it off when advised by the incident commander.
- Mandate that only the incident commander may order the main water control valve be shut down.

Figure 8-18 shows an example of an SOG.

For the most part, as far as water supply concerns, fire department operations for sprinklered buildings and those with standpipes are basically the same.

Some other operations specific to sprinklered buildings are that an activated system will disrupt the thermal balance, making visibility very low to almost zero. A primary search is always conducted. The cooler environment increases the odds of survival for trapped occupants. Each firefighter, or at least each company, needs to have some type of sprinkler wedges or tongs in order to control water flow from activated heads once the fire has been extinguished.

There are also some specific concerns for operations within buildings protected with standpipe systems. The fire department should never utilize any hose that may be in the in-house hose cabinets. This hose is typically not as good quality hose as that found in the typical fire department and it is not tested annually for any weaknesses. An additional major concern is the possible presence of pressure-reducing devices at the standpipe connection outlets. Each company should carry some type of high-rise kit to be utilized in these buildings. There are very basic designs and some that are quite elaborate. **Figure 8-19** shows an example of a high-rise kit.

In general, a high-rise kit should contain at least the following equipment. This equipment must be maintained and be reliable.

- Minimum of 100 feet of $1\frac{3}{4}$-inch or 2-inch hose (more if long hallways are present)
- Gated Y, either $2\frac{1}{2}$ to $1\frac{1}{2}$ inches or $2\frac{1}{2}$ to 2 inches (depending on the size hose line being utilized)
- Correct size spanner wrenches

STANDARD OPERATING GUIDELINES

SUBJECT: SPRINKLER AND STANDPIPE OPERATIONS

It is the philosophy and practice of this Fire Department to use built-in private fire protection systems whenever possible or practical.

Sprinkler System Operations

1. Second-arriving engine on the first alarm assignment shall be responsible to position at the fire department connection (FDC) and secure a water supply.

2. Officer on that engine shall assign one firefighter to locate the main valve and ensure that it is open.

3. Any smoke or fire in the structure requires charging the FDC to 150 psi.

4. FDC shall always be charged by a minimum of two 3" supply lines.

5. The system should remain charged until ordered shut down by the incident commander.

Standpipe System Operations

1. First-arriving engine on the first-alarm assignment shall be responsible to position at the fire department connection (FDC) and secure a water supply. **Exception:** If the fire is on the first floor and the first-arriving engine can attack the fire with hose lines directly off the engine, then the second-arriving engine on the first alarm shall be responsible to position at the fire department connection (FDC) and secure a water supply.

2. Officer on that engine shall assign one person to locate the main valve and ensure that it is open.

3. Any smoke or fire in the structure requires charging the FDC to 100 psi plus 5 psi per floor above the first if using a smooth bore handline nozzle and 150 psi plus 5 psi per floor above the first if using a fog nozzle.

4. FDC shall always be charged by a minimum of two 3" supply lines.

5. The system shall remain charged until ordered shut down by the incident commander.

6. Interior attack teams shall use fire department hose in conjunction with standpipe outlets. (Fire department units shall not use in-house hose lines.)

General

1. Engines should avoid using hydrants that are connected to the private system so as not to steal water supply from the system.

2. Private fire pump installations should be checked early in operations to insure proper operation and their status reported to the incident commander.

Figure 8-18 An example SOG for fire department support of sprinkler and standpipe systems.

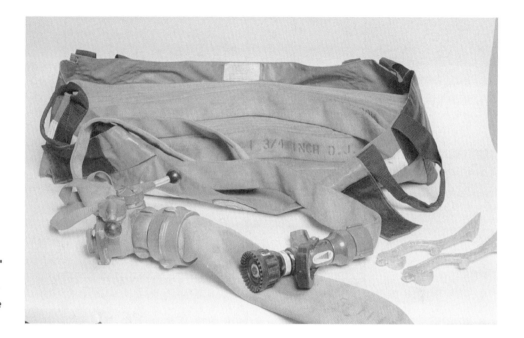

Figure 8-19
Equipment typically carried in a high-rise or standpipe pack.

- Pipe wrench
- Hand wheel for standpipe outlet
- A nozzle that can be used for either a fog stream or a solid stream
- Door wedges
- Typical hand tools and carried equipment such as hand lights and forcible entry tools

The items in this list provide much of the equipment that is needed initially, and they are not overly heavy or cumbersome.

The standpipe connections should always be made on the floor below the fire floor and the hose flaked out within the stairwell. The easiest way to advance the hose from the stairwell would be to flake the hose up the stairs to the floor above the fire floor so as the line is advanced, gravity is assisting rather than fighting against it. This type stretch is shown in **Figure 8-20**.

Support of special extinguishing systems will usually be after the main body of fire has been contained. The systems will have typically discharged, and the fire department must ensure total extinguishment is accomplished, search is completed, ventilation is done, and that the occupants work to get the system back in proper service.

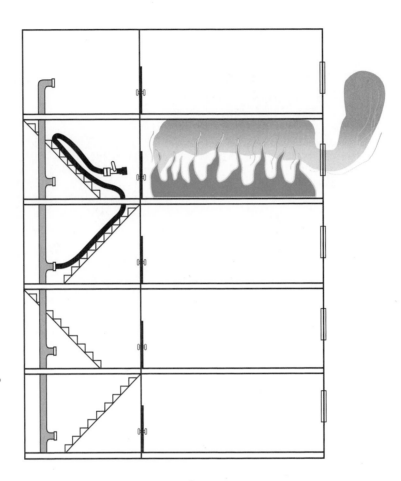

Figure 8-20 *Proper hose line layout from a standpipe connection to the fire floor.*

SUMMARY

Built-in fire protection systems are becoming more popular and more complex due to building owner desires or local mandates. All fire department personnel must have a good working knowledge of the various built-in protection systems available and specifically those that are utilized within the areas they protect. The four types of sprinkler systems all basically work with the same type of valving and controls, even though they have the variations discussed in this chapter. Along with sprinkler systems are standpipe systems, which are similar to the sprinkler system, but with only one or more outlets on each floor. It is important to recognize the different types of standpipe systems, because a class II is not designed for fire department use and should not be used by the fire department. Residential sprinkler systems are also becoming more and more popular; the fire department should know how they work and how to control them, because they may eventually have to work with them. Although there are many types of specialized extinguishing agents, those described in this chapter are the most common ones.

Remember, built-in fire protection systems are an asset to the fire service. They help us do the job more efficiently while providing the general public with a greater degree of fire safety. An additional benefit is that of enhanced firefighter safety.

Built-in fire protection systems must be utilized by the fire department in order to be completely effective. Every fire department must develop guidelines to enhance fire department use of these systems.

REVIEW QUESTIONS

1. Why are built-in fire protection systems on the increase and an important ally of the fire service?

2. List and describe the four main types of sprinkler systems.

3. Name the three main types of main water control valves and how to determine whether the valve is open or closed.

4. List three common types of valving, other than the main water control valve for sprinkler systems, and their purposes.

5. What outside component is a critical part of a preaction sprinkler system?

6. Name three ways, other than fire department supplementation, that sprinkler and standpipe systems receive their water supply.

7. What NFPA Standard describes fire department operations in buildings with sprinkler and standpipe systems?

8. True or False? The first- or second-due engine company to a building protected by a sprinkler system should be directed to the FDC.

9. Explain the three classes of standpipe systems and their designed uses.

10. What are the hazards to occupants and firefighters associated with the activation of a dry chemical extinguishing system.

11. The standard rule of thumb for pump pressure while supplying a sprinkler system with two lines is

    A. 100 psi

    B. 125 psi

    C. 150 psi

    D. 175 psi

12. Who is the person who can order the main water supply valve of a sprinkler system to be closed?

ACTIVITIES

1. Review your departmental SOGs for fireground operations. Do the SOGs address operations for buildings with built-in fire protection systems? If not, how would you change them to include such operational concerns?

2. Preplan a building with built-in fire protection that is within your response district. Be sure to include locations of FDC, where water supply comes from, and initial operations of a fire in the building utilizing the built-in protection system.

3. Review your departments high-rise kit and see if it has the minimum required hose, equipment, and other items.

Chapter 9

AFTER THE INCIDENT

Learning Objectives

Upon completion of this chapter, you should be able to:

- Define an incident termination plan.
- Discuss the three stages of resource demobilization.
- Discuss the purpose and need for formal and informal postincident analysis.
- Describe the critical incident stress management system.
- List four types of critical incident stress debriefing.

CASE STUDY

In what has been termed a career fire for many Kansas City, Missouri, firefighters, the members responded to the city's first general alarm fire in its 117-year history.

The fire in the early morning hours of July 18, 1998, started in the seven-story section of the Midwest Sales Company building. This seven-story section of concrete-reinforced steel made up one-third of the total structure. The other two-thirds was a six-story heavy timber mill-type construction. The seven-story section where the fire began still bore damage from a two-alarm fire nearly 3 months prior. This building is part of a complex of large industrial buildings in an area of the city called the West Bottoms. The West Bottoms is the central industrial district in the city and has been the site of many major fires and other disasters.

At 2:12 A.M. the fire alarm office received a call from an employee in a building across the street, reporting a fire in the Midwest Sales Company building. The initial dispatch sent three engine companies, two trucks, Heavy Rescue 1, and two battalion chiefs.

Upon arrival at the scene, the first engine company reported heavy smoke and fire on the fifth, sixth, and seventh floors on the east side of the building. Hearing the initial report and seeing the size of the fire upon his arrival, the downtown battalion chief called for an immediate second alarm. This alarm brought two engines, two trucks, a battalion chief, and two deputy fire chiefs. In anticipation of further equipment needs, the dispatcher moved four rescues into a nearby fire station.

The fire was rapidly spreading to the west toward the six-story section of the Midwest Sales Company building, a large lumber company, and a six-story grain elevator. The thermal column was carrying a shower of embers skyward and raining them down on the roofs of nearby buildings. Some of the embers were described as the size of baseballs. The radiant heat from the fire was causing tremendous exposure problems on other nearby multistory buildings. The radiant heat ignited the roof of the lumber company and stacks of lumber in the yard.

A third alarm was sounded as the fire continued to spread to other buildings. This alarm brought the four rescue units to the scene that dispatchers had moved to the nearby fire station. As the buildings began to collapse, the radiant heat and exposure problems were reduced. At this point in the fire, there were twenty-two apparatus from the Kansas City, Missouri, Fire Department on the scene. There were additional units from Kansas City, Kansas. Four fire units responded from Johnson City, Kansas, and eventually fourteen neighboring fire departments sent assistance either to the scene or to cover empty fire houses. At the height of the fire, more than forty fire companies and nearly 175 firefighters were deployed on the fireground.

The city's first general alarm fire destroyed nine multistory industrial buildings and kept fire crews busy on scene for six days. The fire was successfully handled by creating divisions on the fireground and assigning a supervisor to oversee the activities within each division. By managing the scene effectively and maximizing the resources as the fire grew in size and complexity, the incident was handled without injuries to firefighting personnel. The incident command system (ICS) was implemented early in the fire and allowed the incident commander and division supervisors to establish control of their areas of responsibility, request resources, and track resource usage.

INTRODUCTION

The mobilization of many resources over a short period of time requires good management skills and the effective use of the incident command system (ICS). As fire crews are responding to the scene, they must be assigned either to a staging area or placed in operation on the fireground. The crews will go to work as a single resource, a group, or in a division and often commit their apparatus on the fireground by laying hose, raising ladders, or establishing master streams.

In a large-scale incident requiring many resources, the resource unit within the planning section will develop a demobilization plan that includes specific instructions for all personnel and resources demobilizing (see **Figure 9-1**). As any incident, large or small, is brought under control, a demobilization plan is used to assure a safe and effective demobilization, or release of resources. The

demobilization
process of returning personnel, equipment, and apparatus after an emergency has been terminated

Figure 9-1   *If the incident commander does not develop an incident termination plan, it can lead to unsafe practices.*

demobilization plan should be consistent with NIMS/ICS criteria and use ICS Form 221 for demobilization checkout.

In addition, many agencies require their personnel to pass through rehab before leaving the incident scene.

This chapter provides information to consider in postincident activities. These postincident activities include terminating the incident, postincident analysis, and **critical incident stress management (CISM)**. Whether the incident is large or small, termination of the incident must be a planned sequence of actions.

critical incident stress management (CISM)
a process for managing the short- and long-term effects of critical incident stress reactions

INCIDENT TERMINATION

Incident termination includes at least three stages: demobilization, returning to quarters, and postincident analysis. The postincident analysis must include the review and update of policies and procedures to ensure improved incident operations in the future.

Demobilization

■ **Note**
Whether the incident is large or small, termination of the incident must be a planned sequence of actions.

Demobilization is the stage in the incident when the incident commander (IC) begins to release operating companies. In a large incident, the IC works with the planning section to evaluate the on-scene resources and compare them with the present activity and needs. Based on the current priorities and resource needs at the incident, the IC begins to establish the demobilization plan and an order to release units/crews. There must be a systematic plan to release of all apparatus and crews from the scene.

■ **Note**
Demobilization is the stage in the incident when the incident commander begins to release operating companies.

The importance of demobilization is ignored in many incident action plans. There are times when equipment is kept on the incident scene too long, leaving areas without adequate fire protection, or times when equipment is returned too soon, allowing the incident to escalate. The transition from the assignment of tactical operations to demobilization must be organized.

As the incident is brought under control, the IC must begin decommitting companies. This must be a preconceived plan that takes into account the needs of the incident and the needs of the community. A number of considerations must be evaluated while forming this step of the overall action plan, including personnel, apparatus, and equipment.

The order of release must be based on maximizing resources necessary to complete the overhaul, salvage, and cleanup of the incident scene. It only makes sense to release first those companies that have been there the longest. In the past, the unit that was first on the scene was the last unit to leave.

■ **Note**
The order of release of personnel must be based on physical or emotional stress.

Many organizations operate using the first-in, last-out approach, meaning that the first-arriving unit, because it usually is the most committed, is the last to leave. This approach is not logical, because fatigue is a significant factor in the injury equation. Often the first-arriving crews are those that have worked the

Figure 9-2 *Firefighter fatigue must be a consideration in the incident termination plan.*

■ **Note**

In long and complicated operations, those crews that arrived first should leave first.

standardized apparatus apparatus that has exactly the same operation and layout of other similar apparatus in a department—for example, all of the department's pumpers would be laid out the same, operate the same, and have the same equipment; useful for situations when crews must use another crew's apparatus

hardest and are at the scene the longest. As depicted in **Figure 9-2**, firefighter fatigue must be a consideration in demobilization.

Many departments are now rethinking this approach. The order of release of personnel must be based on physical or emotional stress. In situations where the crews have been engaged in long and complicated incident operations, those crews that arrived first should leave first. Doing so allows those who have arrived in the height of the operation and have worked under the worst conditions to be relieved first. The replacement of the fatigued crews with personnel in the best physical condition to complete the overhaul and cleanup operations provides a safer termination environment. This will maximize the effectiveness of the crews and provide for personnel safety. Many firefighters have been injured during overhaul because they were too tired to do basic tasks correctly. When personnel get fatigued, they tend to lose their perspective on safe operations due to the desire to get done and get back to quarters as quickly as possible.

One argument with the first-in, first-out approach is that the first arriving company will often strip their apparatus of hose, ladders, and hand tools. Ladders are still lying against the building, hose lines may still be connected to hydrants, standpipes, or inside the building, and tools may still be in use inside. These problems can be overcome when departments have **standardized apparatus** and equipment (see **Figure 9-3**). The crews can then leave their equipment and return on the relief crew's apparatus. In some cases, the apparatus operator returns after rehab to ensure the apparatus and all the equipment are properly returned to

Figure 9-3 *Example of apparatus standardization is shown in these two engines.*

■ **Note**

The mindset of "owning" the fire must give way to the logical thinking of personnel care and conservation.

service. The mindset of "owning" the fire must give way to the logical thinking of personnel care and conservation. The first-in, first-out policy is important during extreme weather conditions, such as extreme cold, heat, and humidity.

Personnel The personnel being demobilized should pass through rehab for a medical evaluation, including vital signs, before leaving the scene. Determine the fatigue factor of the personnel. How long have they been there? Under what conditions have they worked? Are they wet, cold, overheated? Have the members been placed in a highly stressful situation? Will they need to go through CISM?

All these questions must be considered during the decommitting stage of the incident. The health and welfare of the firefighters is important. Often during the overhaul phase, which is sometimes personnel intensive, as in **Figure 9-4**, many

Figure 9-4 *Overhaul can be physically demanding. Remember to keep enough resources on the fire scene to meet the goal of overhaul.*

crews have been released and the remaining crews must do the overhaul without relief.

Apparatus When considering the movement of apparatus from the incident scene, there may be times when the crews are relieved and sent back to quarters on a different apparatus (**Figure 9-5**).

During a major operation, the personnel need to be relieved, but their apparatus will be stripped of tools, equipment, ladders, and hose. The apparatus may be in a key position for overhaul or it may be trapped by hose or other needed equipment.

In some colder parts of the country, there are times when the apparatus is frozen in place. The crews then use reserve apparatus until the apparatus is removed from the ice by the service garage, using steam and propane heaters.

Equipment When equipment is still being used, it can be left at the scene. Hose, ladders, and other important pieces are often left behind. If the department has an established identification system, the tools can be quickly returned to

Figure 9-5
Firefighters may be relieved and have to leave on another company's apparatus.

Figure 9-6 *Color-coding equipment helps to ensure the equipment gets back to the correct apparatus.*

■ **Note**

As the incident commander begins demobilization, input from the branch or division officers, safety officers, and the planning section is necessary to determine the most immediate needs of the incident.

the appropriate apparatus. **Figure 9-6** shows a color-coding system for this purpose.

As the IC begins the process of demobilization, input from the division supervisors, safety officer, and the planning section is necessary to determine the most immediate needs of the incident. Based on the determined needs, the IC can use the division of the fireground for demobilization, the same as it was used to build the fireground. The division supervisors can determine which companies in their branch or division have completed their tasks, which need relief, and which can be returned to service. The companies that have completed their task and have their crews and equipment intact can be returned to service or used as relief. The more efficient and effective the process, the quicker the companies can be returned to service for the next alarm.

During demobilization, the incident commander must also consider reducing the size of the command structure. As branches and divisions are relieved, those officers can be reassigned or returned to their stations.

There is typically a reverse in the command structure at this time. The fire chief or deputy chiefs may now begin to transfer command back to the battalion chiefs or other company officers. As each branch or division is relieved, the command structure decreases in size and complexity. This continues until all units have been returned or only the companies assigned to "fire watch" are left behind. At this point, command is terminated, sounding the end of the incident.

Returning to Quarters

In an incident where the personnel have the potential to be exposed to blood or bodily fluids, they must do as much personal decontamination as possible at the scene before entering their apparatus (**Figure 9-7**). This also includes incidents where the crews are exposed to gasoline, battery acids, greases, and other contaminants. Many contaminants can be carried into the apparatus and back to the firehouse if the crews are not careful and observant.

Once released from the incident, the crews should return to quarters to replenish supplies. They may need to fill air bottles, load hose, and generally clean their tools and apparatus. The chain saws may need gasoline and new saw blades, the axes may need sharpening. The exhaust fans may need to be refueled and cleaned. The hand tools need to be cleaned of plaster, drywall, and tars that have accumulated. The handles and tips on all tools need to be checked for cracks or other damage. The crew members may need to clean and dry their PPE. The crew members should shower, as this is a decontamination process from the elements of incomplete combustion from the fire. Department standard operating guidelines (SOGs) should give direction to this process. **Figure 9-8** is one example of an SOG.

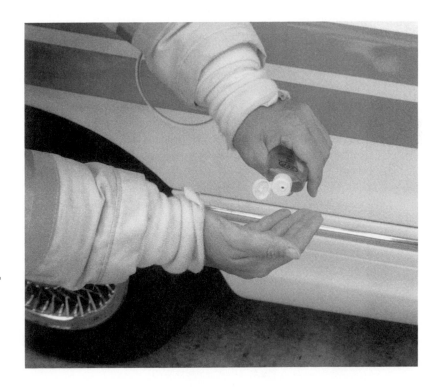

Figure 9-7 *Personal decontamination should be started, in as much as possible, at the scene.*

GENERAL ORDER NO. **359**

Rev. 10/97

PERSONAL PROTECTIVE EQUIPMENT

359.1 **PURPOSE:** To provide a consistent and standard policy on the proper inspection, fitting, and care of PPE for structural firefighters in accordance with NFPA 1500, 1971, 1972, 1973, and 1974.

359.2 **DISCUSSION:**

A. It is to be understood that the ultimate responsibility for the proper care, reports of deficiencies, and wearing of PPE lies with the wearer.

B. The training officer will conduct a visual and physical inspection of all PPE issued to each member of the department on an annual basis, to be done during the month of May. This inspection will be recorded on the Personal Protective Equipment Record form.

C. The shift commanders will conduct a visual and physical inspection of all PPE issued to each member of their respective shift on an annual basis, to be done during the month of November. This inspection will be recorded on the Personal Protective Equipment Record form.

D. Each lieutenant will conduct a visual inspection of all of his/her respective crew members on a monthly basis. The inspection is to be accomplished in accordance with the Department Inspection Guide, General Order No. 360. The inspection is to be noted in the Company Journal Account along with any deficiencies reported to the Training/Safety Office ASAP.

E. Each firefighter should visually inspect his/her PPE daily and report any deficiencies to his/her Immediate supervisor for repair or replacement.

F. Any PPE in need of repair or replacement will be handled by the training/safety office.

G. Any PPE turned in for replacement will either be destroyed or sent out for repair for future use.

H. Damaged or improper PPE is NOT TO BE USED.

359.3 **FITTING:**

A. All new employees will be fitted for PPE by the training/safety office and the manufacturer representative.

B. Current employees with PPE in need of repair or replacement will be fitted by the training/safety office.

359 - I

Figure 9-8 *Many contaminants remain in firefighting PPE, requiring decontamination. Here is an example SOG.*

359.4 **CLEANING:**

A. Soiled PPE expose firefighters to toxins and carcinogens that can enter the body through ingestion, inhalation, and/or absorption.

B. When clothing or equipment becomes laden with particles and chemicals, other problems are encountered in addition to exposure to toxins. These include:

   1. Soiled turnout gear reflects less radiant heat. After materials are saturated with hydrocarbons, they tend to absorb rather than reflect radiant heat.

   2. Gear heavily contaminated with hydrocarbons is more likely to conduct electricity increasing the danger when entering a building or vehicle where wiring may still be live.

   3. Materials impregnated with oil, grease, and hydrocarbon deposits from soot and smoke can ignite and cause severe bums and injuries (even if the material is normally flame-resistant).

C. PPE shall be washed every six months regardless of exposure or lack thereof (excluding that of command officers and staff, which will be laundered on an as-needed basis).

D. PPE shall be cleaned as soon as possible after exposure to contaminants.

   1. During times of normal business hours, the training/safety office will handle this.

   2. During off times, the shift commander shall contact our current laundry facility.

E. Exposed PPE should be hosed down at the emergency scene. Hosing down immediately after the termination of an emergency can remove up to 90% of all contaminants before they have a chance to set in.

F. Under no circumstances should contaminated gear be laundered in the fire stations or taken home to be laundered.

G. Any blood/blood product-contaminated gear should be taken care of according to the Dept. Exposure Control Manual (SOP-357).

Figure 9-8   *(Continued)*

In some cases, the crews are not returned to their station, but to another empty firehouse to fill a void in the response system. In this case, the crews have to do their best to get their equipment and themselves back in service.

In incidents involving train wrecks, plane crashes, serious vehicle accidents, and other significant incidents where there have been deaths and great personal suffering, the crews may not return to quarters but go to a critical incident stress debriefing area. Here, they will have a chance to evaluate their feelings, understand what they have just experienced, and be able to release the stress they may not know they are feeling. There is a separate section on CISM later in this chapter.

Fire officers and firefighters should all be alert for signs of critical incident stress within the crews. If they recognize any of the signs or symptoms, they should notify their supervisors to begin CISM.

The company officer or supervisor must determine whether the crew can be placed into service. It is their duty to ensure that the crews are ready for their next response. While cleanup is in progress, look for signs of fatigue and reassess the crew members for injury or illness that can be attributed to the incident. For example, in 1999, a Philadelphia firefighter died in the station after suffering an injury on the fireground. Company officers must ensure that the apparatus, equipment, and personnel are ready and capable of going back into service. After the apparatus and crew members have been cleaned and are ready for the next alarm, the company officer should sit down and discuss the incident. This is just an informal talk to discuss how tasks can be accomplished better the next time. It is a time for each member to have input and improve communication in an informal setting. This is the first step in the postincident analysis.

postincident analysis
a critical review of the incident after it occurs; should focus on improving operational effectiveness and safety

■ **Note**
The critique, sometimes called a *critical review*, is the postincident analysis (PIA).

■ **Note**
The term *critique* carries with it the negative connotation of criticism; therefore, the term *postincident analysis* is favored by this text.

■ **Note**
The purpose of the postincident analysis is to focus on reinforcing the most appropriate and effective strategies while discouraging inappropriate or impulsive strategies.

POSTINCIDENT ANALYSIS

A postincident analysis (PIA) should be conducted on any incident that multiple units have responded to and operated, including fires, hazmat, technical rescues, multiple casualty incidents, vehicle accidents, and entrapments. The critique, sometimes called a *critical review,* is the postincident analysis. It is a step-by-step approach to examining how the incident developed and how the different tactical operations at the incident worked in tandem to accomplish the goals and objectives set by the incident commander (see **Figure 9-9**). The components of the PIA should focus on:

- Personnel
- Equipment
- Resources
- Operational effectiveness

Figure 9-9 *The postincident analysis should be performed after incidents and focus on personnel, equipment, resources, and operational effectiveness.*

The term *critique* carries with it the negative connotation criticism; therefore, postincident analysis is the terminology favored by this text.

Purpose

The purpose of the PIA is to focus on reinforcing the most appropriate and effective strategies while discouraging inappropriate or impulsive strategies. The lessons learned in a PIA should be carried back to the planning section of the department. The information discovered can be used to review and update outdated or ineffective policies and procedures.

The PIA and planning phase are as much a mind set as they are an activity. The PIA activities are relevant to every member of the department, not just the chiefs and the crews that were on the scene. This process can have a far-reaching effect on every member and on future operations of the department. Because there is a decline in the number of actual fires, this transfer of information becomes invaluable. The smaller the department, including both paid and volunteer departments, the more valuable.

The PIA must look at the entire operation, the big picture. This overview includes all operations and activities that took place at the incident and considerations concerning environmental impacts, mutual aid, community resources, and additional response demands. Concerns with understanding human limitations in certain operations must be evaluated. The ability or inability of personnel to know and understand how they must perform at an incident is vital to the overall effectiveness of the organization.

Informal PIA

There are two steps of the postincident analysis. The first is the informal company-level discussion that takes place after the incident. At this discussion, the crew members should be free to openly discuss their ideas as to how and why they performed their specific duties. This is a good time for each member to openly discuss his or her individual actions and those of the crew as a whole. The company officer should take notes during this session to take to the formal PIA.

The fire service has witnessed an overall reduction in the number of serious fires during the past decade. With this reduction in actual working fires, many of today's new officers and firefighters have less opportunity to develop their skills through actual experience. The ability to relay information during the PIA is one way to fill this gap. The PIA gives senior officers and firefighters a chance to explain what they saw in reading a building, the fire and smoke movement, and the actions they took. This allows the members of the unit to discover problems and develop solutions.

The company-level PIA allows the members to express what went right in the operation, what tactics worked and how. This process should be a positive experience meant to improve the members' thought processes and decision-making abilities. By analyzing the activity elements of the incident and putting these elements into a complete picture, the crew can see how their specific activities affected the overall incident operations.

The company-level PIA should occur as soon as the crew has had the chance to get the equipment ready for the next alarm and they have gotten themselves cleaned and ready to respond. It can also occur at the scene if the crew is sent for rehabilitation, briefly sitting on the back step of the engine, or even as shown in **Figure 9-10**, while hose is being reloaded. This allows the members to sit and discuss the events while they are fresh in their minds. After a major incident, the company-level PIA should be held regardless of the time. A few minutes discussing the lessons learned in this operation may be just enough information to prevent an injury or save a life on the next response. If it is late at night and the crew is fatigued, hold the PIA to 10 or 15 minutes and then address the issue again the next morning or at the beginning of the next shift during company training time. In a volunteer fire department, members could be asked to return to the station the next day or evening to conduct the company-level PIA. Be careful: Delaying the company-level PIA may give some members time to develop defensive behaviors or mount attacks against other members, thus compromising constructiveness.

Use the company-level PIA after every response, from the minor investigation of an odor to a working fire. Each incident provides an opportunity to learn and communicate ideas. Every activity is open for discussion, from the response to the scene to the return to quarters. Have each member of the crew provide input during the session, from the apparatus driver to the rookie. We can all learn and we all have ideas and information to provide. If time allows and there

■ **Note**

If it is late at night and the crew is fatigued, hold the PIA to 10 or 15 minutes and then address the issue again the next morning or at the beginning of the next shift during company training time.

Figure 9-10 *A quick, informal PIA may be held while readying the engine for the next alarm.*

is interest in the PIA, play "what if" games. What would you have looked for, how would you have vented, entered, searched? Make the session a positive, constructive learning experience. Leave criticism and blame out of it!

Formal PIA

The formal PIA is a scheduled event for those required to attend and participate. Those in attendance should include the crews that responded to the scene and carried out specific tasks, those that filled incident command positions, and those that supplied support activities. There often are formal investigations completed by the fire marshals office or Bureau of Fire Investigations (see **Figure 9-11**). An individual from the investigative team should participate.

This individual must provide the postincident pictures and analysis to show why the smoke, fire, and construction features acted the way they did based on their investigation. We have all been to fires and said, "This worked before. Why didn't it work on this one?" The post-fire investigation team can often answer these questions. Other people present will be representatives from the training bureau, fire operations, and the communications office. If there are policy or procedural plans that need to be changed, these people will need to know what, how, and why changes are required.

The news media should be contacted as soon as possible to obtain raw footage of the incident. Most news agencies will provide copies of the raw

■ Note

Those in attendance should include crews that responded to the scene and carried out specific tasks, those that filled incident command positions, and those that supplied support activities.

Figure 9-11 *If a cause and origin investigation was conducted, the investigator should attend the formal PIA. (Photo courtesy of Craig Brotheim.)*

footage for free or at nominal cost. This request for raw footage must be made the day of the incident. Once the footage is edited, the raw, unused footage is generally discarded. The raw footage allows the viewing of the fire in the right time sequence, if the media was on the scene (**Figure 9-12**).

When preparing for the formal PIA, choose a place large enough to accommodate the group. The formal PIA should be held in a large room or the training academy where the presenters can use slides, overheads, video and audio tapes, and other visual aids. Using the visual aids, the presenters should describe their response to the incident, what they thought about, what the preincident plan information provided them en route and upon arrival at the scene.

The presenters need to detail their size-up upon arrival and their initial actions. Detail what worked and what did not work at the incident. If actions or operations were performed outside the realm of the SOGs, the reasons for the actions and the results achieved should be described. This information could lead to updating the SOGs.

Start with the initial companies and move through the scenario in a time line as more alarms are sounded and other companies arrive. Allow each branch or division officer to present information regarding his or her activities and understanding of the operation. Discuss each and every significant part of the operation from the dispatch to the incident being placed under control.

Figure 9-12 *Often news agencies provide copies of raw footage, which can enhance the PIA.*

Overall, the formal PIA must focus on the entire incident and not dwell on any single aspect. Many incidental activities must come together for the incident to be handled effectively. Ignoring these incidental activities can lead to a lack of focus on the needs of the support positions. It can lead to underestimating the need for additional resources or specialized apparatus.

Reviewing and Updating Procedures

The SOGs that have been established by an organization to carry out a response are often developed with a minimal knowledge base. A combination of past experiences, broad generalizations, and common sense seems to dictate the planning function. Little, if any, attention is paid to current circumstances and change. There is a major difference in the necessary response to a single-family house fire as compared to an unsprinklered high-rise full of elderly and infirm residents, but some SOGs do not differentiate in their response to these significant incidents.

The objective of the PIA is to examine how the incident was handled and what SOGs need to be created or changed for the next operation to be improved. The PIA highlights both the positive and negative outcomes of the incident. This information will broaden the knowledge base from which SOGs can be developed or updated. These changes can then be implemented by changing the SOGs

■ **Note**

The PIA is a positive
event for the
organization as it
provides an objective
platform from which
the department can
affirm strengths and
discover weaknesses.

and, as necessary, equipment. The training division can then broadcast the lessons learned and the policy or procedural changes so that others can benefit.

One final note: The PIA is a positive event for the organization as it provides an objective platform from which the department can affirm strengths and discover weaknesses. All personnel must share what was right and what was not so right so that others may learn from the experience. The lesson someone learns may very well help him or her to save a life one day.

CRITICAL INCIDENT STRESS MANAGEMENT (CISM)

In today's society, it is nearly impossible to live without stress as a part of every-day life. Most people can deal with this stress and suffer no ill health effects, but a constant bombardment of high stressors can cause negative health effects, which leads to negative lifestyle changes, such as illness, job burnout, loss of productivity, and other problems. Emergency response personnel are expected to function under highly stressful situations on a regular basis; therefore, they are prone to the stress-related problems.

There is always the uncertainty of what the next call will bring, the interruptions of daily routines, and the dangers associated with the job. Emergency responders are expected to perform any time human tragedies are involved in circumstances that bring hazards associated with rescues and firefighting. They are always concerned with the circumstances and fears of doing something wrong that may prolong or aggravate human suffering.

Some of these incidents are usually high in stress levels, but in some cases it is the combination of many smaller events that build up in terms of emotional strain and stress. These incidents leave a lasting impression on responders that can affect their ability to function both on the job and in their home life. These events are called critical incidents.

critical incidents
incidents that have a
high potential to
produce critical
incident stress, for
example, children, a
family member, or a
coworker being
involved in incidents

Many emergency response personnel learn to deal with these stressors as "part of the job." In the past, responders were expected to handle these events on their own and were often criticized if they displayed weakness or emotion toward these critical events. Fortunately this is no longer the case for most emergency services. Critical incident stress management is now a part of most organizations, and written guidelines have been developed. One such guideline is shown in **Figure 9-13**.

Many events can lead to critical incident stress such as the following:

- Sudden or traumatic death or severe injury to an infant or child
- Prolonged rescue efforts with mass casualty or severe suffering of the victim
- A responder knowing the victims
- Traumatic death or severe injury of a co-worker whether or not in the line of duty
- Suicide of a co-worker

TEAM ACTIVATION

If initial notification is via the state phone line (1-407-373-CISD):

Operator will obtain county and region of incident and notify appropriate regional coordinator, who will evaluate the need and activate appropriate local team.

If initial notification is to the regional coordinator:

Regional coordinator will evaluate the need and activate the appropriate local team.

If initial notification is to the local team:

The local team will initiate team SOPs. Notify the regional coordinator and request assistance if needed.

OPERATIONAL GUIDELINES

PEER SUPPORT NETWORK

The basis for the entire CISD process is the peer. The local team coordinator should encourage each agency to identify individuals to be trained as CISD peers. This group of trained personnel will then be able to make initial contact in the event of a critical incident or be available to provide support to an individual who is experiencing an acute or cumulative stress reaction. The peer support network integrates with the CISD team, providing peer members to the team or by acting as a liaison to the interagency CISD team.

ON-SCENE SUPPORT SERVICES

Support services and interventions may be utilized during a critical incident. These services may be provided at or near the scene of operations. In most cases, these services will be provided by peer support personnel, although mental health team members may be requested and required if the situation warrants.

On-scene support may consist of the following types of services:

1. One-on-one counseling to those emergency service workers showing obvious signs of distress as a result of their participation in the incident.

2. Advice and counsel to the incident commander or his or her liaison on the topics of stress management and, specifically, issues related to the critical incident.

3. Demobilization of personnel being released from the scene.

4. Until more appropriate agencies arrive, control of victims, survivors, and families in order to ensure that the work of emergency services is not impeded.

Figure 9-13   *Critical incident stress management (CISM) programs are a necessary consideration in postincident activities. (Courtesy of CISD of Florida Region V.)*

PEER SUPPORT PERSONNEL ENGAGED AT THE SCENE

Any peer support member who is dispatched to the critical incident as a member of an emergency services organization is primarily responsible for operating with that organization. For example, peer support/firefighters who accompany their units to the scene will serve in the capacity designated by their supervisors. This holds equally true for all emergency services personnel involved at an incident.

While performing assigned duties, it may be possible for the emergency service personnel to observe the scene for situations that may increase the potential for stress affect. It may also be possible, while performing emergency duties, to observe personnel for obvious distress. While these are not the primary functions of emergency service workers at this time, appropriate disclosure of their observations may provide insight to command officers. If the need to make recommendations to command becomes obvious or if the peer suspects that the potential is unusually high for the development of acute stress reactions, the peer may suggest to the commander that he or she consider calling the CISD team.

Even if the incident commander designates the peer's function as CISD on-scene support, the peer support personnel may request additional team support to be dispatched to the scene. The rationale for this action is:

1. To keep the team coordinator advised of the activity and ensure continuity.
2. It may be inappropriate for the peer to provide services to his or her own units.
3. It may prove to be too "draining" for the peer who is or has engaged in emergency operations to carry out on-scene support activities.
4. It may not be in the emotional best interest of the peer member to provide services in this situation.
5. The task may be too involved for one or two team members to handle effectively.
6. The peer support member's unit may be released from the scene before on-scene support activities are completed.

PEER SUPPORT PERSONNEL DISPATCHED FOR CISD SERVICE

CISD team members dispatched to the incident scene by the team coordinator may rendezvous so they may go to the scene as a unit. Every attempt should be made to have team members escorted to the scene by an emergency service agency to permit easy access to the incident scene. If this is not possible, the team members may take the minimum number of vehicles required to transport the team to the scene.

Figure 9-13   *(Continued)*

ON-SCENE TEAM LEADERSHIP

Once on the scene, one team member may act as team leader and report to the command post. This member should advise the IC or his or her liaison of the number of support personnel and request direction from the officer. The team leader should advise the officer of where the team will be located and what they will be doing. The team leader should act as a liaison between command and the team throughout the incident when possible.

Therefore, any recommendations and observations of any team members should be made to the team leader, who will, in turn, report to command. It may be ideal if the team members can arrange "report times" to offer information to the team leader and so that the team leader will not interrupt the IC more frequently than necessary.

ADDITIONAL ON-SCENE CONSIDERATIONS

1. All CISD team members, acting on behalf of the program, may wear CISD ID at all times while on-scene, travelling to the scene, and so forth.

2. Team members should be appropriately dressed with protective clothing, including proper shoes/boots, protective helmet, and so on, to enter the internal perimeter.

3. No team member should go inside the internal perimeter unless requested to do so by a command officer.

4. The team leader should keep track and know where all team members are during the operation.

5. Except circumstances, the team leader will be the liaison between command and the team.

6. The team leader should assign tasks to team members as required.

7. CISD operations should be set up in a safe area, out of the inner perimeter and out of view of the scene.

8. Communications between team members, especially between command liaison and CISD operations, should be established using portable radios or cellular phones if possible.

9. Food and drink should be obtained for the rehabilitation area if not provided by command.

10. Warm, dry clothing should be provided for severely affected personnel who are taken out of service.

Figure 9-13   *(Continued)*

■ **Note**
Before a critical incident occurs, it is important to have a process in place for both stress defusing and debriefing.

Before a critical incident occurs, it is important to have a process in place for both stress defusing and debriefing. The critical incident stress management (CISM) is a system that provides the necessary means to accomplish this need. This program includes training that prepares supervisors and peers to recognize signs, symptoms, and situations that may indicate the need for stress debriefing (**Figure 9-14**). Each organization should establish procedures for the times when debriefing is needed and the method for accessing the system. The system should be implemented automatically for any of the stressors listed previously, or whenever called for by an incident commander or company supervisor.

■ **Note**
CISM allows emergency responders the opportunity to vent their feelings about an incident that has had a profound and powerful emotional impact.

The ability to discuss an incident after the call with peers, supervisors, or critical incident stress debriefing (CISD) personnel allows a positive form of communication to alleviate the stress and anxiety of the incident. It allows emergency responders the opportunity to vent their feelings about an incident that has had a profound and powerful emotional impact. Without CISM, the responder may be more likely to develop negative physical, behavioral, and psychological reactions.

The emergency responder may begin to develop generalized feelings of anxiousness and apprehension. This anxiety may create unwanted physical reactions: fatigue, sleeplessness, changes in eating habits, and/or body aches. The behavioral changes that can occur include changes in activity levels, difficulty in concentration, forgetfulness, nightmares, flashbacks, isolation, and other manifestations. The psychological symptoms may include fear, guilt, sensitivity, depression, or anger.

Figure 9-14   *Fire department personnel, particularly supervisors, must be alert for signs of critical incident stress, clearly not a problem in this case.*

No one is immune from the anxiety and stress of constant emergency situations. We try to deal with these anxieties based on past experiences and our own psychological compositions. Some people a more prone to these stressors than others. In some extreme cases, a person suffering from critical incident stress reactions may turn to alcohol or drugs as an escape. Still others may just become fed up with the stress and resign from the service. There are also the tragic tales of those who could not handle the stress and chose to commit suicide.

The organization's safety and health program should include CISM to help emergency responders deal with the issues associated with critical incidents, and hopefully avoid the unwanted physical, emotional, cognitive, and behavioral responses listed previously.

Types of CISM

peer diffusing
the concept of using a trained person from the same discipline to talk to emergency responders after a critical incident as a means to allow them to talk about their feelings about the event in a nonthreatening environment

There are four different types of CISM both at the scene and after the event. The first is **peer diffusing** that occurs at the scene. Peer defusing involves an informal discussion about the event by a group of peers who experienced the event together. As the incident begins to de-escalate, the responders get together with their peers and discuss the incident and the actions they took. This can be seen on most major incidents as the emergency responders are replacing equipment and preparing to leave the incident.

The second type is on-site defusing. This is sometimes used during long-term operational incidents of major proportions, such as plane crashes, train wrecks, or natural disasters. In this defusing scenario, the responder is pulled from the operational area and sent to rehab branch or division. While in rehab, the responder is evaluated by a member of the local CISM team for signs and symptoms of distress. If the person appears to be suffering anxiety and stress, he or she may be removed from the incident site for further evaluation.

The third type in the CISM system is the defusing after demobilization from a major incident. In this process, the emergency responders are sent to a neutral area, such as an empty firehouse, where they have a buffer from the highly stressful incident scene. This allows them the opportunity to get some food and drink and relax before returning to their own station. The session will last from 30 to 60 minutes and provide a brief defusing session during which the emergency responders are briefed on signs and symptoms of critical incident stress.

■ Note
All members involved in the incident should try to attend the session. The peer counselor should have had training in this type of formal debriefing.

The fourth session is the formal debriefing that commonly occurs sometime after the incident, maybe several days. It allows the members who operated at the scene to get back together and discuss their roles at the incident and the overall response. This session has two specific requirements: First, all members involved in the incident should try to attend the session, and second, the peer counselor should have had training in this type of formal debriefing session. This is a group session led by a CISM team of specially trained staff. It allows participants to focus on the event and to express their feelings.

Many CISM teams are regionalized as part of the mental health community in conjunction with the emergency response agencies. Regionalization is the best approach because most emergency response agencies do not have the appropriately trained staff, and each debriefing would involve a peer from the same organization. The regionalization of the CISM team allows mental health professionals who are nonpartisan and properly trained and experienced to conduct the sessions.

In summary, the critical incident stress debriefing function of the CISM provides a supportive environment for the participants to recognize that they are having normal feelings and reactions. It provides them with the signs and symptoms of stress and anxiety so they can look out for themselves and their co-workers. It reassures them that they are part of a caring and supportive organization. This contributes to the prevention of job burnout, isolation, and feelings of inadequacy by members. It also strengthens their coping skills for future incidents.

The safety and health of emergency responders and their longevity in the response system depends on the support of a good critical incident stress management system.

One final note when referring people to a debriefing area: The critical incident stress management plan must be developed to take into account people other than the emergency service responders who have responded to the incident. This may include tow truck operators who must pull cars apart to get to victims; crane and heavy equipment operators; city, county, and state employees; special service agencies; and the police officers, as shown in **Figure 9-15**. Anyone who has operated in a task function at the incident can feel the stress and anxiety

Figure 9-15   *CISM must take into account other agencies that were involved in the incident. (Photo courtesy of Palm Harbor Fire Rescue.)*

of the event. They must be part of the debriefing if they show any signs or symptoms of stress. Many of these people are not trained on how to deal with deaths and severe injuries and this may be a first-time event for them. It can be emotionally overwhelming, and the fire service has an obligation to support them.

SUMMARY

Understanding factors relating to the termination of an incident are as important as the other areas of study in strategies and tactics. Incident termination requires an organized plan for demobilization. This plan must provide an orderly transition for returning resources to their stations, while still filling gaps within the response system. The demobilization of the resources must include input from the planning section and the fireground branch or division officers. The safety officer will be involved in the plan to ensure those who have high fatigue factors leave first. The first-in, last-out approach may be a contributing factor to injuries that occur during the final stages of an incident. This is why the first-in, first-out approach should be adopted to ensure the health and well-being of the firefighters.

Another important component that follows the incident is a PIA. The purpose of the PIA is to reinforce appropriate and effective strategies while discouraging the inappropriate and impulsive strategies. The PIA should be both informal and formal. Each provides information for changes in policies and procedures for the organization. The information gathered during the PIA can cause positive changes in both procedures and equipment.

Critical incident stress management is yet another postincident consideration. Although incident-related stress can come from response to a variety of incidents, studies have shown that incidents of a particular type are more stressful than others. The fire department should have a procedure in place for CISM. There are a number of approaches and steps in the CISM process.

The management of an incident begins with the preplan information and continues throughout the incident until the last units return to service. This process is and should be part of the incident strategy.

REVIEW QUESTIONS

1. During an incident, what plans must be developed to return resources to their stations?

2. Which of the following is not one of the three stages of incident termination?

    A. Demobilization

    B. Returning to quarters

    C. First-in, first-out

    D. Postincident analysis

3. True or False? The incident safety officer should be involved in the demobilization planning.

4. Describe the first-in, first-out concept.

5. True or False? The purpose of the CISM process is to focus on reinforcing the most appropriate and effective strategies while discouraging inappropriate or impulsive strategies.

6. True or False? The informal PIA may be held at the scene while the crew is in rehab.

7. Which of the following is a consideration in the demobilization phase?

   A. Personnel

   B. Apparatus

   C. Equipment

   D. All of the above are considerations.

8. List three events that can be called critical incidents.

9. When and how do you prepare for critical incident stress management?

10. List and briefly explain the four types of critical incident stress debriefing sessions.

ACTIVITIES

1. Look through your department guidelines and find guidelines for incident termination. See when they were updated last. See whether they promote the first-in, last-out theory.

2. Obtain from the National Fire Academy, NFPA, or a local resource a copy of a post-incident analysis for a large incident. Review the contents and information provided. See whether it lists the lessons learned.

3. Obtain information from your local critical incident stress management team. When and how can they be contacted?

Chapter

10

BASIC BUILDING CONSTRUCTION

Learning Objectives

Upon completion of this chapter, you should be able to:

- Identify the five basic types of building construction.
- Learn how to identify each basic building construct method.
- Discuss the basic benefits and inherent strengths and weaknesses of each type of construction.
- Discuss the hazards of each construction type when involved in fire.

CASE STUDY

The 150-year-old, three-story, wood-frame building was once a prominent hotel that catered to the elite citizens and visitors of this small river town. But over the past several decades after the hotel closed its doors, the building had seen one failed business after another. The building had been neglected and most firefighters referred to it as a fire trap. The fire prevention officers made the decision to recommend the building as a non-entry structure. The wood-frame structure was showing signs of structural problems from years of neglect. There were no tenants, as the area was seeing signs of commercial decline.

It was 3:00 A.M. when a police officer was cruising through the rundown area and saw smoke pouring out of many windows on the first floor. Upon investigating further, he saw an active fire through the windows in the area of the wide sweeping stairs to the upper floors. He immediately called his dispatcher to notify the fire department. The first engine company was more than 4 minutes away. The police officer moved his vehicle out of the way and walked around the building to see if anyone was trapped inside. He watched the fire continue to grow at a fast pace and move upward to the second floor.

Upon the arrival of the first engine company the flames were already throughout the first floor with heavy smoke pouring from the second and third floors and cockloft. The lieutenant immediately called for a second alarm and then had his crew try a quick attack with the stang nozzle on top of the engine. This tactic had little to no effect on the fire. As the lieutenant met face to face with the police officer, he learned that no one was seen in or around the building. The lieutenant took a quick look at exposures from three sides and decided to lay a line to a hydrant and begin exposure protection. As the second engine arrived, the fire was already coming out of second- and third-floor windows. The fire was burning fast and furious throughout the building. The first truck arrived and began setting up the ladder when fire blew through the roof and the entire building erupted into flames. The ladder crew captain established command and set a collapse zone to keep the crews from being buried if a collapse occurred.

The large stars on the front of the building were a clue to the captain that the building had spreaders to provide support to the structure. He knew that with this amount of fire the building couldn't stand much longer.

The battalion chief arrived and took command from the captain. It was now 10 minutes into the fire, and the building was fully involved. The battalion chief took a quick look around the area to ensure the personnel were well out of the collapse zone and had just returned to his vehicle, when suddenly and without any warning the front wall bowed outward toward the street. There was a loud cracking sound and then the building just dropped, pushing the front wall straight out into the street toward the firefighters.

A quick Personnel Accountability Report (PAR) determined that everyone was accounted for and no injuries had occurred.

The officers knew the construction of the building was wood frame with Type I spreaders holding it together. They knew how the building went up, what was holding it together, and how it would fall. By knowing the building construction, they knew how to run a safe fireground. The building was lost, the exposures saved, and most important, no one was injured when the collapse occurred.

INTRODUCTION

Throughout this book each chapter has a discussion of the construction type for each building fire being described. The purpose of this chapter is to provide a basic explanation of the five building types, the methods used to identify them during size-up, and the safety concerns with each construction type when involved in fire. Every fire science student and future officer must take more in-depth classes on building construction to study how different loads, forces, materials, and structural elements affect the stability of each construction type.

There are different methods for classifying building construction. The National Fire Protection Association (NFPA) has set a standard categorization for each type:

Type V: Frame (wood)

Type IV: Heavy timber

Type III: Ordinary

Type II: Noncombustible

Type I: Fire-resistive

In your response area there may be only a few types of construction for you to work with. In older response areas, such as the old river towns and urban areas, there is a mix of multistory wood-frame buildings, large heavy timber factories, and a mix of every building type possible. In newer cities, there is basically newer construction and most of the buildings fall within Type III to Type I construction. But, it is important for firefighters and officers to have a basic understanding of how construction type, design, alterations, and materials influence the buildings' reaction to fire (see **Figure 10-1**).

Let's start with Type V, wood-frame construction, because it is very common throughout the country.

■ **Note**

Every fire science student and future officer must take more in-depth classes on building construction to study how different loads, forces, materials, and structural elements affect the stability of each construction type.

■ **Note**

But, it is important for firefighters and officers to have a basic understanding of how construction type, design, alterations, and materials influence the buildings' reaction to fire.

Figure 10-1 *Not all buildings are built alike.*

TYPE V FRAME (WOOD)

A wood-frame building is one in which structural materials are wooden. The interior floor supports, interior walls, and roof supports are made of wood. The main point to remember in the performance of wood as a structural member is surface-to-mass ratio. Mass is the key factor in fire performance of wood. Wood carries its load by mass. The larger the beam, the heavier the load it can carry and the longer the beam can burn before losing its structural integrity.

The second-most important point to remember in wood-frame construction is the connection method used to hold the structural members together. The connection methods can become the weak link in the manner in which wood structural members will react and maintain structural integrity in a fire.

Whenever you are on the road in your district, stop at construction sites, whether it is new construction or buildings undergoing rehab, and look at how the structural members are connected. If we know how buildings go up, we know how they will come down.

Types of Frame Buildings

Balloon Frame This type of frame construction was common throughout building construction from the 1800s through the end of World War II. With the invention of cheap machine-made nails, it became possible to build multistory wood-frame buildings faster and cheaper. It involves a wood stud framing system in which the studs run continuously for the full building height and there is no inherent

Figure 10-2 *Example of fire spread through balloon-frame construction.*

firestopping provided between floors. In frame buildings with basements (cellar), there is an open channel that runs from the cellar to the attic. The floor joist channels are usually open to vertical stud channels, creating an open area for fire to run throughout the interior. The entire structure is one open void space involving the entire building (see **Figure 10-2**).

The balloon-frame structure can often be identified by looking at the doors and windows. All wall openings, doors, and windows are usually stacked vertically.

We have watched basement fires run up the stud channels straight into the attic, requiring quick dual-action attacks on the cellar and the attic at the same time. Interior room fires can spread into the stud channel and run the exterior stud channels without visible fire extension.

One very windy, cold day, we responded with units to a large, two-story farmhouse that had a small fire in the basement. As we were moving hose lines into the kitchen to gain access to the basement stairs, we were stopped by an odd noise. There was a roaring sound overhead. After checking around, we found the fire had burned a hole through an exterior wall. The hole was only about six inches in diameter, and you could see the wind blowing the fire in the hole like a blow torch. The wind was pushing the fire through a floor joist channel above our heads and throughout the frame house. The fire pushed by the wind had free run throughout the structure. What began as a small basement fire gained access to the wall stud channel and spread quickly vertically and horizontally through the structure.

The hazard associated with these structures is the ability for the fire to spread vertically and horizontally throughout the structure without visible indication. The structural voids occur in supportive structural assemblies, such as the floors, walls, and roofs. Once ventilated, the fire is drawn throughout the walls, floors, and attic. There can be early collapse of the attic and floors in these structures as the fire destroys the mass of the wood near the connection points.

Platform Frame The platform frame, sometimes referred to as *western framing,* is the most common type of new frame construction in use today. The wood stud framing system is set on a foundation or slab and built to story height. The wall studs are attached to a single board on the bottom called the sill plate. The studs are then attached at the top with double boards. This creates an inherent firestop within the framing structure. The second-story floor joists are placed on top of the first-floor walls and the subflooring is laid in place. The second-floor walls are then erected on the subfloor. Once this platform is completed, a third floor can be erected in the same manner. This type of construction can be used for up to three-story buildings. The platform framing method creates a compartmentation on each floor and an inherent firestopping between each floor.

This type of construction began sometime after the mid-20th century and continues today (already stated above). You can sometimes identify these buildings by offset windows and doors. Remember, in the balloon-frame construction, the doors and windows were stacked vertically.

Post and Beam The post and beam is a framing system using post (vertical) member and beams (horizontal) members to create a load-bearing frame connected by rigid points. The nonbearing wall panels are attached to the frames. The interior framing may be left exposed for aesthetic purposes. The nonbearing wall panels provide the lateral support for the post and beams. The barn is a typical post and beam construction. Barns have a plank and beam floor and roof system. The minimum dimensions for the roof planks are 2 inches, with beams set 6–8 feet apart. The minimum dimension for the post and beams are 4" × 4".

The post and beam framing is used in barn and farm buildings and in some commercial building construction. It is easy to identify these buildings as the interior walls are usually exposed. In commercial buildings, it becomes more difficult to identify this type of construction. Updated preplans and company inspections will assist crews in identifying these structures. These buildings are usually structurally sound during a fire, but when collapse occurs, it can involve the entire structure. Barn fires are usually fully involved to the point that you can see the entire post and beam structure before collapse occurs.

Plank and Beam A framing method that resembles the post and beam, but uses much larger beams, is being used today for large residential structures, commercial buildings, churches, and places of assembly. Plank and beam construction uses boards laminated together to form large beams. The floors are typically thick tongue and groove planks. This type of construction reduces the concealed spaces in the building. The large laminated beams are left uncovered and become part of the interior décor. The large, rigid laminated frames are covered with plank roofing in many churches and assembly halls. These structures are easily identified upon entering as the large laminated frames are a visible part of the interior. The benefit of these large beams is that wood carries its load by mass. The hazard is that the large beams and exposed finished plank surface of the ceiling become more fuel for the fire.

Truss Frame A very common and controversial construction method uses lightweight truss construction. The truss frame is an engineered construction in which the entire structure is tied together into a unitized frame. The stud walls are load bearing and built in the typical platform-frame method. The difference is that the floors and roof are built with lightweight wood trusses as opposed to heavier wood beams.

This type of construction can be used for buildings up to three stories. The wooden roof trusses can be set on top of buildings much taller. The integral part of the truss system is that each member is dependent upon the next (see **Figure 10-3**).

These buildings are easy to identify as most residential and commercial construction today uses some form of roof or floor trusses. Watch the buildings under construction in your district and you will see many trusses used. The benefits for the construction industry are the lightweight members are less expensive, easier to work with, and are pre-engineered to span large openings.

Figure 10-3 *Example of truss construction.*

■ Note
The hazard is that wood carries its load by mass. The truss is constructed of the lightest weight and smallest dimension lumber as possible, typically 2" × 4".

The hazard is that wood carries its load by mass. The truss is constructed of the lightest weight and smallest dimension lumber as possible, typically 2" × 4". They can span large, open spaces without support from below. This creates the potential for large areas of collapse. A student of building construction should study the many truss structure fires that have taken dozens of firefighters' lives during the past decades.

Another type of truss is the timber truss. It is built with larger framing members and used when large openings need to be covered. They are used in supermarkets, bowling alleys, theaters, and other large, open-floor buildings. The most dangerous of these trusses is the bowstring, identified by the characteristic bow shape of the roof. These roofs have been the death of many firefighters as they can collapse without warning. When they collapse, they can push the exterior load bearing walls outward. The typical rule of thumb is that if there is heavy fire involvement of the building and/or attic area of the bowstring truss roof, *do not* put anyone on top, under, or at the sidewalls of the bowstring truss roof.

■ Note
Do not put anyone on top, under, or at the sidewalls of the bowstring truss roof.

Variables in Fire Performance of Wood

The condition of the wood plays a part in the structural integrity of a building. If the building is old, it may contain dry rot or termites. Fungus and decay also play a part in the loss of structural integrity. Laminated wood members have different fire performance with some delaminating when exposed to heat or fire.

Defects in the wood members, such as warpage, shrinkage, and torn or lifted grain, can weaken the members. The loading can cause the wood members to bend and deform. The biggest concern is the condition of the connection methods. Remember that the weak link in any construction method is how structural members are connected together.

When working with truss construction, the metal fasteners used to hold the truss members together are a concern, because heat causes them to fail. These are used in most of the lightweight wood trusses used in construction today. The metal fasteners are referred to as gusset plates, gang nailers, or staple plates. The metal staple penetrates the wood less than $1/4$" deep.

Wood Treatments

Wood can be treated to resist moisture, insects, and fire. Years ago, builders adopted a practice of placing fire-retardant plywood on firewalls and roofs. However, the woods treated with phosphates and sulfates to protect them from fire caused another problem. When used in roofing systems, the plywood begins to deteriorate due to high humidity and temperatures in attics and cocklofts. This deterioration of the plywood caused it to become brittle, enough to collapse under the weight of a person walking on the roof.

■ **Note**
Throughout this discussion of wood-frame and wood structural members, one must remain cognizant that size does matter.

Size Matters

Throughout this discussion of wood-frame and wood structural members, one must remain cognizant that size does matter. As mentioned previously, wood carries its load by mass. The more mass, the larger the load. Wood loses its mass during a fire, and at some point the loss of mass will be greater than the load it is carrying and collapse will occur.

TYPE IV HEAVY TIMBER

■ **Note**
Heavy timber construction, or what some refer to as *mill construction*, is constructed with very large dimension lumber.

Heavy timber construction, or what some refer to as *mill construction,* is constructed with very large dimension lumber. The heavy timber construction is capable of sustaining massive loads commonly found in old industrial buildings (see **Figure 10-4**). The large wood members are capable of spanning great distances allowing the open floors for industrial machinery. The large wooden columns are greater than 8" × 8" and can be used for buildings up to eight stories. The floors are typically 3 inches or greater in thickness and are builtup using 1-inch tongue-and-groove construction laid crossways. Large sections of wood more than 6 inches thick can have more fire resistance than exposed steel.

The exterior walls of the heavy timber construction are made of masonry. Some large buildings will have firewalls built to separate the structure. A firewall consists of a free-standing masonry wall that penetrates the roof structure and will confine fire. The openings between sections of buildings separated by firewalls will have fire doors that drop down or close when subjected to heat.

Figure 10-4 *Example of Type IV—heavy timber construction.*

These structures are identified by the size and age of the industrial building. They possess a certain degree of fire resistance based on the sheer size of the timbers used in the construction.

The hazard in heavy timber structures is the fact that the walls and ceilings are typically unfinished and lack any real fire protection. The fire can quickly expose the connection point for the timbers. The connection becomes an important factor in how long the building will stand.

One cold, wintry day, a six-story heavy timber building burned in the middle of downtown in a large city in Ohio. The fire started on the third floor and quickly spread to the three floors above. Before fire crews could begin an attack on the fire, flames began spreading the length of the building. The fire moved past fire doors and closures and entered other sections of the building. The fire spread through the buildings for two city blocks and ultimately destroyed three six-story heavy timber buildings!

TYPE III ORDINARY

■ **Note**
Ordinary construction is composed of masonry load-bearing walls with wood-joisted floors and a wood roof.

Ordinary construction is composed of masonry load-bearing walls with wood-joisted floors and a wood roof. The building materials for the load-bearing walls consist of brick, concrete block, or brick *and* concrete block. The walls can vary from 6" to 30" thick. In multistory buildings, the walls are thicker at the bottom to support the load from the stories above (see **Figure 10-5**).

Figure 10-5 *Example of Type III— ordinary construction.*

These buildings are common throughout most cities. They are used for hotels, motels, office buildings, retail, and other commercial and light industrial purposes. They typically run one to three stories tall, although there have been some built to heights of ten stories.

The floors are constructed of wood joists with at least 1-inch-thick tongue-and-grooved boards. The roofs can range from wood joists to parallel chord trusses and triangle trusses.

In many buildings the floor and roof joists have what is called fire-cut joists. The joists have a 30-degree cut on them. The longer side of the joist sits in a pocket in the load-bearing wall. During a fire, the floor and roof structure is designed to collapse downward and save the exterior walls from being pushed outward and collapsing. The ordinary constructed building will have the load-bearing walls along the longest part of the building. This will allow the floor and roof joists to span the shortest distance possible.

Spreaders

In some older ordinary construction buildings you might see a star, "S", channel, circle, or other decorative device on the exterior of the building. These are called *spreaders*. A spreader is any device used to spread a load between two or more structural members.

- *Type 1 spreader.* A rod or cable, runs parallel to the joists and ties the walls together to increase the stability of either or both walls.
- *Type 2 spreader.* Tensile members, rod or cable, run perpendicular to the joists which tie the first three or four floor joists to the wall. This may be part of an original design of heavy timber building, or it may be added to a building to support a weak wall.

The ordinary constructed building is easy to identify mainly due to its common nature. The typical corner gas station, drugstore, or supermarket is made of ordinary construction. The buildings are structurally sound and will generally stand up to the effects of room and content fires. The fire will stay inside the buildings because the concrete block and bricks are poor conductors of heat and will not allow fire to spread to nearby buildings unless there is a wall failure.

The hazards are in the potential for the walls to collapse during heavy fire load fires. Failure of part or all of a wall can occur during large fires. The wall will typically begin to bulge before collapse, thus providing some warning. Another warning of possible collapse is smoke and/or fire pushing through cracks in the walls.

TYPE II NONCOMBUSTIBLE

The noncombustible construction method employs building materials that will not add to the fire development. The materials are not combustible and will prevent the fire from becoming a structure fire. When you watch wood-frame or ordinary construction buildings burn, you can see that the structural elements of the building begin to burn. In the wood-frame building, the entire structure can be consumed. In the ordinary construction building, the roof and floors become involved in the fire. This translates into what we call a *structural fire.*

The noncombustible building does not employ building materials that add to the fire. The building materials can range from steel to concrete. The steel-framed building with metal-clad siding and steel deck roofing is common in many small-town firehouses. They are typically prefabricated buildings brought to the job site and erected (see **Figure 10-6**).

The typical modern corner drugstore is constructed with concrete block and topped with steel parallel chord trusses and a flat, metal deck roof. There are few interior wall coverings, and the building itself is noncombustible. The roof structure can be flat or peaked. In either case, the structural material used is metal-framed truss. The most common method for building the roof structure is with metal decking on the metal trusses. The roof itself is a buildup of different materials to keep the weather out of the building. The buildup begins with an insulation material laid on the metal decking. Then alternating layers of hot tar and roofing felt are applied until the roof is sealed. The roof then has light gravel or rocks laid on it for further protection. There are many new, soft and pliable materials for roof covering. They give you an uneasy feeling when you walk on

Figure 10-6 *Example of Type II— noncombustible construction.*

them because they are spongy. When walking on these or any flat roof, walk along the side next to the wall as it is better supported and less likely to collapse under your weight.

TYPE I FIRE RESISTIVE

The fire-resistive building is built with steel, concrete, and other fire-resistive or fire-rated materials. Fire-resistive buildings are built with structural components that will not burn and will resist the effects of fire for a long period of time. The invention of steel and the use of steel for a building material began the era of fire-resistive buildings. Buildings could be built higher and stronger than ever before (see **Figure 10-7**).

However, steel used as a structural material will begin to elongate and can fail during fires. The typical rule of thumb for the expansion of steel is 1 inch for every 10 feet at approximately 800°F. Because of the effects fire has on steel, it is usually covered with a fire-protective material. The steel member can be encased in concrete or covered with brick or a fire-rated drywall. The drop ceiling can be used to protect overhead steel beams and steel trusses and wall coverings can be used to protect beams. There are spray-on coverings consisting of a cementatious material that can be applied to steel structural members. All of these methods serve to protect and prevent the steel from elongating due to heat.

Many fire-resistive buildings are constructed with reinforced concrete and steel bar joists for floors and the roof. The concrete can be reinforced with steel rebars in a prefabricated manner or the concrete can be poured in place over and

Figure 10-7 *Example of Type I—fire-resistive construction.*

around the rebar. The rebar provides the concrete with lateral strength. There are other items that can be added to concrete to provide additional strength.

Truss Construction

There are many different styles of trusses and truss materials used in the building of each of the five types of construction. For this chapter, we identify the types of trusses and the benefits and hazards of those trusses (see **Figure 10-8**).

The truss is used mainly because of the benefits of lightweight members used interdependently of each other to carry a load. They consist of lightweight building materials arranged in a group of triangles in which loads applied will be distributed throughout the other members. Because the triangle is inherently stable, any load applied to the lightweight member will actually strengthen the entire triangle as long as each member maintains its structural stability. The top chord is compressed and the bottom chord is in tension.

The top and bottom members of the truss are called the **chords**. The inside members are called the **web**. The trusses are tied together with connecting members called **ties** and the connections are **panel points**.

The most dangerous truss is the **bowstring**. It is easy to identify by its curved top chord. These trusses are common in bowling alleys, skating rinks, and other large buildings requiring a long, uninterrupted span.

The most common truss for roof construction is the **peaked roof truss**. This truss is found in most of today's homes and commercial buildings. The truss is triangular in shape to provide the peaked roof. The trusses are placed close together—16" to 24" on center. They are held in place by the plywood roofing

chords
top and bottom members of the truss

web
inside members of the truss

ties
connecting members, such as gusset plates, that hold the members together

panel points
metal plates, sometimes referred to as gusset plates, that have teeth that enter the wood member to hold them together

bowstring
the most dangerous truss is the bowstring. It is easy to identify by its curved top chord; these trusses are common in bowling alleys, skating rinks, and other large buildings requiring a long, uninterrupted span

peaked roof truss
this truss is found in most of today's homes and commercial buildings; the truss is triangular in shape to provide the peaked roof

Figure 10-8   *A truss roof.*

parallel chord truss
the top chords and the bottom chords run parallel with each other with the web between them

interstitial space
opening between the top and bottom chords of a parallel chord truss

material and then covered with shingles, terracotta tiles, metal, or other roofing covers.

The **parallel chord truss** is used for both roofs and floors. The top chords and the bottom chords run parallel with each other, with the web between them. The top and bottom chords can be either steel or wood. The webs can also be steel or wood. We have seen parallel chord trusses with wood chords and steel webbing. The trusses used in Type 2 commercial structures will typically be all steel.

In most residential structures the common practice is to use all wood for the parallel chord trusses. The problem with the parallel chord truss in residential structures is the void spaces created by the parallel chord truss. This **interstitial space** becomes the location for electrical wiring, HVAC, and other utilities to be placed. In one development, there were more than a dozen three- and four-story residential dorms built on the college campus. By code, the buildings were fitted with sprinklers. The fallacy to this protection system is the interstitial space created between each floor by the wooden parallel chord trusses were not sprinklered, and this is where all the utilities and recessed lighting were placed.

Truss Problems

The truss is designed to allow the minimum amount of building material to carry the designed load. The truss is engineered by the building industry to provide a less expensive way to carry loads for both the roof and the floors. While there is concern for structural stability against natural elements and forces on a building, it appears that structural stability during a fire is less of a concern.

■ **Note**
The weak point of the truss construction is the same weakness we find with any construction method: the connectors. The truss is interdependent upon each and every member.

The weak point of the truss construction is the same weakness we find with any construction method: the connectors. The truss is interdependent upon each and every member. This means that every part of the truss is important. If one element of a truss fails, it can cause other members to lose their structural stability and lead to failure of the entire truss.

In wood trusses, the lightweight gusset plates are very important components that hold the chords and the webs together. In some instances, these gusset plates are damaged during the construction of the building and not repaired. These problems are then hidden when the roof deck and the drywall are put in place. In other cases, the metal gusset plates pull out of the wood members when heated and cause failure of the truss. The last concern is one to always remember: *Wood carries its load by mass.* When a lightweight wooden truss burns, it will quickly lose its mass and its structural stability. The steel bar parallel chord truss is used for both roofs and floors. In unprotected steel trusses, the steel can begin to loose strength at 800°F and fail around 1,000°F. The steel bar joist truss will also begin to elongate while heating and can push masonry walls to collapse. The steel trusses will also begin to sag and can collapse under the weight of firefighters trying to ventilate the structure.

Fire officers and firefighters are greatly encouraged to read all they can about trusses and truss construction as it exists in every town in the United States. For more information on trusses, see http://www.truss-frame.com.

SUMMARY

The ability to understand the inherent safety features, and the dangers, associated with buildings comes from knowing building construction principles. In essence, knowing how a building is built will allow fire personnel to know how they will act during a fire.

In this chapter, the reader was introduced to the five types of building construction as identified by the National Fire Protection Association, NFPA and included clues to identify the types during size-up. Depending upon the age of your response district, you may find each and every one of the five types of buildings in your response district. It is important for every firefighter to have a basic understanding of the construction types in his or her district and how the building materials can influence the spread of fire and potential collapse.

Type I Fire-resistive buildings have the most inherent fire safety factors associated with both the building material and the building techniques. Many of today's office buildings are built with fire-resistive building components such as those found in type 1 construction.

Type II Noncombustible has building materials that will not add fuel to the fire, but can suffer from the effects of the fire. These buildings often have exposed steel trusses and iron beams that can elongate and fail during fires. Walk into many of your new corner drug marts or large box stores and you will see the exposed steel trusses and beams.

Type III Ordinary construction comprised of masonry load-bearing walls and wood-joist floors and roofs are safe, but they have building materials that can add fuel to the fire. An inspection of older districts will reveal ordinary constructed buildings with roofs built with wood rafters, wood trusses, and—the most dangerous—bowstring truss roof. There are parallel chord trusses and peaked trusses.

Type IV Heavy timber or mill construction can be found in many older factory towns, old barns and storage buildings. They have an inherent built-in fire protection in the size of the beams and columns. Remember that wood loses its strength by losing its mass.

Type V Wood-frame buildings add to the fire in that nearly every structural support will burn. These types of buildings become structural fires as the entire structure can burn and collapse. Many firefighters have lost their lives in these structures that range from single-family homes to multistory multifamily apartment buildings. There are many types of wood-frame building practices. It is your responsibility to know and understand the types of wood-frame structures in your first- and second-due district.

This chapter represents a short introduction to building construction features and their inherent safety and dangers. It is recommended that firefighters continue learning all they can about building construction. The National Fire Academy has two courses in building construction:

H103 - Principles of Building Construction: Noncombustible

H104 - Principles of Building Construction: Combustible

Know your building construction and the types of buildings in your response district.

REVIEW QUESTIONS

1. Name the five building construction types and their NFPA designation.

2. What does wood carry its load by?

3. Why is it important to know how a building is constructed?

4. Name the important factor to know about balloon-frame construction.

5. Why is platform construction a safer method for wood-frame construction?

6. Why is the use of lightweight wood-frame construction a concern for firefighters?

7. Explain the danger of a bowstring truss roof building. Explain where it is unsafe to place firefighters at the fire scene.

8. Name the two types of spreaders and how they differ in building construction.

9. What are the main building materials used in a fire-resistive building?

10. Name the elements that make up a lightweight wood truss and how they work to carry the load through the building.

ACTIVITIES

1. Identify the various types of construction in your response area. Try to find one example of each of the types.

2. Photograph various buildings in your response district and identify the hazards with the construction of each building.

3. Photograph buildings under construction in your response district and identify the strengths and weaknesses found in the building construction.

4. If a wood-frame building three stories or more in height is required to be sprinklered, why are there garden apartments with three floors found without sprinklers?

5. Photograph buildings in your district that have spreaders and explain how you can identify which type they represent.

Chapter

11

ONE- AND TWO-FAMILY DWELLINGS

Learning Objectives

Upon completion of this chapter, you should be able to:

■ Identify the common types of construction for one- and two-family dwellings.

■ Describe the inherent life safety problems in these dwellings.

■ Describe the hazards associated with firefighting in these structures.

■ List the strategic goals and tactical objectives applicable to fires in these structures.

CASE STUDY

The calm of the afternoon is broken by the sound of a prealert tone. The dispatcher announces the call for a structure fire in a single-family dwelling. As the crews head toward their apparatus, the engine company officer on the first-due engine starts painting a mental picture of the types of houses in that district: Size-up has begun. The house is in the older, downtown section of the city, where the typical homes range anywhere from 50 to 100 years old. They are basically wood frame with some balloon-frame construction.

As the engine company approaches the scene, the crew observes smoke drifting across the street. The engine arrives on the scene and stops short of the one-and-a-half story house, leaving the front open for the ladder truck coming from the other end of town. This gives the engine company officer a view of only two sides of the structure. Smoke is pushing through openings around the windows on both floors and pouring from the eaves. As the engine company officer climbs out of the engine, he is passed by the firefighters pulling a 1¾-inch line off the attack engine. The fire has not vented and the smoke is a thick, grayish-brown color. The firefighters walk up the porch and kneel down at the front door to adjust their face masks, helmets, and gloves. They force the door open and are met by a rush of thick smoke and heat that covers them from top to bottom on the porch. They drop to their knees and begin crawling into the structure, dragging the hose with them.

As the firefighters enter the house, they notice that the smoke has not lifted. It appears to be coming out of every hole and opening in the house. They slowly make their way across the living room toward the back of the house, bumping into furniture and going around obstacles. They continue to move through the house in zero visibility, feeling their way across the floor toward the kitchen and the back of the house. The heat is intense, but there are no signs of fire on this floor.

The battalion chief arrives on scene and parks behind the first-due engine. As the chief walks toward the building, he sees smoke enveloping the house. He quickly radios orders to the truck company to proceed to the roof for vertical ventilation as soon as it arrives on the scene. The battalion chief then takes a position in the middle of the street in front of the house. He has only seen two sides of the house, the front where the crews entered and the side from where he approached. The smoke has enveloped the house and continues to push from the entire structure.

The second-due engine arrives on the scene, where the driver stops and lets the crew off in front of the house. As the crew gather their tools and equipment, the driver drops a supply line at the first engine and proceeds to the hydrant. The lieutenant from the second engine meets with the battalion chief for an assignment. The battalion chief sends one crew member to break out windows around the house for horizontal ventilation. A long pike pole is needed to reach the windows on both floors. The lieutenant and other crew member start dragging a backup 1¾-inch line off the attack engine toward the front door to stand by.

Inside, the engine crew cannot find any fire, but the heat buildup is getting worse. They find a stairway and mistakenly assume the fire is upstairs. It only makes sense because they could not find any fire on the first floor. As they move up the stairs, they feel in front of them and to both sides for any potential victims. When they reach the top of the stairs, they feel more heat but still cannot find the glow of the fire. They begin to wet down the heated gases over their heads, fearing a flashover may occur. There are no flames visible, no rollover, nothing to indicate they are getting close to the fire. They are becoming confused. Smoke is everywhere. The heat is intense, but there is no fire visible. Why? They push on across the second floor, looking and wondering where the fire is hiding. Their facepieces vibrate as they begin to run out of air and are forced to retreat back down the stairs and out of the structure. They are physically whipped after 10 minutes of looking for fire and finding nothing.

The ladder crew has arrived on the scene and the operator has set the aerial ladder to the roof. The lieutenant and a firefighter move up to cut open the roof and begin ventilating the structure. Smoke pushes forcefully from the hole as they are cutting, stalling their saws. They have to move out of the smoke to restart them. As they continue to make their cuts, the smoke is pushing more forcefully out of the vent hole. Somewhere inside this house is a deep-seated fire waiting for just enough air to quickly grow and take total control of the structure.

A firefighter from the second-due engine has been breaking windows around the house on both the first and second floors to enhance the horizontal ventilation. During the trip around the house, the firefighter observes a well for a basement window at the rear of the house. As the firefighter completes the ventilation and returns to the front of the house for the next assignment, he reports the window well for the basement. The battalion chief looks at him in disbelief. A basement?

The battalion chief regroups his personnel. This house has a basement. We forgot to look for a basement. Did anyone do a 360° survey?

Now, armed with sufficient information to make a logical, planned attack, the crews are sent back into the structure to find the basement stairs and make an attack. Once inside, the crews are met at the basement door with a raging basement fire that is quickly extending up the interior walls and pipe chases. Fire begins to vent from the hole in the roof. The combination of both horizontal and vertical ventilation is now providing the fire with all the air it needs to grow quickly and consume the structure. The fire is running from the basement up the concealed spaces to the upstairs and the attic. As the fire takes control of each floor and the attic, the battalion chief orders everyone out of the structure.

A second alarm is sounded and multiple lines are placed in service to battle the fire. The precious time lost in not knowing the fire's location has doomed this house. The failure of the first-arriving engine company officer and the battalion chief to do a 360° walk of the structure for a proper size-up and to locate the fire has allowed the fire to take control of the entire structure.

Implementing tactical operations without knowing a fire's location and how to battle the fire is comparable to not showing up at the fire scene at all. The building construction must be identified, along with any special structure features. The smoke and fire behavior must be read and understood. The operational priorities must be known and implemented in a logical sequence for an effective fireground operation. In this case, they were not, and the result was a house with a total fire loss and firefighters whose lives were needlessly jeopardized.

INTRODUCTION

■ Note

The most common structure fire response in the United States involves the one- and two-family dwelling.

The most common structure fire response in the United States involves the one- and two-family dwelling. According to the National Fire Protection Association (NFPA), one- and two-family dwellings are defined as residential structures in which not more than two families reside. In some cases however, more than two families may actually occupy these structures. The dwelling units may be stand-alone structures as seen in most suburban housing developments (**Figure 11-1**) or they may be attached units, such as those found in inner-city row houses or suburban townhouses (**Figure 11-2**). It is common to find attached one- and two-family dwellings in many small, older towns. One- and two-family dwellings also include manufactured homes. For this chapter, we consider these structures as being in the same category as detached single-family dwellings, duplexes (**Figure 11-3**), and townhouses.

Figure 11-1   *A typical single-family dwelling.*

Figure 11-2 *Attached one-family dwellings.*

Figure 11-3 *A two-family dwelling or duplex.*

There are many different designs for these structures. Each offers its own inherent building construction and safety concerns. There are one-story ranches that range from as small as 400 square feet to as large as 25,000 or more square feet. There are two- to three-story colonial styles, the Queen Anne and Victorian houses. There are split- and tri-level homes and townhouses. They can have attached or unattached garages. They may have basements or be built on a slab or crawl space. These features add to fire problems. The most common construction style is wood frame and masonry. The exterior covers are wood, vinyl, asbestos shingle, brick facade, aluminum, and others. These sidings can help to contain fire, as with brick and asbestos, or spread the fire, as with the vinyl and wood sidings.

In the NFPA full report *Fire Loss in the United States During 2004* residential home fires accounted for 78% of the total structure fires and more than 82% of the fire deaths. Somewhere in the nation, a fire department responds to a residential structure fire every 77 seconds. Of these residential structure fires, one- and two-family structures account for 57% of the fires and 69% of civilian fire deaths, **Figure 11-4**.

Nearly half of residential fire deaths occur in fires that start from 11:00 P.M. to 6:00 A.M. The peak night hours are from 2:00 to 5:00 A.M., when most people are in deep sleep. Most occupants succumb from the effects of smoke and the carbon monoxide gas it contains while sleeping. This is why smoke detectors are so important in these structures in order to provide early warning and allow sufficient time for escape.

Because of the numbers of fires in these structures, the fire service handles many fires in single- and two-family dwellings, but we sometimes become complacent with our training efforts toward them. Because they are a common cause for our fire responses, we sometimes take them for granted as being easy fires. This should not be the case, because according to the 1998 USFA FF Fatality report, 40% of fireground firefighter fatalities occurred in residential structures.

■ **Note**

In the NFPA full report *Fire Loss in the United States During 2004* residential home fires accounted for 78% of the total structure fires and more than 82% of the fire deaths.

DEATHS (1,503 CASES)

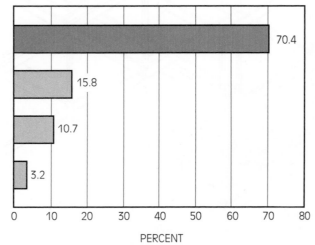

Figure 11-4 *Graph of total fire deaths in residential occupancies. (Courtesy of USFA.)*

CONSTRUCTION

Incident commanders, company officers, and firefighters must have a working knowledge of building construction in order to understand the effects fire will have on different types of homes and businesses. The construction ultimately affects the tactical objectives chosen for the incident. It is equivalent to understanding fire dynamics. On-scene personnel can make educated assumptions on the direction the fire will travel, the amount of heat that will be generated, the typical length of building stability under fire conditions, and the length of time firefighters should be left in buildings before evacuation should be considered.

The most common building construction method for one- and two-family dwellings are type V (wood frame) and type III (masonry). The wood-frame constructed homes typically fall within one of three framing methods: post and frame, balloon frame, or platform frame.

Post and Frame

The post and frame (some called plank and beam) home has a few large wooden members in place of many small wood-frame members, as shown in **Figure 11-5**. The larger wooden members are usually 4" × 4" posts that serve as the main structural members. The smaller, 2" × 4" studs are used to fill in between the

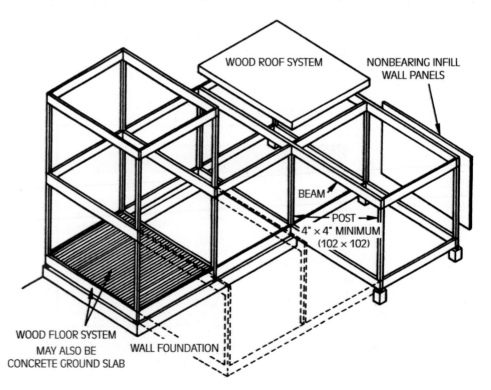

WOOD ROOF SYSTEM

NONBEARING INFILL WALL PANELS

BEAM

POST
4" × 4" MINIMUM
(102 × 102)

WOOD FLOOR SYSTEM
MAY ALSO BE
CONCRETE GROUND SLAB

WALL FOUNDATION

Figure 11-5 *An example of post and frame construction. (Courtesy of USFA.)*

larger framing members. This construction method can be found in both old and new homes. Many associate this method of construction with log cabins. In today's larger, modern homes, with their open air concepts and "great rooms," builders are once again using this building method. This does not mean that the entire house is constructed in this manner, but plank and beam construction can be found in some rooms, entryways, and other portions of the structure.

Balloon Frame

The balloon-frame constructed home (**Figure 11-6**) uses standard 2" × 4" studs. The difference in this construction feature is that the exterior framing members

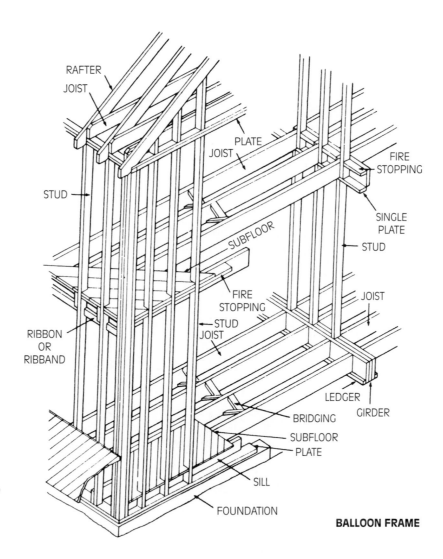

BALLOON FRAME

Figure 11-6 *Balloon-frame construction. (Courtesy of USFA.)*

■ **Note**
In houses built prior to 1940, these framing members form an open channel from the foundation to the eave line.

fire stops
pieces of material, usually wood or masonry, placed in studs or joist channels to slow the spread of fire

run two or more stories high. They start at the foundation line and run to the eave line. In houses built prior to 1940, these framing members form an open channel from the foundation to the eave line. This allows a fire to run these channels from the bottom to the top of the structure. The joist channels for the floors may also be open from one end of the home to the other, rendering the entire home open for fire to run every channel. If the house has a basement, these channels are open and susceptible to fire spread from the basement straight to the attic. When fighting a fire in this type of structure, the incident commander must consider this movement of the fire from the basement or first floor to the attic. This is very important if the fire is to be stopped.

The balloon-frame constructed homes built after 1940 usually have fire stops in the walls. An example of a fire stop is shown in **Figure 11-7**. These consist of 2" × 4" wood members placed in the joist channel. This placement helps to limit the spread of fire up these channels and out across the floor channels. Fire

FIRE STOP

Figure 11-7   *An example of fire stopping.*

stopping is critical in balloon-frame constructed homes. All of the fire stopping must be in place correctly for it to be effective.

Platform Frame

In platform-frame construction, each floor is built as a box. The framing in each level is set on a sill plate and runs the height of one floor, as shown in **Figure 11-8**. The subflooring is laid on the joists with the bearing and nonbearing walls placed on top of that floor. The framing is then capped with 2" × 4" horizontal studs, thus capping that wall section. This forms a box. The second-floor joists are then set on top of the bearing walls. The subfloor is laid into place and

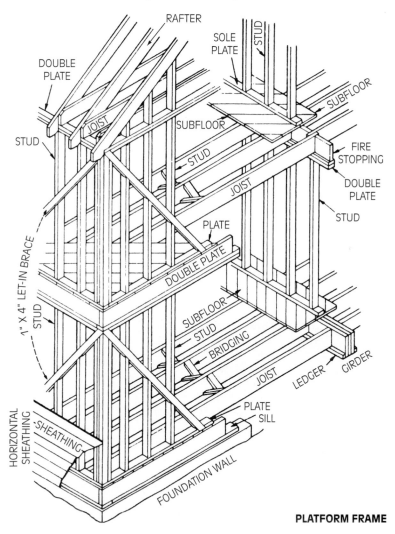

PLATFORM FRAME

Figure 11-8
Platform-frame construction. (Courtesy of USFA.)

the second-floor walls are built on this floor. This construction method creates inherent fire stopping by creating a substantial barrier that prevents fire spread vertically and horizontally through the walls and floor. This method creates individual compartments capable of confining the fire to that compartment even when the fire enters a confined space. With platform-frame construction, the room and contents fire can be held to one compartment or a portion of the home.

The platform construction is found in most contemporary construction standards. There are many variations in construction styles and methods. The split-level, bi-level, and tri-level houses are all variations of platform construction. It is nearly impossible to categorize all of them. To be effective, you must study the types of homes and building construction methods in your jurisdiction. Study the old home styles, the new home styles, and visit those under construction and renovation. This is the only way you will learn the inherent dangers and safety features these homes present.

■ Note
To be effective, you must study the types of homes and building construction methods in your jurisdiction.

Ordinary Construction (Masonry)

The difference between the masonry home and the wood-frame homes just discussed lies in the bearing walls. In this home-building method, the bearing walls are cement, concrete block, or other masonry product (**Figure 11-9**). The interior walls are still constructed with wood studs, floor joist, and roof trusses. This home building style is very popular in the southern states where hurricanes are prevalent. The construction methods used are platform and/or post and beam.

Figure 11-9 *Ordinary construction.*

HAZARDS ENCOUNTERED

■ **Note**
The most common danger is the spread of smoke and fire throughout these structures due to their openness.

In all of the building construction methods presented there are certain inherent fire dangers. The most common danger is the spread of smoke and fire throughout these structures because of their openness. Heat, smoke, and fire travel along the path of least resistance. In the case of the one- and two-family dwelling, the most common paths are the hallways and stairwells. They are usually open with no built-in fire protection measures. The simple task of closing doors assists in confining a fire in these structures. You must also consider the travel of fire through duct work, utility shafts, air shafts, and voids.

The exterior of the house can assist in spreading the fire. The wood, vinyl, and tar coverings found on the exterior assist spreading the fire up the side of the structure and into the soffits and the attic. In the brick, concrete, stucco, and asbestos sidings, the fire spread does not occur. This limits the spread of the fire into the attic from the siding. Fire spread can still occur if the fire comes out of a window and travels into the soffit or into the attic, as shown in **Figure 11-10**.

In the balloon-frame construction, if a fire enters the exterior wall, it has a free run up the wall into the attic. A basement fire can quickly send fire up these channels. Upon arrival at a fire involving a balloon-frame home, the incident

Figure 11-10 *Smoke and fire travel throughout a single-family dwelling. (Photo courtesy of Greg Sutton.)*

■ **Note**
There is a chance that fire can also drop down.

commander must consider extending lines to the fire floor and the attic. There is a chance that fire can also drop down these channels. A fire on the second floor or attic can travel downward in the form of burning embers. Fire crews must be constantly alert to signs of fire and smoke spread, such as scorching or blistering paint.

The platform frame and the masonry structures are built in a method that creates individual compartments that contain the fire to room and contents. The real danger of fire spread lies in the open halls and stairways. The increasingly popular use of great rooms and vaulted ceilings in many newer style homes allows fire to spread more quickly.

Hazards are inherit in the construction of these structures as well (see **Figure 11-11**). Lightweight wood truss members for the roof or floor assemblies can fail as a result of fire exposure in the attack. There are various types of roof coverings common on one- and two-family dwellings, such as heavy barrel tile, asphalt shingles, wood shake, slate, and many others. A new product on the market is a stone-coated steel roofing that is made from 26-gauge structural-grade steel coated with a heavy resin and other protective layers. An extra-thick coating of ceramic-fired earth stone is then applied. The different profiles can emulate contemporary clay tiles, traditional asphalt shingles, and even natural cedar shake (see **Figure 11-12**).

These roofs have the highest strength rating for hail, hurricane, and other storm damage; however, they are nearly impenetrable. A fire crew may not be able to ventilate a roof with this type of covering; worse yet, they may not know

Figure 11-11 *Tile roofs add significant weight to the roof members. They are also difficult to ventilate.*

Figure 11-12 *Wood shake shingles add to fire load and fire spread.*

what type of roof covering they are up against until the post-fire investigation. In many cases, the use of positive pressure ventilation is the method of choice in residential structures. Placing firefighters on roofs that are weakened from fire is not advisable. Remember that many roofs are made from lightweight structural members, and their design strength is only to carry the snow, wind, and rain loads. When the roof structure is compromised by fire and the weight of firefighters, collapse can occur.

Residential one- and two-family homes are generally not subject to periodic inspections and requirements of fire prevention codes for proper storage and housekeeping. This results in the possibility of just about anything being stored within the structure (**Figure 11-13**), including flammable liquids and gases, explosives, and poisonous chemicals. This must always be a consideration at these fires.

The normal contents of a house add to the **fire load**. The amount of furnishings varies widely. Some living rooms may have a sofa and a chair, whereas others might have sofa, chairs, and tables. Around holidays, the fire load may be increased because of presents or decorations. Bedroom contents range from single beds to multiple bunk beds, again increasing the fire load.

There is almost always electric and water in the structure; however, gas service may vary. If so, is it natural gas or liquid petroleum gas (LPG), as shown in **Figure 11-14**?

The contents of a residential structure will probably be unknown on the initial response; however, all possibilities must be considered when applying the strategic goals.

■ **Note**

Just about anything can be stored in a one- or two-family home.

fire load
all the combustible parts of contents of a building

Figure 11-13   *Various flammable products in many forms may be found in one- and two-family dwellings.*

Figure 11-14   *An LPG cylinder behind a house should be noted during size-up, indicating some type of LPG service to the house.*

STRATEGIC GOALS AND TACTICAL OBJECTIVES

The strategic goals and tactical objectives for fighting fires in one- and two-family structures involve the items presented in the model in Chapter 5. As with any fire, the incident priorities must be applied and the following topics need to be considered.

Firefighter Safety

In the NFPA report *Firefighter Fatalities in the United States–2005* there were twenty-five firefighters recorded as fireground deaths. Almost all of the eighteen structure fire deaths occurred in residential properties. The most significant factor is that fires in one- and two-family dwellings killed eleven firefighters. We commonly refer to fires in the one- and two-family structures as "routine fires." For this reason, tactical objectives leading to the goal of firefighter safety must be employed at these incidents. These objectives include accountability, the assignment of a safety officer, the use of a rapid intervention crew, and operating under an incident command system. Fire spread in some types of these structures can be fast and, in the case of balloon construction, may go unnoticed. Life safety is the top priority in these structures. Firefighters may often be involved in aggressive search and rescue operations. Therefore, procedures must be in place to monitor fire growth and spread to prevent the firefighter from becoming trapped or caught in a flashover.

Search and Rescue

■ Note
Although the entire structure must be searched, certain areas may present a higher likelihood of finding a victim.

Unless the occupants of the home are accounted for, an interior primary search must be started, assuming the mode of operation is an offensive fire attack. Although the entire structure must be searched, certain areas present a higher likelihood of finding a victim. For example, paths of travel, below windows, behind doors, in hallways, and in bedrooms present a high priority. The use of interior search teams and exterior vent-enter-search (VES) teams must be considered to reach these areas. Search and rescue should always be assigned to one of the first-arriving units.

Evacuation

In two-family dwellings such as duplexes, evacuation may be necessary. Occupants of one-half of the duplex may not be aware of the fire in the other half, particularly if the fire vents itself out a window and smoke does not travel into the uninvolved half. In this case, a simple evacuation of the attached uninvolved half must take place.

Exposure Protection

Depending on the distance between structures, exposure protection can be a significant concern. This is particularly true when the structure is well involved and a defensive operation is ordered. Hose streams must be placed to prevent fire extension to surrounding structures or other properties. Furthermore, in duplexes and attached row or townhouses, exposure protection must be considered early.

Confinement

The confinement of the fire also depends on the mode of operation. In an offensive attack, hose lines should be positioned to confine the fire from the unburned side. Hose lines must also be in a position to protect interior paths of egress and stairwells. This not only supports the confinement effort, but also supports the means of egress for search and rescue purposes and firefighter safety.

In defensive operations, a strategic goal is to place hose streams in a position to prevent fire spread beyond the structure of origin (**Figure 11-15**). As can occur in row or townhouse structures, firefighters are often playing catch up when it comes to defensive attacks. The incident commander must remember that it takes time to get hose lines placed and in operation—often a longer time than it takes for the fire to spread between structures. If it appears that the fire will not be confined to the structure of origin, then work toward saving the block of origin.

Figure 11-15

Defensive operation in single-family dwellings should be aimed at keeping the fire in the building of origin.

Extinguishment

Once the fire is confined, efforts are made toward extinguishment. Extinguishment requires the operation of hose lines or master streams, depending on the magnitude of the fire, of sufficient flow to absorb the heat being produced. A direct attack on the burning materials may be indicated for a fire in which flashover has not occurred. Once flashover has occurred and the fire is in the free-burning stage within a compartment, an indirect attack may be indicated.

■ **Note**
A direct attack on the burning materials may be indicated for a fire in which flashover has not occurred. Once flashover has occurred and the fire is in the free-burning stage within a compartment, an indirect attack may be indicated.

Ventilation

The location of the fire within the structure dictates the ventilation necessary. In many instances, roof ventilation is not necessary at these types of fires, unless the fire has extended into the attic. The ventilation effort must support the fire attack as well as the search and rescue operation. Horizontal ventilation, both natural and mechanical, is applicable and effective in these type fires.

Overhaul

Overhaul is necessary to ensure that the fire is completely extinguished, particularly in balloon-frame construction where a basement fire can easily extend through the walls from the basement into the attic. By design, many of the furnishings in these structures add to the overhaul needs. Overstuffed furniture, beds, carpets, and clothes need to be properly overhauled to ensure that they are not smoldering. It is often best to remove these materials to the outside of the structure if they have been involved in the fire. In some cases, the fire investigators may want to start their investigation during the overhaul process.

Salvage

■ **Note**
Salvage can be accomplished by covering the materials, boxing and removing them, or sometimes just by getting them off the floor and away from standing water.

Although salvage is important at all fireground operations, fires in residential structures present the greatest challenge and the greatest opportunity for firefighters. Clothes, pictures, heirlooms, furniture, jewelry, and insurance papers are found in homes. The overall operation must include consideration for removing when possible or otherwise protecting those items that are not affected by the fire. This can be accomplished by covering the materials, boxing the materials and removing them, or sometimes just by getting them off the floor and away from standing water. Although most homeowners have insurance to cover these losses, many do not. Further, some of the losses in residential fires are sentimental items that are irreplaceable.

SPECIFIC FIRES

The following is a look at some of the common fires found in one- and two-family dwellings and the specific hazards, strategic goals, and tactical objectives.

Basement Fires

Hazards Encountered Basements are often the final destination of many unwanted items in the house. For this reason, the biggest hazard encountered in basements is the owners' ignorance of the items that are there. These items can be anything from normal household furniture, as may be the case in finished basements, to a hazardous materials storage area. Along with this hazard may be the sheer amount of the owners' belongings that are packed into the basement.

A second hazard encountered in basement fires is the lack of entrance and egress points. Most basements have one entrance, as depicted in **Figure 11-16**. This one entrance is now also the only exit. If this exit gets cut off, then firefighters operating in the basement are in serious danger. It is important during size-up to locate any alternative entrances into and potential exits from basements. These may include interior stairwells or ground-level windows.

Strategic Goals and Tactical Objectives The strategic goals and tactical objectives for basement fires closely relate to the general ideas that were presented earlier in this chapter; however, the specific differences are introduced here.

Firefighter safety In addition to all the other safety issues thus presented, one of the first safety considerations for the incident commander at a basement fire is the safety of the firefighters entering the basement. Many of the safety issues closely relate to the hazards encountered. For example, in many cases, basement windows are not large enough or low enough for a firefighter to escape without assistance if conditions deteriorate. The Radio Intervention Crew (RIC) must note the location of the basement windows. Are they regular glass plate windows or glass block windows? Some are covered with plywood or metal plates to prevent break-ins. Others are covered during remodeling of the residence. Some basements have no windows at all.

Search and rescue Although basements may be the dumping ground for all sorts of unwanted items, they may also be the living areas for people. Many homeowners finish the basements and make them in-law quarters, recreation rooms, or studio apartments. Therefore, with this type of basement, search and rescue must be done as soon as reasonably possible. A good size-up for basement fires includes gaining input from homeowners or neighbors, if possible, as to the content of the basement and the likelihood of occupants in the basement. Further, as the firefighting operations begin, crews need to conduct a primary search on the upper floors. The number of floors of the building determines the proper method for search. Basement fires are dangerous to all occupants of the building; however, the

Figure 11-16
Entrance and egress from basements may create tactical and safety problems.

■ **Note**

The crews must maintain control of the stairs to prevent fire spread.

first floor should be searched first. The crews must maintain control of the stairs to prevent fire spread. In the balloon-frame-constructed house an additional hose line must go to each floor of the house. The heat and smoke will push up the stairs and throughout the house. This type of hose line placement allows companies to search the entire house and maintain a safe means of egress.

Evacuation Evacuation for basement fires generally consists of evacuating the house. Decisions must also be made as to whether adjacent homes need to be evacuated. This will certainly be the case if the buildings share a wall or the fire building has become unstable as a result of the basement fire.

Exposure protection Exposure protection for basement fires is generally the same as for all one- and two-family dwellings.

Confinement Confinement of a basement fire comes from locating the fire and making a quick attack. Determine where in the basement the fire appears to be located. Determine whether an exterior entrance exists. A walk around the structure will assist you in making this determination. A primary clue to a basement fire is smoke from bottom to top of the structure that is not lifted when the front door is opened and ventilation is started. If the smoke cannot be lifted from the floor level with ventilation, then begin the search for other clues that will lead you to finding the existence of a basement.

In an unfinished basement, the fire will be easier to find due to the openness of the area. If the basement is finished, it becomes more difficult to find the fire due to the compartmentalization. The rooms will confine the fire, making it difficult to find and extinguish. The fire must be contained, while protecting the stairways.

Extinguishment The basement fire is inherently difficult to fight because you have to advance hose line down into the fire itself. The engine company needs to take a hose line to the top of the basement stairs. After making sure there is enough hose to make it to the bottom of the stairs, the hose should be advanced quickly, but controlled, for a descent into the basement. Once in the basement the fire can be extinguished in the usual manner. If ventilation is poor, then the use of smooth bore nozzles should be considered for greater reach and to lessen the interruption of the thermal layers of the fire.

Ventilation Ventilation is more difficult in these situations. Outside ventilation may actually make conditions worse inside the basement. If exterior windows can be vented without jeopardizing the interior crews, this action must be taken. If vertical ventilation is required, place the hole as near as possible over the top of the stairway, especially in a two-story structure. The hole over the stairs helps to remove smoke from the second floor that would otherwise mushroom out through the hall and open bedroom doors. If the basement is not venting well, consider cutting a hole in the first floor under a window, about 1 foot from the wall, then remove the window.

Overhaul Overhaul in basement fires consists of checking every floor and the attic of the house. Overhaul ties in directly to knowledge of building construction and general building features. The attic, for example, is the first area checked if incident command suspects a balloon-frame home. Furthermore, fire will spread through walls and floors. Overhaul must be thorough.

Salvage Many of the items found in basements hold sentimental value to the owners or occupants. Care should be given to protect these items from unnecessary damage. If the basement floods due to hose lines, then the owner's possessions should be moved to upper floors if it is a safe option. Otherwise, meeting with the owner will help to determine the best place for these most precious possessions.

First-Floor Fires

Hazards Encountered One of the biggest hazards encountered in first-floor fires is the potential speed of fire spread. The first-floor fire can be located in the kitchen, living room, family room, bedroom, or bath. The type of room does not matter as much as where the room is located in the house and the ease of travel of the fire outside that room. Most bedrooms and bathrooms have doors that can be closed by the occupants to contain the fire. The family room, living room, and kitchen, as shown in **Figure 11-17**, are usually open to the entire house and allow the rapid spread of smoke, heat, and fire throughout the structure.

A second hazard is the possible complacency of firefighting crews. A first-floor room and contents fire may not be taken as seriously as it should. This is a fire that many firefighters feel comfortable fighting. This type of complacency can kill firefighters. Remember the potential for collapse.

Firefighting forces must be aware of the possible instability of the structure. Determine the building's height. This may not be as obvious as it seems. Many times, structures are built on the sides of hills, creating a building that looks like a single-story structure from the road, but may in fact have a full basement. A basement must be identified early in the incident and the danger of floor collapse passed on to interior crews.

Figure 11-17 *Large open interiors and great rooms provide open access to many combustibles.*

Strategic Goals and Tactical Objectives One of the key elements in setting strategic goals and tactical objectives in first-floor fires is determining the exact location of the fire. The incident commander can determine where the fire is most concentrated by making a walk around the structure. Look through windows and doors for the signs of the fire. Look for fire venting from windows. The travel of the fire can be determined by looking at the concentration of smoke coming from the windows and eaves. The intensity of the fire can also be determined. Is the smoke drifting from the structure lightly, is it being pushed by a flame front, or is it coming out forcefully in a cyclonic effect under tremendous pressure? Are flames exiting the structure through a window or traveling throughout?

Determining where the fire is located, how fast it is traveling, and the direction of travel helps the incident commander to apply the following goals and objectives specific to these type fires.

Firefighter safety Up to this point, important issues that help to ensure firefighter safety have been presented; however, specific safety issues need to be addressed in first-floor fires.

One of the most critical elements to firefighter safety is a good size-up to determine hazards. For example, if an incident commander notices on a walk around that there is a basement or the house is built with a stem wall and crawl space, as shown **Figure 11-18**, this information must be passed on to firefighters so they will make sure to check for a floor before entering a structure. This may seem obvious to firefighters if they are moving onto a roof or second floor from a ladder, but not so obvious on the first floor.

Figure 11-18 *Note the space under the house, an avenue for fire spread and a safety hazard for firefighters.*

Search and rescue If the structure is likely to have occupants, then search and rescue is absolutely crucial in first-floor fires. The first floor should be searched first and all occupants accounted for. A team must also be deployed to search the second floor. Ideally, these tasks would be completed simultaneously; however, if not possible, the first floor is first, followed by the second floor.

Evacuation and exposure protection Evacuation and exposure protection as related to first-floor fires are typical to all one- and two-family structures.

Confinement and extinguishment While firefighters are extending hose lines, the incident commander will have completed the walk around the structure and may be able to tell the crews where to find the fire. The crews should bring the first line in the front door and proceed directly to the fire area. While the engine crew is placing the line between the fire and the remainder of the house, an additional crew should begin search and rescue procedures. The first hose line should be used to contain the fire and allow search and rescue to begin. There is always the potential for failure of the first hose line. Therefore, a backup hose should be deployed. A backup line provides protection for the first-in hose line crew and the search and rescue crew.

Ventilation A window that has failed allows natural ventilation of the fire, which obviously assists the incident commander in finding the fire. If the structure remains tightly closed, the smoke buildup may obscure the actual fire area. It allows the smoke and fire to spread throughout the structure, following the path of least resistance. The smoke and fire travels throughout the first floor and up any stairs, duct work, or open pipe chases affecting the second floor and attic. Ventilation should be considered for a quick and effective attack. In most small, one-story ranches, vertical ventilation is not necessary. Properly placed positive pressure ventilation with the right windows removed opens the structure and allows the smoke to lift and be removed. This creates better working conditions for both the hose line crews and the search and rescue crews. In the two-story structure with smoke and fire conditions that require vertical ventilation, the ventilation hole should be placed over the fire area if safe to do so. If egress is a concern, it should be placed over the stairway, allowing the smoke to be drawn up the stairs and out of the building. This placement makes the second-floor areas more tenable for search and rescue operations.

Overhaul and salvage Overhaul is used to determine that all the hot spots have been located and the potential for rekindle has been eliminated. A thermal imaging camera is an excellent tool to use during overhaul. Salvage is the process by which we save and protect irreplaceable keepsakes, clothing, furnishings, and other items. The officer in charge of salvage should meet with the occupant to determine the importance of items before they are carelessly tossed out a window into the yard.

■ **Note**

What looks like a two-story house in the front may be a four-story house from the rear because of the sloping terrain.

■ **Note**

Because firefighters are working on upper levels, rapid intervention teams must be aware of the differences in building construction in a two- or more story building and prepare themselves accordingly.

Upper-Floor Fires

Hazards Encountered The second- or higher floor fire consists of the same room and contents type of fire found in a first-floor fire. The mattress, box springs, and bedroom furniture produce significant smoke and heat. The smoke on a second floor does not have a way to escape unless a window fails. This type of fire situation is more difficult for the fire officer to evaluate during the walk around because it is not as easy to see into the building, which creates a hazard that is specific to these type structures. The 360° survey is always necessary to look for occupants in danger, smoke and fire conditions, and other peculiarities with the structure. As with first-floor fires, the incident commander must be particularly aware of elevation changes. What looks like a two-story house in the front may be a four-story house from the rear because of the sloping terrain.

Strategic Goals and Tactical Objectives

Firefighter safety The biggest concern for firefighter safety when fighting fires in upper floors is the lack of egress. The way into the building is not always the best way out of the building. If firefighters enter the building via interior stairs, then a ladder should be placed to a window and the firefighters made aware of the location of the ladder (**Figure 11-19**). Similarly, if firefighters enter upper floors by means of a ladder, then a second ladder should be raised and the interior stairway should be protected.

Figure 11-19 *When firefighters are operating on upper floors, ladders must be placed to provide a secondary means of egress.*

Because firefighters are working on upper levels, rapid intervention teams must be aware of the differences in building construction in a two- or more story building and prepare themselves accordingly. They should be ready to search and rescue firefighters on the second floor and in the basement, in building or floor collapse scenarios. The proper tool choices must be made based on the rescue possibilities, including having ground ladders placed to the upper floors.

Search and rescue The search crews will have a hard time searching the hallways until the fire is confined. Firefighters must remember that the top of the stairs and the hallway are places to find victims overcome by smoke while attempting to escape. In some cases, as discussed in firefighter safety, crews find it easier to go off ladders or porch roofs into second-floor windows to perform search and rescue, as shown in **Figure 11-20**.

The idea is to enter the window and check directly under the window sill. If time is needed inside to search thoroughly, the vent, enter, and search (VES) firefighter should immediately shut the door to the bedroom to seal it from the hall. This buys the VES firefighter time to conduct a more thorough search with less danger from flashover from the hallway.

Evacuation and exposure protection Evacuation and exposure protection for upper-floor fires is typical of any one- and two-family dwelling fire.

Confinement and extinguishment The first hose line into the front door is stretched up the stairs to begin to confine and extinguish the fire. Fire officers must consider the length of the hose lay (**Figure 11-21**). The 150-foot preconnect may not be enough to reach all portions of the second floor. All building features

Figure 11-20 *Access may be possible from porch roofs to second-floor areas.*

Figure 11-21 *When making a hose stretch to upper floors, consideration must be given to the length of the hose lay.*

should be considered. Much of the information is based on initial size-up. How large is the building? How far away is the apparatus placed? Can the preconnect stretch across the yard, up the stairs, and to the back of the house? If there is any doubt, then more hose should be pulled.

The second hose line is a backup line and should be deployed to protect the stairway, as discussed earlier in this chapter.

Ventilation If positive pressure ventilation is indicated and ordered, it must be coordinated very carefully. If the fan is turned on before the appropriate windows have been opened, the smoke, heat, and fire can be blown back down onto the hose line crews. In some areas of the country, it is common to have a firefighter take out as many windows as possible with a long pike pole, providing some re-lief for the inside crews. The problem is that the windows will not vent a lot of smoke and heat if the drapes and blinds are not completely removed. Opening the window first, then starting the positive pressure ventilation is typically a more useful ventilation method. Keep a close watch for any indication that the fire is being driven into the attic, walls, or back on the interior crews.

■ **Note**

If the fan is turned on before the appropriate windows have been opened, the smoke, heat, and fire can be blown back down onto the hose line crews.

Overhaul and salvage Overhaul is accomplished as it is for other types of one- and two-family dwelling fires. All the hot spots are located and the potential for rekindle is eliminated. A thermal imaging camera is an excellent tool to use during overhaul. Salvage is the process by which we save and protect irreplaceable keepsakes, clothing, furnishings and other items. The officer in charge of salvage should meet with the occupant to determine the importance of items before they are carelessly tossed out a window into the yard.

Attic Fires

Hazards Encountered Much like basement fires, the unknown element of the attic fire is the most hazardous condition that is encountered. Attic fires can involve both finished and unfinished attics. The finished attic probably has better access than the unfinished. The finished attic usually has the same fire load as the typical bedroom.

If an attic is unfinished, then nobody knows what is stored in it. Even the owner many times does not realize the amount of junk hidden away in the attic. The storage and fire load varies with the size and age of the house. Typically, the longer someone has lived in the house, the greater the fire load in the attic.

Other hazards encountered in attics include low sloping roofs, unstable floors, electrical wires, lack of safe entrance and egress points, extreme weight, and the potential for rapid collapse.

Low sloping roofs on houses can cause firefighters to become disoriented in attics and lose their way. They can be led into tight corners. Further, roofs with fiberglass or asphalt shingles can have nails protruding into the attic space that rip and tear bunker gear easily.

■ **Note**
Most unfinished attics do not have flooring throughout the attic. Firefighters are forced to balance themselves on trusses or beams.

Most unfinished attics do not have flooring throughout the attic. Firefighters are forced to balance themselves on trusses or beams. Firefighters can easily lose footing and fall through the attic floor.

Electrical wires and other utilities such as plumbing and cable generally run through the attic space, causing tangle hazards for firefighters. In parts of the country where homes do not have basements, it is common to find air-conditioning units and hot water heaters in the attic, adding weight. Finally, attics do not have safe entrance and exit points, creating a nightmarish trap for firefighters working there.

Strategic Goals and Tactical Objectives

Firefighter safety When firefighters are placed into attics to fight fire, everyone on the fireground must work for the safety of those firefighters. Ventilation has to be performed quickly and properly, hose lines have to be stretched to the proper length and moved as a team, and everyone on the fire scene has to be aware of signs of structural instability.

Incident commanders should not hesitate to provide for secondary means of entrance or egress by having firefighters pull ceilings and place attic ladders in them. Do not count on or use a scuttle hole as the only entrance to an attic. Placing a ladder to the attic provides this secondary means of escape.

Search and rescue If attics are unfinished, then one can make a reasonable expectation that no one is in the attic. Therefore, heroic firefighting efforts that threaten the lives of firefighters should not be attempted in unfinished attic fires. If the attic is finished, then it should be searched with the same methodology as bedrooms in one- and two-family dwellings.

Evacuation Evacuation of the structure is absolutely necessary if the attic is burning. Unfinished attic fires are impinging directly on the structural elements, and therefore collapse potential is greater than in fires where rooms with drywall provide a level of protection against fire damage.

Exposure protection Exposure protection for attic fires is essentially the same as for all one- and two-family dwellings.

Confinement Confining the fire to the attic space will not be difficult if the fire is located and extinguished quickly. If firefighters cannot locate the seat of the fire in a reasonable time, then the incident commander should consider evacuating the building. Roof collapse is inevitable.

Extinguishment Extinguishment requires the first hose line to be taken through the best access point of the house to the entrance of the attic. These fires must be attacked from below to be extinguished. Firefighters are required to get into the attic area and advance the hose line under the heat and smoke conditions. If the fire has broken through the roof, resist attacking it from outside, providing it is safe to employ an offensive attack. The process of lobbing water streams onto a roof to extinguish an attic fire from the ground below is often ineffective. The water will not get down into the base of the fire. An attack from above the roof with a ladder pipe or tower ladder can help to knock down most of the fire. As always, experience and knowledge of building construction and fire behavior help firefighters make the correct fire attack decision.

Another method of extinguishing attic fires is the piercing nozzle. This nozzle can be forced up through the ceiling and into the attic space, down through the roof, or in through a side wall. The nozzle creates an effective broken stream of water. This nozzle does not completely extinguish the fire, but used in conjunction with proper ventilation, it allows firefighters much easier access to the attic, with greatly improved visibility.

Ventilation To adequately ventilate an attic that has significant fire, vertical ventilation is the choice of many experienced incident commanders. However, the incident commander must consider the stability of the roof before committing firefighters to vertical ventilation. If the roof is deemed safe, get a crew to the roof as quickly and safely as possible to open a large enough hole to remove the smoke and heat. In some cases, the attic will have a small window that can be used in ventilating until a hole is cut in the roof. The size of the attic and the amount of fire present dictates the ventilation methods used.

Many fire officers do not consider positive pressure ventilation (PPV) in attic fires; however, this tactical method is viable. An argument could easily be made that heavily involved attic fires create unstable roofs. Therefore, if there is not a life safety issue with victims, then incident commanders should not create one with firefighters. In these cases, PPV is certainly the safest option and can be successful if performed correctly. It must be emphasized that PPV still requires an exit for the smoke in the attic area. If firefighters cannot go onto the roof, an exit can be made by removing attic vents from the safety of a ground ladder.

Attached Garage Fires

Construction The current housing stock in the United States offers at least three common styles of attached garages on the single-family home. In most of the single-story or single-story homes with a basement, the garage is an extension of the structure. The garage roof follows the same elevation and is integrated with the roof over the living quarters (**Figure 11-22**). The roof truss system for both the garage and the house are the same; therefore, the attic space is open over the entire house and the garage, with no firebreak between the two areas. The interior entry door from the garage goes into the kitchen or utility room of the house.

The second style can be found in most two-story houses. The garage roof is separate from the living area roof and ties into the side of the house as shown in **Figure 11-23**.

This style provides an attic over the garage with no interconnection between the garage attic and the living quarter's attic. The tie into the side of the house provides an inherent firebreak between the garage attic and the living quarters.

Figure 11-22 *A typical garage attached to a one-story dwelling.*

Figure 11-23 *A garage attached to a two-story house.*

The interior entry door from the garage goes into the kitchen or utility room of the house.

The third style can be found in either one- or two-story houses. The houses typically also have a basement. The garage is under part of the living quarters of the house (**Figure 11-24**). The ceiling of the garage is the floor for the living quarters above. In many cases, the area above the garage has the bedrooms. The interior entry door from the garage goes into the basement of the house. This style may also be seen in split-level houses, but the door to the house does not enter into the basement, but into a kitchen, hallway, laundry, or utility room.

Depending on the jurisdiction and the use of various fire codes throughout the country, the interior wall of the attached garage, next to the living quarters, may be required to have fire-resistant drywall. The other walls may be left with the studs exposed and no drywall. The ceiling of the attached garage should have drywall. If the ceiling of the attached garage is the floor of the living quarters, as found in the third style, it should also have fire-resistant drywall.

Hazards Encountered Garages tend to be like basements regarding storage. Everything the homeowner wants out of the house ends up in the garage. This means that garages can be hazardous due to the type of materials in them, such as gasoline, yard and pool chemicals, and pesticides (see **Figure 11-25**). Garages can also be hazardous because of the sheer number of items in them. Some garages can be packed so full that entry with hoses and SCBA is nearly impossible.

■ Note

Garages can be hazardous because of the type of materials in them, such as gasoline, yard and pool chemicals, and pesticides. They can also be hazardous because of the sheer number of items in them.

Figure 11-24   *A garage situated under living areas.*

Figure 11-25   *Garages can be full and have a variety of materials and products stored in them.*

Overhead doors can also be hazardous. There are two common types of garage doors, with either door being made of wood, aluminum, or light steel. The first type is the standard roll-up door that rides on tracks on each side of the door. The door has four to six hinged panels and rolls up out of the way. The second door style is a solid door that raises up and slides back along the ceiling of the garage. The hazards come if the door closes with firefighters inside. It is imperative that either of the door styles be braced in the open position during firefighting operations to prevent them from accidentally closing and trapping firefighters.

■ **Note**
It is imperative that either of the door styles be braced in the open position during firefighting operations to prevent them from accidentally closing and trapping firefighters.

Strategic Goals and Tactical Objectives

Search and rescue Although garages are not typically living areas, they still need to be searched. In many areas of the country, garages have been converted into family rooms or bedrooms. From the exterior, the garage door stays intact but the room is actually inside the garage. Searching the garage may have to come after the fire is extinguished if the garage is heavily involved with flames. A search of the house is also necessary, especially if the conditions inside the house are smoky.

Evacuation When a garage is heavily involved, it is important to evacuate the house. A garage fire can easily extend into the house due to various construction characteristics or compromised fire protection features. Further evacuation of surrounding homes may be necessary if hazardous materials are discovered in the garage.

Exposure protection Exposure protection initially focuses on protecting the house from the garage fire, which is usually accomplished through timely confinement and extinguishment.

Neighboring houses may also need protection. In many areas, houses are built very close together. Even suburban areas are seeing a boom of gated communities where homes are built 5 to 10 feet away from each other. In cases such as these, garage fires may be a threat and the property needs to be protected appropriately with hose lines until the garage fire can be extinguished. This is especially true if wind is blowing in the direction of an exposure.

Other exposure considerations are items such as cars in the driveway.

Confinement The main objective of a garage fire is to confine the fire to the garage and keep it from extending into the house. This is best accomplished by using proper hose line placement and extinguishment tactics.

Extinguishment There are two different schools of thought regarding attacking a garage fire. One recommended practice is to attack the fire from the unburned side of the fire, as normal. This involves making entry into the house and attacking through the interior door that leads to the garage. In order for this attack to work properly, the garage must also be ventilated. This tactic ensures that the fire will not extend to the house but it does have some limitations. For example, smoke, heat, and fire can unnecessarily be introduced into the interior of the structure and damage the living area.

Another school of thought involves fighting the fire from the garage side. When this is done, the fire officer must consider and address the integrity of the interior entry door to the living quarters. In this method the standard fire attack ideology of attacking from the unburned side is not used. In a working fire with the garage door closed, the fire crews must raise or cut through the garage door to make an attack. If the door must be cut with a saw, cut about 6 inches in from each side to avoid the hardware and rollers. If there is a side entry door from the outside, that door can be used, but the best method is to open the overhead garage door.

This tactic is based on the assumption that the living quarters side of the attached garage has fire-resistant drywall. The interior entry door to the living quarters must be fire-rated in order to use this method. A team should be sent inside the living quarters to check the structural integrity of the interior entry door and close the door if necessary. In the third style of an attached garage, crews must also be sent to check above the fire.

There are also some limitations to this tactic. Because one- and two-family dwellings are not normally inspected for fire code violations, the integrity of the fire wall may be compromised. For example, a homeowner may choose to change from drywall to peg board. This can be done easily, without a permit and without the fire department ever having knowledge of it.

Ventilation   It is rare that vertical ventilation is needed in a garage fire, but good horizontal ventilation is important. Opening the main overhead garage door is the best form of horizontal ventilation. The garage may also have a window or door on the side or rear. The garage door itself may also have windows in it. These can be used for ventilation if the overhead door cannot be opened.

Overhaul   Overhaul for garage fires may be very labor-intensive depending on the amount of items stored in the garage. In heavily stored garages, it may be necessary to remove many items to ensure that overhaul is complete.

Questionable overhaul is not acceptable. If the fire is extinguished completely before the fire reaches the house, the owners or occupant may choose to stay in the house. If this is the case, quality overhaul cannot be overemphasized. The fire can not be allowed to rekindle.

Salvage   As the old saying goes, "One man's junk is another man's treasure." What firefighters see as junk may in fact be irreplaceable keepsakes for the homeowner or occupant. The incident commander should attempt to meet with the occupant to determine what is important and salvage the items as carefully as possible.

Manufactured Mobile Homes

Manufactured homes (formerly called "mobile" homes) are transportable structures that are fixed to a chassis and specifically designed to be towed to a

Figure 11-26   *Typical single-wide mobile home.*

residential site. They are not the same as modular or prefabricated homes, which are factory-built and then towed in sections to be installed at a permanent location. (see **Figure 11-26**).

The federal government regulates the construction of manufactured housing. Since 1976, manufactured homes have been required to comply with U.S. Department of Housing and Urban Development (HUD) manufactured housing construction and safety standards, which cover a wide range of safety requirements, including fire safety. Post-1976 manufactured homes bear a label certifying compliance.

Hazards Encountered   Three times as many people die in mobile home fires, proportionately, than in single- and two-family home fires. For every 1,000 fires that break out, twenty-one victims will die in mobile homes, whereas fewer than seven will die in single- or two-family dwellings. The primary cause of the problem is that fire spreads rapidly through mobile home contents, while the structure itself intensifies heat and smoke buildup. In addition, most mobile homes have fewer safe exits than a traditional home.

The furnishings placed in a manufactured home are the same as those found in most other residential homes. With today's furnishings we are experiencing higher heat and faster fire spreads. When contained inside the manufactured home, these items create a fast-spreading fire that quickly engulfs the inside of the manufactured home.

In the past, it was rare to save the structure of many manufactured home fires. In most cases, the smoke column was seen long before the arrival of the first-due engine. In one case, an older mobile home park (1950–1970 vintage trailers) was located just blocks from a firehouse, but even with that close proximity, the fire grew so fast that trailers were destroyed before the firefighters' arrival.

Strategic Goals and Tactical Objectives Upon arrival at the mobile home, as with any structure, the first concern is life safety. Knowing that these dwellings account for many fire deaths, the first-due officer will want to quickly determine any rescue concerns. A quick fire knockdown from the exterior may be a good first tactical move based upon the extent of the fire. This action will accomplish two initial concerns: (1) It will begin efforts for exposure protection, and (2) it will allow crews to enter the manufactured home to complete fire control.

The close proximity of manufactured homes in these parks can create immediate exposure concerns once the fire begins to vent from the windows. Firefighters may not be able to commit their apparatus close to the scene because of cars parked on the roadways and the potential for fire spread to exposures. In some cases, manufactured homes sit literally feet from each other, creating immediate exposure hazards.

Firefighter Safety The fast spread of fire throughout a manufactured home should be considered before deploying crews inside the structure. The initial incident commander (IC) must complete a rapid size-up and know when it is safe to enter the structure. Roof collapse is a danger in manufactured homes, as the structural stability is rapidly compromised by fire.

Search and rescue If a person is reported trapped or missing, a primary search must be conducted as rapidly as possible. In some manufactured homes there is a door on the back side of the structure approximately in the middle or back third of the structure. Depending on the location of the fire and the reported location of the victim, this door may be an option.

Evacuation and exposure protection As stated earlier, the rapid spread of a fire through a manufactured home and the close proximity of other homes must be considered. The evacuation of the residents on each side of the burning structure should be taken into account, along with setting exposure protection lines in place.

Confinement and extinguishment The initial size-up will be used to determine whether a quick exterior knockdown of fire is required or whether the crews can begin an interior attack. In many manufactured home parks, there are propane cylinders attached to the structure. The IC should consider this when conducting the initial size-up. If propane cylinders exist, they must be turned off and protected from excessive heating.

Ventilation The use of horizontal ventilation by removal of windows and doors is the easiest way to vent a manufactured home. At no time should a person be sent to the roof to try vertical ventilation. The roof of a manufactured home will not support a firefighter if it has been compromised by an intense fire.

Overhaul and salvage Overhaul and salvage operations at a mobile home fire are very similar to that of any single-family home. Remember that most mobile homes sit on metal frames and the potential is there for someone's foot to break through the floor.

SUMMARY

The one- and two-family dwellings present a multitude of possibilities in size and complexity, depending on where they are located. It is important for firefighters and incident commanders to observe the construction types and styles of dwellings in their response districts. This knowledge of the structures in their district will help them to formulate tactical objectives to meet the necessary strategic goals when the fire occurs.

To be effective, firefighters must know the various construction types, the hazards associated with each, and how construction type affects fire spread.

As with all fires, each strategic goal must be considered and those that are applicable must be used. Fires in one- and two-family dwellings result in the highest percentage of firefighter and civilian deaths each year. Therefore, the goals of firefighter safety and search and rescue must be a strong part of the overall action plan.

The residential structure fire may involve rooms and contents or components of the structure itself. In general, the fires can be described by five areas: basement, first floor, second or upper floor, attic, and attached garage. Each requires particular tactical objectives in order to meet the strategic goals.

REVIEW QUESTIONS

1. Describe the need for a walk around the fire building during initial size-up.

2. How is the location of a fire in the structure determined during the initial size-up?

3. What percentage of the population of the United States lives in one- and two-family dwellings?

 A. 60   B. 80   C. 75   D. 90

4. List the three principal wood-frame building construction methods found in residential dwellings.

5. What time of day do most deaths occur in residential structures?

6. Where should the first hose line be placed for operations at a basement fire?

7. In balloon-frame construction, what is the inherent construction feature that an incident

commander must consider during a basement fire?

8. True or False? Platform wood-frame construction is a safer building method because each room is built as a separate compartment to contain a fire.

9. What kind of building material(s) is used for the load-bearing walls in ordinary construction?

10. List three paths of fire spread in a residential structure.

ACTIVITIES

1. Review your response district. What are the primary construction types and features of the residential structures in your district.

2. Review a postincident analysis from a residential structure fire. How were the strategic goals met? What may have been done to better meet these goals?

3. Review your fire departments current standard operating guidelines (SOGs) for response to residential fires. Write a list of recommendations for updating these SOGs based on the new information you have learned from this and other chapters regarding the strategic goals.

Chapter 12

MULTIPLE-FAMILY DWELLINGS

Learning Objectives

Upon completion of this chapter, you should be able to:

- Identify the common types of construction for multiple-family dwellings.
- Describe the hazards associated with firefighting in these structures.
- List the strategic goals and tactical objectives applicable to fires in these structures.
- List specific tactical objectives applicable to fires in these structures.

CASE STUDY

On July 4, 1998, at approximately 2310 hours, fire units were directed to respond to a fire reported on the third floor of a four-story, twenty-eight family, H-type multiple-family dwelling. The first-arriving engine company notified the dispatcher that they had a working fire on the third floor, B wing, with fire showing out three windows in the front. The engine company then proceeded to a nearby hydrant, away from the front of the building, allowing access for the ladder companies. The engine company hooked up to a hydrant and started stretching the initial attack hose line toward the fire building. The first-arriving ladder company, which shares a station with the first engine and arrived simultaneously with the engine, began carrying out common ladder company functions for fires in multiple dwellings. The ladder operator placed his aerial ladder to the roof to allow a firefighter to get to the roof to perform vertical ventilation.

The officer and his forcible entry team proceeded to the fire floor to begin ventilation, forcible entry, and search (VES) of the fire apartment and adjoining apartments. An outside ventilation firefighter used the front fire escape in an attempt to get into the fire apartment from the exterior. While these operations were underway, the second- and third-arriving engine companies assisted in getting the first hose line in operation and stretching a second backup hose line. The second-arriving ladder company officer notified the units on the fire floor that they were proceeding to the floor above the fire for VES purposes. As the initial hose line advanced into the fire apartment through heavy smoke, firefighters discovered the apartment had undergone major alterations. As they proceeded down what should have been the main hallway, they ran into a wall. They backed the hose line out in an attempt to find a way to the front of the apartment. At this time fire was showing out five windows on the third floor and auto extending to the fourth floor. The officer of the second ladder company was now reporting fire in two of the seven apartments on the fourth floor with possible extension into the cockloft. The incident commander transmitted second, third, and fourth alarms in rapid succession. The firefighters on the roof performing cutting operations reported an extensive cockloft fire and began cutting a trench to save the A wing of this multiple dwelling. Firefighters were ordered out of the building, and after offensive operations failed to extinguish this fire, a defensive attack began.

This case study demonstrates the complexity of fire suppression operations in a multiple-family dwelling.

■ **Note**
Multiple-family dwellings are occupancies that house three or more families.

INTRODUCTION

For the purpose of this chapter, multiple-family dwellings are occupancies that house three or more families. These buildings come in all shapes and sizes and present a severe life hazard at all hours of the day due to the number of occupants. They also present serious life safety risks to firefighters because of the size and complexity of interior layouts. These structures can house many families. A simple four-story multiple-family dwelling with seven apartments per floor, as mentioned in the case study, contains twenty-eight apartments. With an average of four persons living in an apartment, emergency responders are faced with a building housing more than one hundred occupants. In large urban areas, these buildings are not unusual to find. The construction and occupancy of these multiple-family dwellings usually dictate fire department operations. It is incumbent that fire companies responding to multiple-dwelling fires have standard operating guidelines (SOGs) dependent on the type of fire and structure they confront. For example, a brownstone multiple-family dwelling is usually a one-building fire, with minimal attention needed in exposures, whereas a row-frame multiple-family dwelling of wood construction not only presents a severe life hazard to the buildings occupants, but also poses a severe life hazard in adjoining exposures and beyond. The location of the fire in the building is also an indicator of how operations should be conducted. A fire on an upper floor necessitates a quick upper floor search and rescue and possible roof cutting operations, while occupants below the fire are usually out of harm's way. Conversely, a lower-floor fire endangers occupants on all floors.

Apartment houses and other multiple-family dwellings present varying challenges to firefighting forces. These may be described as tenements, garden apartments, and row-frame apartments. Also the era in which the structure was constructed has an impact on the firefighting strategies, as older buildings may not have the same built-in fire safety features as new building.

There are variations to the types of apartment buildings in the following descriptions. Nothing can be written that accounts for every type of building that may be in a community. Effective preplanning and knowledge of response area are critical components in applying fireground strategies and tactics.

In general, multiple-family dwellings can be classified into older apartment houses, newer apartment houses, fire-resistive multiple dwellings, row-frame multiple dwellings, brownstone multiple dwellings, and garden apartments.

■ Note
Effective preplanning and knowledge of response area are critical components in applying fireground strategies and tactics.

STRATEGIC GOALS AND TACTICAL OBJECTIVES

The strategies and tactics presented here are general to all multifamily buildings. There is detailed discussion later in the chapter on construction features, hazards encountered, and strategic goals and tactical objectives as they apply to the individual building types.

Firefighter Safety

Many of the same basic elements of firefighter safety can be applied to all structure fires that have been presented thus far such as accountability, the assignment of a safety officer, the use of rapid intervention crew (RIC), and operating under an incident command system. These principles are applicable here as well; however, due to the size and construction features of these buildings the importance of firefighter safety cannot be overstressed. Firefighters may be operating on upper floors which requires the placement of the RIC in a position from which they can respond quickly to a firefighter in need. This might mean a position inside the fire building but away from the fire area.

In terms of accountability, because of the potential size of the building (**Figure 12-1**) and the number of entrances, there must be an accounting of who is in the building and where they are at any given time. This can be accomplished through the assignment of accountability officers at entry and exit points.

For the most part, the buildings are very large and could have undergone renovations, producing a maze for firefighters to operate in. The use of ropes for search crews that might not have a hose line may be indicated. Ground ladders should be raised to various window heights and in multiple locations in order to provide alternative means of egress.

Figure 12-1 *In a large building, an accountability officer should be placed at entrance points.*

■ **Note**
Search and rescue requires a great commitment of resources because of the area and number of people that may be affected.

Search and Rescue

Search and rescue requires a great commitment of resources because of the area and number of people that may be affected. The incident commander (IC) should be ready to assign these resources and call for additional resources early in the incident. Victims often require rescue by ground or aerial ladders, which creates further demands on available resources. Searches must be conducted systematically with those in the worst danger taken care of first. The person at the third-floor window screaming may not be the one in the most danger. In ideal situations a simultaneous search is conducted of all areas of the building; however, in reality many fire departments do not have the personnel or resources available to perform this task. General priorities for a multiple floor search are as follows:

■ **Note**
The incident commander (IC) should be ready to assign these resources and call for additional resources early in the incident.

- Fire floor
- Floor above the fire
- Top floor
- The rest of the building

■ **Note**
The IC must consider the resources needed for a total evacuation of occupants as compared to shelter in place.

Evacuation

Evacuation of the remainder of the building on fire or adjacent buildings has to be considered. The IC must consider the resources needed for a total evacuation of occupants as compared to shelter in place, which is described further later in the text. For example, in a fire-resistive multiple-family dwelling with a contained fire on the ninth floor, the occupants on the first floor should remain in their apartments. Conversely, a first-floor garden apartment fire with potential fire spread to the cockloft requires a total evacuation of the entire building.

shelter in place
a form of isolation that provides a level of protection while leaving people in place, usually homes or unaffected areas of large buildings

Exposure Protection

Exposure protection needs are varied. In a well-involved building fire, there is a need to protect external exposures of similar, close buildings. In a fire that involves several apartments, internal exposures must be considered. For example, in an H-type apartment building (**Figure 12-2**) with a well-involved fire in wing A, the goal for exposure protection should focus on preventing fire extension into the B wing. Likewise, in cockloft fires in row-frame and garden apartments, the exposure protection goal must be directed at preventing spread throughout the cockloft to other buildings.

■ **Note**
In a well-involved building fire, there is a need to protect external exposures of similar close buildings. In a fire involving several apartments, internal exposures must be considered.

Confinement

Hose lines should be placed to confine the fire to the room, apartment, or floor of origin. It is critical to place streams to protect stairwells and other means of egress (**Figure 12-3**). Sometimes, if the hose line is not yet in place, simply

Figure 12-2 *A typical H-style building.*

closing an apartment door can slow the fire spread and help to confine the fire. Hose lines going to the floor above and into adjacent apartments should be deployed early to further be prepared for fire confinement.

Extinguishment

Once confinement is accomplished, the extinguishment operations begin. With the fire loading present in some of these buildings, particularly in basement storage areas, consideration must be given to the size hose line required. Application of the fireground fire flow formula to determine the correct amount of water flow required to extinguish the fire can serve as a guide in determining the proper size hose. Consideration must also be given to the ability and amount of available staffing to maneuver larger hoses to upper floors.

Ventilation

The location of the fire in the building dictates the ventilation location. In multiple dwellings with interior stairwells, ventilation will most likely be necessary over the stairwell if the fire is of any magnitude. Fires where smoke is not creating a problem in the stairwells may only require ventilation on the involved apartment or floor. Both mechanical and natural ventilation are applicable in these buildings. Positive pressure ventilation is particularly applicable to stairwell ventilation.

Figure 12-3 *Hose placement in a multifamily dwelling. In this case, fire could continue vertical spread up the stairwell if it were not protected.*

Overhaul

Overhaul is always indicated because of unprotected vertical openings between floors and unprotected horizontal openings between apartments. The search for hidden fire should focus in areas where these openings exist. Like single-family homes, overstuffed furnishings and bedding require overhaul and should be removed from the building after the fire. As with all fires, the needs of the fire investigator should be considered during overhaul.

Salvage

Most of the materials found in these occupancies are personal and require that the firefighters do everything possible to protect them. Unlike a single-family dwelling fire, a fire in a multiple dwelling can affect many families. Water from extinguishment on upper floors seeks the path of least resistance on its way to the ground and can damage many apartments and property along the way. Incident commanders should consider the travel of water and smoke throughout the building and, as resources become available, assign firefighters to salvage operations in these areas.

SPECIFIC FIRES

Many of the tactical objectives that have been presented thus far in the chapter can and should be applied also to the specific fire types that are discussed here; however, each fire maintains its own unique problems and special considerations which are addressed here.

Older Apartment Houses

Construction   Apartment houses built around the 1900s were between four and six stories in height, approximately 30 feet wide and 80 feet deep (**Figure 12-4**).

These buildings, though not very big, present a severe life hazard. These older buildings were built exclusively of fire-resistive materials consisting of brick exterior and wood interior. There are anywhere from two to four apartments per floor.

Hazards Encountered   Fire stopping in older apartment houses is limited at best, and the building contents and construction features lead to very hot and smoky conditions, with excellent chances of fire spreading to the apartments and floors above. **Figure 12-5** shows examples of avenues of fire spread. Other possible avenues of fire extension are:

- Pipe recesses
- Utility shafts (plumbing, electrical, etc.)
- Air/light shafts
- Dumbwaiter shafts
- Renovation/alterations
- Auto extension from floor to floor

Some inherit construction features are hazardous to the firefighters during the operations. Fire escapes are used as a secondary means of egress for building tenants. Stairways and landings in these buildings are made of wood and are very

Figure 12-4   *A typical older apartment building.*

narrow, causing difficulties in advancing hose lines and getting firefighters above the fire floor. If occupants are fleeing down this interior stair, an extreme life hazard to occupants and firefighters is present. Basement access to these older buildings is via a staircase found under the main stairway inside the building (**Figure 12-6**).

Because the stairs to the basement are old and the risers normally are open and without any fire resistive qualities, they are susceptible to collapse under a firefighter's weight. Under fire conditions, in the basement, these stairs must be used with balanced judgment. Another feature in these types of older buildings are air and light shafts located between buildings as shown in **Figure 12-7**. A fire originating in a shaft or an apartment fire extending to the shaft can expose adjacent apartments, the floors above the fire apartment, and the adjoining building.

Strategic Goals and Tactical Objectives

Firefighter safety In addition to what has been covered previously regarding firefighter safety, there are specific issues surrounding older apartments that can

■ **Note**

Because the stairs to the basement are old and the risers normally are open and without any fire-resistive qualities, they are susceptible to collapse under a firefighter's weight.

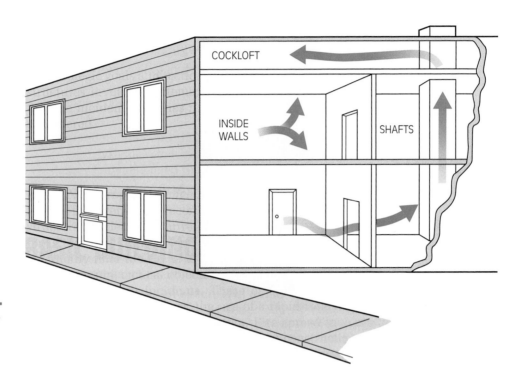

COCKLOFT

INSIDE
WALLS

SHAFTS

Figure 12-5 *Example of fire spread in an older apartment complex.*

■ **Note**
Getting the initial hose line between the building's occupants and the fire may save more lives than trying to perform multiple rescues.

■ **Note**
A visible life hazard upon arrival, such as people hanging out of windows or coming down fire escapes, may indicate an even greater life hazard within the building that is not visible from the street.

keep the fireground safe. One of these issues is the age of the building. Old apartment buildings typically have hazards that come from the lack of fire protection built into the concealed spaces, such as fire stops. Also, the building may have been deteriorating due to age and usage prior to the fire, which creates a very unstable structure under the attack of flame. Incident commanders must be aware of these hazards and use the knowledge to keep firefighters operating in a safe environment.

Search and rescue Regardless of the number of personnel responding to a fire in an older apartment complex, the basic principles of firefighting apply, which is to preserve life and confine and extinguish the fire. At times, resource constraints may not make it possible to perform these tasks simultaneously. At that point, the incident commander has to decide whether to remove trapped occupants or get a hose line in operation. It is important to note that getting the initial hose line between the building's occupants and the fire may save more lives than trying to perform multiple rescues. Remember that a visible life hazard upon arrival, such as people hanging out windows or coming down fire escapes, may indicate an even greater life hazard within the building that is not visible from the street. If a decision is made to perform rescue operations without simultaneously stretching an attack hose line, time will not be on your side and you will ultimately be confronted with a rapidly advancing fire.

Figure 12-6
Basement entrance to an apartment building may create access or egress safety issues.

■ Note

For a basement fire, it is crucial that the first hose line be stretched through the front door to the cellar stairs to protect the interior hall for fleeing occupants.

Evacuation Evacuation of older apartment buildings is imperative. Smoke and fire can move anywhere in the building and threaten the lives of occupants who assumed they were remote from the fire.

Exposure protection The tactics for exposure protection presented under general multiple dwelling can be applied to older apartment buildings.

Confinement and extinguishment It is recommended that a hose line be stretched via the interior stairway to the fire apartment to ensure protection of this vital stairway. For a basement fire, it is crucial that the first hose line is stretched through the front door to the cellar stairs to protect the interior hall for fleeing occupants. If possible, after occupants have left the building, an attempt can be made to descend these stairs to extinguish the fire.

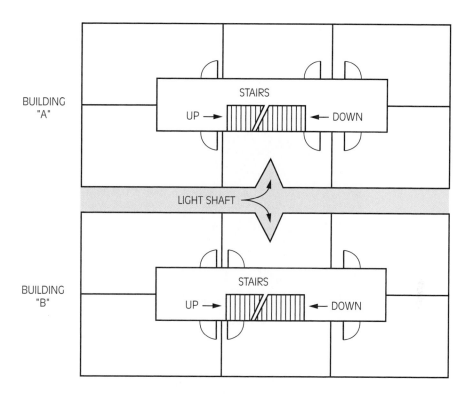

BUILDING "A"

STAIRS

UP → ← DOWN

LIGHT SHAFT

BUILDING "B"

STAIRS

UP → ← DOWN

Figure 12-7 *A shaft may be present, contributing to fire spread, as in this example of a light shaft between similar buildings.*

bulkhead/scuttle
the opening from a stairway of ceiling to the roof

The firefighters inside will also commence VES operations in their areas of responsibility and apprise the incident commander of fire conditions as well as the status of all primary and secondary searches.

Ventilation Another vital function at older apartment fires is to have a division proceed to the roof of the fire building to ventilate the stairway. By venting the stairway **bulkhead/scuttle**, particles of combustion are released to the outside atmosphere (**Figure 12-8**). This procedure prevents smoke banking down from roof level to upper floors and makes it easier for the engine company to advance its attack line and perform searches on the fire floor and floor above. The company assigned to this critical position must gain access to the roof by way of an adjoining building or via an aerial or tower ladder. This position is never attained by going up the interior stairs of the fire building in this type of structure. There is only one stairway in this building and a firefighter can easily become trapped or overcome if the bulkhead or scuttle are padlocked from the inside. After opening up the door to a roof bulkhead, a firefighter should reach in and probe the stairs leading to the roof for possible overcome victims who attempted to flee to the roof. After checking the stairs, the company should continue vertical ventilation at roof level, such as opening up skylights and ventilation covers. This team should also examine shafts and the rear and sides of the building for

Figure 12-8
Bulkheads can be used to ventilate stairwells.

persons in distress, fire extension, and location of fire escapes. This information should be communicated to the incident commander as soon as possible. It is also important for firefighters operating inside the fire building to know that vertical ventilation has been accomplished, after which the roof division can proceed down the fire escape to assist in searching the floors above the fire. For a fire on the top floor or where fire is suspected to have entered the cockloft and the building has a scuttle, the scuttle cover should be removed and the returns opened to examine the cockloft. The benefits of vertical ventilation are summarized in the following list:

■ Note

For a fire on the top floor or where fire is suspected to have entered the cockloft and the building has a scuttle, the scuttle cover should be removed and the returns opened to examine the cockloft.

- Relieves upper portions of heat and smoke
- Allows for a thorough search above the fire floor
- Creates a vertical flue for smoke and heat removal
- Limits horizontal fire spread
- Increases victims chances of survival

After a hose line is in position and ready for the fire attack, horizontal ventilation can begin. Horizontal ventilation entails removing windows and glass in the fire apartment. Under heavy smoke and heat conditions, these windows

■ **Note**

By maximizing the opening, you are also increasing the survival time of overcome victims by expelling products of combustion and allowing fresh oxygen to enter.

should be broken and window sashes removed in the event a firefighter has to use a window to escape and to maximize ventilation to relieve heat and fire gases. By maximizing the opening, you are also increasing the survival time of overcome victims by expelling products of combustion and allowing fresh oxygen to enter. If the fire is small and localized, horizontal ventilation can be accomplished by removing window blinds and curtains and simply opening the windows. For maximum ventilation, open windows two-thirds from the top and one-third from the bottom, as shown in **Figure 12-9**. The benefits of horizontal ventilation follow:

- Allows for interior hose lines to rapidly attack the seat of the fire
- Minimizes heat and fire rolling over the attack team
- Allows for a thorough search of the fire apartment
- Expels products of combustion from the fire area
- Introduces fresh air into the fire apartment
- Increases survival time of unconscious victims

■ **Note**

Horizontal ventilation must be coordinated with the incident commander so that it can be coordinated with the attack hose line.

Horizontal ventilation must be coordinated with the incident commander so that it can be coordinated with the attack hose line. Remember that premature ventilation could result in a backdraft.

Positive pressure ventilation in older apartment buildings, used in conjunction with horizontal ventilation, is still a viable option. These ventilators are usually set up at the door to the fire apartment as the attack team gets ready to

Figure 12-9 *A double hung window can be used for ventilation. It should be opened two-thirds at the top and one-third at the bottom.*

advance its hose line. Positive pressure must be used with good judgment so as not to push fire into uninvolved areas or where victims may be trapped.

Overhaul Overhaul on older apartments must focus on finding the hidden fire. Because of the varied construction features, the fire may be in places that are not typically considered. The attic should always be checked for extension. Also the floor above the fire should be thoroughly checked.

If thermal imaging cameras are available, they should be put into operation to check for hot spots. Thermal imaging cameras reduce damage done to a structure during the overhaul phase and still provide for a complete overhaul. They allow the operator to actually look through drywall to see hot and cold spots.

Salvage Older apartments may have families living in them with years of keepsakes and memorabilia. After the fire has been brought under control, incident commanders should direct a person to find out what is of value to the victims and retrieve it for them if possible. This act of kindness and extra care will reap many benefits for the department.

Newer Apartment Houses

Construction Newer-type apartment houses (**Figure 12-10**) built later in the 1900s are higher, wider, and deeper than the older apartment buildings previously described.

Figure 12-10 *A newer-style apartment house, built later in the 1900s.*

throat
the component of a building that connects two wings

The biggest structural feature was the introduction of unprotected steel I beams. These steel I beams are used as both vertical and horizontal components that support sections of these buildings. Because of the use of steel, these structures can take on various configurations such as **E**, **U**, **H**, and double-**H** buildings. In most cases these buildings have at least two wing sections and an area called the throat that separates the wings. Apartments are generally located in the wing sections, while the stairs and elevators are located in the throat, but it is not unusual to find an apartment in the throat section. There could be many apartments located on each floor in each wing. The design features make these buildings safer than the older tenements; however, as described in the case study, this building has great potential to burn. There are no interior stairs leading to the basement, which is a big advantage from a firefighting point of view. The basement ceilings may have fire-resistive construction. Access to the basement is through exterior courtyards and alleys. This feature takes away the concern of a basement fire racing up the interior stairs, trapping occupants. The building's interior stairways are of fire-resistive construction (metal treads with marble stairs and landings) surrounded by fire-resistive walls. Usually interior apartment walls are fire stopped at each floor. Apartment doors are fire-resistive, limiting fire in the hallway from getting into apartments. Unlike older apartment buildings that have only one stairway, newer buildings may have multiple stairways depending on the design. An example floor plan of a newer apartment building is shown in **Figure 12-11**.

The following are types of stair layouts:

- *Isolated stairs.* These stairs serve a certain part of the building. They are a single set of stairs. In many cases, fire-resistive apartment buildings are

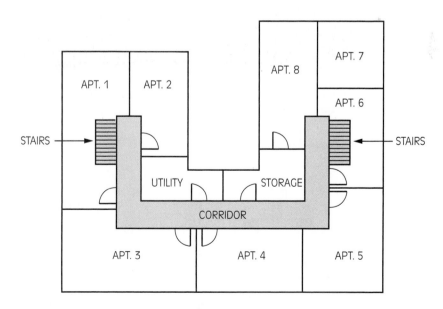

Figure 12-11 *A typical floor plan in a new apartment building.*

divided into sections. Each section has its own address, with its own entrance. One isolated enclosed stair would serve that address.

- *Wing stairs.* Wing stairs are located in each wing of a building. There may be more than one stairway in that wing (for example, front and rear stairs), but you cannot go from one wing to another unless you are in the lobby.
- *Transverse stairs.* Transverse stairs, as shown in **Figure 12-12,** are located on opposite sides of a building with multiple wings. The greatest feature of these stairs is that you can transverse from one wing to another on each floor. These stairs are a tremendous asset to firefighting operations. In many instances there are smoke stop doors in the middle of each floor. These stairs assist and ease evacuation, and hose lines can be stretched to any floor in the building from any stairway.

Hazards Encountered Although the design of these newer apartment houses were more fire protection-conscious—for example, fire stopping between floors and

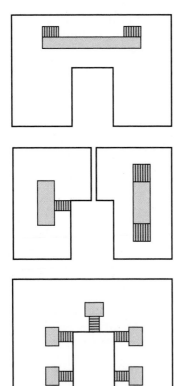

TRANSVERSE STAIRS

LOCATED AT POINTS REMOTE FROM EACH OTHER, BUT A PERSON CAN GO FROM ONE STAIRWAY TO ANOTHER VIA PUBLIC HALL ON ALL FLOORS OF THE BUILDING.

WING STAIRS

THESE STAIRS, ONE OR TWO IN A BUILDING, ARE LOCATED IN EACH WING. THERE IS NO CONNECTION TO THE OTHER WING. THE DIAGRAM REPRESENTS HALF A BUILDING SHOWING EACH TYPE.

ISOLATED STAIRS

USUALLY HAVE INDIVIDUAL ENTRANCE. FLOOR LANDINGS ARE LIMITED. NO CONNECTION TO OTHER.

Figure 12-12
Different stair Layouts.

■ **Note**

The extensive use of steel I beams allows a path for fire travel.

apartments and noncombustible hallways and stairs, there are many features that firefighters must be aware of. There are numerous shafts found in these buildings such as for elevators, trash chutes, and dumbwaiters. The extensive use of steel I beams allows a path for fire travel. Although vertical beams may be boxed in by Sheetrock, concrete, or sprayed with fire-retardant materials, this construction cannot be relied upon to prevent vertical fire travel. A fire entering this vertical flue requires opening up the walls and ceiling where the beam terminates. The base of the beam must also be examined for fire dropping down. Unprotected I beams (**Figure 12-13**) may also be used to support wood floor joists

Figure 12-13

Unprotected steel I beams in an apartment building.

and could assist the fire traveling horizontally. As with older buildings, pipe recesses, utility shafts, and poke-through construction also create excellent channels for vertical fire extension.

Strategic Goals and Tactical Objectives

Firefighter safety Items of concern in newer apartment buildings are the possible difficulty in stretching needed hose line to the fire origin and performing large search and extinguishment operations, with the resulting fatigue of firefighters. These issues can be resolved by having adequate resources on the scene and a good rehabilitation plan.

Search and rescue Search and rescue techniques are the same for most apartments; however, with large apartments there is a possibility for substantially more victims. Apartments should be marked as they are checked so no apartment units are missed. Marking can be any style as long as everyone on the scene is familiar with its meaning.

Evacuation In most cases, evacuation of the entire building will be required, especially if the fire has a potential for extending out of the area of origin.

Exposure protection Exposure protection strategies and tactics are typically the same as for all other multifamily dwellings.

Confinement A key to confining the fire is locating it and getting the hose line to the right area of the building in a reasonable amount of time. In multiple dwellings with multiple wings and stairs it is critical to determine which stairway serves the fire apartment before a hose line and crews are committed. Stairs may also wind around elevator shafts, necessitating a possible stretch via the buildings exterior using a utility rope. This hose line can be brought in on the floor below the fire and then stretched up the interior stairs to the fire floor.

Extinguishment Extinguishment is closely related to confining the fire. Once the fire is located and the hose is in place, the fire can be extinguished by normal tactics. A second hose line should always be stretched to back up the first line and for possible use on the floor above the fire or in an adjoining apartment. In many cases, these hose line stretches are very long and arduous, requiring two or three companies to assist in getting the first line in operation. If a well is present in the building stairs, the hose line should be placed in the well. A well is a space around the stairs where a hose line can be stretched from (**Figure 11-14**). One 50-foot length of hose at the base of the stairs can usually reach approximately five floors.

For a fire in the basement, the first hose line is stretched to one of the building's exterior basement entrances, to the seat of the fire. In many larger apartment buildings in urban areas, store fronts may be located on the first floor. The first hose line should be stretched to extinguish the fire in the store, with the second hose line backing it up. If the first hose line can extinguish the store fire,

■ **Note**
Apartments should be marked as they are checked so no apartment units are missed. Marking can be any style as long as everyone on the scene knows the meaning.

■ **Note**
A well is a space around the stairs from which a hose line can be stretched. One 50-foot length of hose at the base of the stairs can usually reach approximately five floors.

Figure 12-14 *Hose can be placed in the well of the stair case, allowing for less hose to be used for an upper-floor stretch.*

the second hose line can be advanced to the floor above the fire to check for extension.

Ventilation As with the older apartment buildings, roof ventilation may be indicated. In newer buildings, the roof division can use an adjoining stairway in the fire building, if it is isolated from the fire area; however, if there is an adjoining building, it is still the preferred method of getting to the roof. The roof division performs the same functions as in the older buildings. In cases of top-floor fires, the cockloft may have to be examined for fire extension. Firefighters can pull ceilings on the top floor to examine the cockloft. If fire has extended into the cockloft, roof cutting operation should commence immediately. A fire traveling in the cockloft could easily extend to the adjoining wings, causing extensive damage to the top floor. A large ventilation hole should be cut over the fire. This hole may

■ **Note**

Firefighters can pull ceilings on the top floor to examine the cockloft. If fire has extended into the cockloft, roof cutting should begin immediately.

limit fire travel. Top-floor ceilings must be pulled and hose lines on the top floor quickly put into operation. In the event of an extensive cockloft fire, trenching operations should take place to isolate the fire from remote sections of the building. Again, this is a last resort, and performed after a sufficient vent hole is cut over the main body of fire. The trench should be started approximately 20 to 30 feet from the main ventilation hole, and firefighters must always be assured of a second way to get off the roof. A hose line should be stretched to the roof to protect firefighters performing this difficult operation. Even if a trench cut is made, there are many instances when a rapidly moving cockloft fire has passed the trench cut before it has been pulled, resulting in complete loss of the top floor.

If the apartment building is large and the entire building needs to be ventilated with emphasis on protecting stairwells, then roof ventilation is indicated. If the fire has been contained to a bedroom in a second-story building, with light smoke in the breezeway or hallway, then PPV is the ventilation tactic of choice.

Overhaul All floors above the original fire floor must be examined for fire extension, with emphasis placed on the first floor above the fire floor.

Salvage Salvage for these building is essentially the same as presented earlier in this chapter.

Fire-Resistive Multiple Dwellings

Construction Fire-resistive multiple-family dwellings are well-constructed buildings, and fire is usually contained to the fire apartment, unless the apartment door has been left open by fleeing tenants. Each apartment is usually served by two fire-resistive enclosed stairs on each floor. Because the building is of fire-resistive construction, loss of life is kept to a minimum; however, fires that do occur in these type buildings produce great heat and are particularly punishing for the firefighting forces trying to extinguish them. An aggressive interior attack using one, and many times two, $2\frac{1}{2}$-inch hose lines to advance down a hallway where an apartment door has been left open is crucial to controlling a fire in these buildings.

Hazards Encountered An upper-floor fire, where the windows have self-vented, will be fanned by wind conditions outside making extinguishment extremely difficult. On December 18, 1998, three New York City firefighters lost their lives crawling down a hallway on the tenth floor of a fire-resistive apartment building, searching for the fire apartment. The fire apartment door was left open by a fleeing occupant, and when the apartment windows self-vented, a fireball swept the hallway, killing them instantly. Five days later on December 23, 1998, four civilians were killed in midtown Manhattan when they left their apartments while fleeing from a fire in an apartment on a lower floor. These civilians were found in the stairways, killed from smoke inhalation. The tragedy here is that had they remained in their apartments, they would not even have been injured. Lack of public education played a definite role in this tragedy.

Strategic Goals and Tactical Objectives

Firefighter safety Because many of these fire-resistive multiple-family dwellings are equipped with standpipe systems, it is relevant to present some firefighter safety considerations. Some standpipe systems may be connected to a city water main, whereas some may have gravity tanks on the roof. One of the safety concerns when using a standpipe is to make sure there are sufficient firefighters with enough hose proceed to the floor below the fire and make all the necessary hose line connections. There is no justification to hook up on the fire floor, even if firefighters are in an enclosed fire-resistive stairway. It is also generally not recommended to hook up to an auxiliary standpipe connection in a hallway on the fire floor. A firefighter with a radio should remain at the valve wheel to control water pressure in the hose line. It will take several firefighters to advance this hose line down a hallway and into the fire apartment, and all efforts should be made in this direction. In all cases, the standpipe system needs to be augmented by fire department engines to ensure adequate water pressure and flow.

Another safety concern in this type building is rapidly spreading fire. This may not seem possible in so-called fireproof buildings, but there have been instances where fires have been started in enclosed stairs (in garbage or mattresses) and the paint on the walls caused the fire to accelerate and turn into a ball of fire, incinerating the stairway from the fire floor to the roof in seconds. Civilians have been found dead in hallways many floors above the point of origin. If a fire has started in a stairway, no firefighters should enter this stairway above the fire until the situation is controlled.

A final point to be made is that units taking elevators to get to the fire floor should stop at least two floors below the reported fire floor and take the stairs the rest of the way (**Figure 12-15**). Taking an elevator to the fire floor can end in tragic consequences to firefighters and is not an acceptable practice (see Chapter 6 case study).

!Safety

Taking an elevator to the fire floor can end in tragic consequences to firefighters and is not an acceptable practice.

Search and rescue The floor above the fire generally does not become an exposure problem in fire-resistive multiple dwellings; however, this floor must still be searched. The greater emphasis for search and rescue must be made in the fire apartment, the public hall, and the adjoining apartments, particularly if apartment doors have been left open.

Evacuation At least one stairway needs to be designated as an evacuation stairway. It should be one that firefighters are not using to stretch hose lines or connect to standpipes. Additionally, before engine companies advance their lines onto the fire floor, the incident commander must make sure that all floors above the fire and the attack stairway have been evacuated and are clear of civilians.

Exposure protection Exposure protection should focus on the adjoining apartments on the same floor. Also, exposure to upper floors from open or broken windows in the fire apartment is a consideration. If flames are impinging on upper floors, then hose lines need to be placed into the upper floor apartments as soon

Figure 12-15
Firefighters must stop two to three floors below the fire floor when using elevators to access upper floors. (Photo courtesy of Palm Harbor Fire Rescue.)

as possible to prevent a multifloor fire. These exposure hose lines should be placed simultaneously or after initial hose lines are deployed to the fire floor, depending on staffing levels.

Confinement In order to confine the fire, efficient tactics are required. Responding engines should take positions on fire hydrants near fire department connections (FDCs) and supply the system. Depending on staffing levels, the first-arriving engine companies should ensure that the first line gets put into operation. As engine companies are hooking up the attack lines, ladder companies can attempt to locate the fire apartment.

Extinguishment The actual extinguishment of fires inside apartments is generally the same as any dwelling fire once the hose line has been stretched to the fire apartment. There are, though, some items that are unique to these buildings. For example, many of these buildings have more than one stairway. If this is the case, fire attack should be made from one stairway and the other used for rescue or evacuation, as shown in **Figure 12-16**. After initial hose lines are stretched, the backup hose line should be stretched from at least two floors below the fire floor, from the same stairway as the initial attack line. This tactic removes the chance of lines accidentally opposing one another. One stairway can be used as the attack stairs, while others can be used for evacuation.

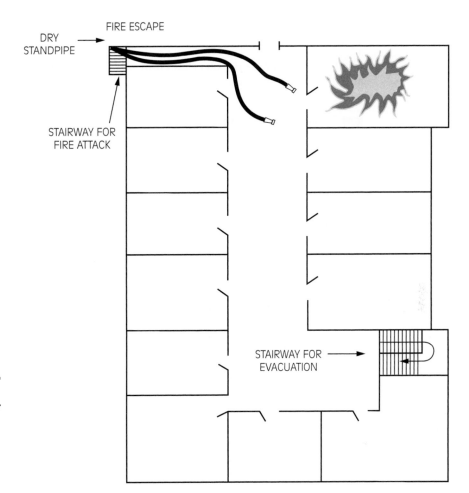

Figure 12-16 *Hose placement on upper floor of an apartment building using one stairway for fire attack and one for evacuation.*

Another example is incinerator fires. In most cases these buildings have incinerators or compactors for the disposal of refuse. A shaft running the entire height of the building is used by occupants, with closets on each floor. A shaft built for an incinerator was designed to burn, whereas a compactor shaft, unless converted from an incinerator, was not. If a large fire is in a compactor shaft, a hose line must be stretched one floor above the burning material and directed into the shaft to extinguish the fire. A heavy smoke condition usually occurs in the public halls at this type of fire and searches of all floors must be thorough.

Ventilation In most cases vertical ventilation is not a major consideration; however, incident commanders do want to ensure that the roof bulkhead is opened so smoke entering the stairs has a way out of the building. In some cases, the fire floor may be quite remote from the roof, making roof ventilation unreasonable.

Horizontal ventilation using PPV is the most efficient means of ventilating individual apartments.

Overhaul and salvage Overhaul and salvage performed in fireproof apartments and condominiums is common to all dwelling fires.

Row-Frame Multiple Dwellings

Another type of multiple-family dwelling found around the country that seriously tests our firefighting forces is the row-frame building, constructed solely of wood and built in rows with as many as twenty buildings in the row, as shown in **Figure 12-17**.

Construction These row frames built at the turn of the twentieth century are as high as five stories, with one or two apartments per floor running from the front of the building to the rear. There are air/light/dumbwaiter shafts present in most of these buildings and fire can travel in all directions. There are no brick walls between these buildings and the prevalent feature in these structures is the common cockloft that runs the entire length of the block. **Figure 12-18** shows an example of this problem.

Hazards Encountered A general concern in row-frame fires is the rapidity with which fire spreads and the great potential of structural collapse under fire conditions. This type of fire taxes fire department resources and quick water on the

■ Note
A general concern in row-frame fires is the rapidity with which fire spreads and the great potential of structural collapse under fire conditions.

Figure 12-17 *Row-frame multiple dwellings.*

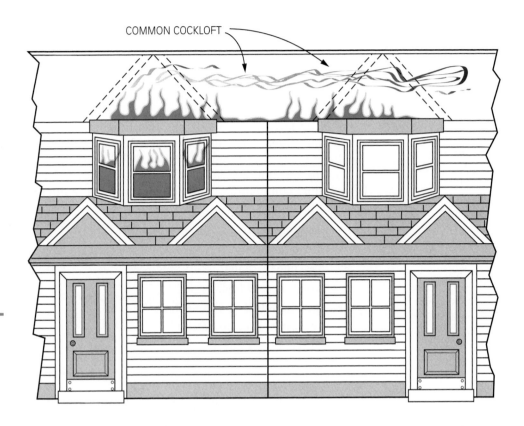

COMMON COCKLOFT

Figure 12-18 *Open cockloft between dwellings. Truss floors can create cocklofts on each floor.*

fire and speed of operations may very well dictate whether you lose one building or an entire block of buildings. The life hazard in this type of structure is great, and resources must be deployed with this in mind.

Other ways that fire spreads in these type structures include:

- Utility shafts
- Air/light shafts
- Interior walls
- Building exterior (wood)
- Common cockloft
- Common cornices
- Common cellars
- Cellar meets that abut one another

Strategic Goals and Tactical Objectives

Firefighter safety In addition to all the other safety issues that have been presented, one of the greatest threats to firefighter safety in row-frame apartments is underestimating the need for adequate personnel early in the incident. Additional resources must be requested promptly and sufficient firefighters with saws must begin roof operations immediately.

Roof operations present their own unique firefighter safety issues. Firefighters proceeding to the roof to perform operations should make every attempt to use an aerial or tower ladder to get to this position. If the interior of a building must be used to get to the roof, use a building that is several buildings away from the fire building. Firefighters should exercise caution using an adjoining building to get to the roof because of the possibility of fire traveling in the cockloft.

Another important aspect of firefighter safety is good communications with the incident commander. Units must notify the incident commander as soon as possible with the following information:

- Location of fire
- Fire extension
- Presence of shafts
- Exposure problems
- Need for additional hose lines
- Progress of roof operations
- Status of searches

This information helps the incident commander have a good understanding of fire movement and status of personnel.

If fire escapes are provided on the building, firefighters operating on them should always be aware that under heavy fire conditions, the fire escapes may pull away from the building walls and collapse.

Caution must be used by firefighters pulling top-floor ceilings to expose fire. If adequate roof ventilation is not provided, built-up gases in the cockloft can ignite, engulfing firefighters on the top floor.

Search and rescue Entry, ventilation, and search of all floors is critical and search of floors above the fire is extremely dangerous and punishing. Except for these issues and the other safety items that were presented earlier in this chapter, search and rescue is generally the same as most multifamily dwellings.

Evacuation If the fire is in the cockloft, it can move through the building very quickly. It is important to get far ahead of the fire and evacuate people to safety. Incident commanders cannot wait until the fire is threatening building occupants. Consideration must be given to where the fire will go next and where it may go after that and move the people in its path.

Exposure protection To protect exposures, the same tactics should be used that were used to determine evacuation plans. Find out where the fire is moving and protect the buildings that can be protected.

Confinement To confine the fire in the place or apartment of origin, a smaller hose line ($1\frac{1}{2}$-pocket or $1\frac{3}{4}$-inch) is recommended due to its mobility and

■ **Note**
Any hose lines stretched into row-frame dwellings must have sufficient hose to cover the entire building.

■ **Note**
In the event of a rapidly spreading cockloft fire, companies may have to skip a building or two to get ahead of the fire.

trench cut
a cut for ventilating a confined fire in which the cut is the entire distance of the roof; useful in confining a fire in a cockloft that is spreading horizontally

maneuverability. Additionally, hose lines stretched to the top floor of the fire building and the top floors of the adjoining exposures must be stretched immediately to get ahead of the fire.

Extinguishment The location of fire in row-frame dwellings dictates where the first hose line is stretched. As with some of the previous multiple dwellings discussed, if the fire is in the basement, the first hose line must protect the interior stairs. If this line can advance down the interior stairs, it should do so while a second line is stretched to protect the interior stairs and cover the floors above. If the first line cannot advance down the stairs to the basement, the second hose line is stretched to an exterior basement entrance.

For a fire on the upper floors, the first hose line is stretched to the fire, while a second hose line is stretched to back it up and possibly cover floors above the fire. A third hose line should be stretched to the floors above the fire or to the top floor of exposure B or D. Any hose lines stretched into row-frame dwellings must have sufficient hose to cover the entire building. In the event of a rapidly spreading cockloft fire, companies may have to skip a building or two to get ahead of the fire. Firefighters advancing hose lines into the top floor of row-frame structures should ensure that they are equipped with tools to pull ceilings.

Ventilation Top-floor fires are extremely dangerous and roof ventilation is critical. If ventilation is performed quickly enough and correctly, it will stop the fire from progressing. In these cases a **trench cut** (**Figure 12-19**) is often necessary.

Figure 12-19 *Trench cuts may be necessary to stop the fire in a common cockloft. This process is personnel-intensive.*

Overhaul and salvage The overhaul and salvage for these type buildings is essentially the same as for most multifamily dwellings.

Brownstone Multiple Dwellings

Construction As shown in **Figure 12-20**, these structures are built of non-fire-resistive construction; however, the exterior is constructed of stone and masonry, thus the name brownstones. These buildings were built in the late 1800s and early 1900s. They are three to five stories in height.

When they were originally built, they were occupied by only one family that was usually financially well off. Today they are usually occupied by three or more families or used as single-room occupancies. Single-room occupancies are individual rooms in an apartment rented out to individuals. These individuals share the same bathroom and kitchen.

These buildings were originally constructed with a high stoop that leads to the second-floor parlor. This is also the main floor. The ground floor, the entrance to which is under the high stoop, is also known as the first floor or basement. Below the first floor is the cellar. Today you can find an apartment on each floor.

Hazards Encountered As with row-frame multiple dwellings, brownstones take up entire blocks. The major differences are that the building's exterior is

■ Note
Single-room occupancies are individual rooms in an apartment rented out to individuals.

Figure 12-20 *A brownstone multiple dwelling.*

noncombustible, and brick walls separate each building. These walls extend through the cockloft, usually limiting fire to the structure of origin. There have been cases where the dividing walls fail and fire does extend, but they are rare. High ceilings on the second or parlor floor may require the use of 10-foot pike poles to pull ceilings. As with the row-frame buildings, there are several ways in which fire could extend vertically, including:

- Utility shafts
- Dumbwaiter shafts
- Pipe recesses
- Air/light shafts
- Hot air systems
- Sliding pocket doors

Pocket doors (**Figure 12-21**) are doors that slide into the wall. When these doors are closed, it creates a void for vertical fire extension. Fire can also extend horizontally via the cockloft, common cellar beams, and common cornices.

Figure 12-21 *An example of a pocket door.*

Strategic Goals and Tactical Objectives The strategic goals and tactical objectives examined here are specific to brownstones or have unique considerations that were not discussed in other types of multifamily dwellings.

Exposure protection The major emphasis is on the fire building, with less staffing needed in the adjoining exposures.

Confinement and extinguishment Once again, speed of operations is critical. Determining fire location is also critical in formulating an offensive attack. A 1½-inch or 1¾-inch hose line is generally sufficient for fire operations. If the fire is located in the cellar, the first hose line is stretched through the first-floor (basement) entrance to the stairs leading to the cellar. This line protects the stairs, and firefighters may advance down the stairs to confine and extinguish fire in the cellar, if practical. If the fire is located in the basement, which is the ground-floor level, the first hose line is stretched to the second floor to protect the stairs. This hose line can either advance down the stairs or a second hose line can be stretched to the first-floor entrance to extinguish the fire. The latter method is preferred. For a fire on the upper floors, a hose line is stretched up the front porch for a direct attack on the fire.

Ventilation Companies initiating VES have to pay particular attention to the top floor, where fire gases will be banking down. The roof division must ensure that scuttles and skylights are vented and that the rear and shafts are examined. For top-floor fires, the roof division needs to open up the returns and examine the cockloft for fire extension. It should also be prepared to operate a saw for roof cutting operations, if it becomes necessary.

Garden Apartments

Construction The final type of multiple dwellings to be discussed are garden apartments (**Figure 12-22**). These multiple dwellings began to be built in the 1940s. In general, these building are constructed of wood flooring, wood studs, and a wood roof. The are covered by siding or brick veneer and may be two- to four-stories high. They may be large, up to several hundred feet in length. The name garden apartment comes from the fact that the buildings are surrounded by landscaping and greenery as opposed to concrete. Building codes vary; however, there is some restriction on the area in between fire walls.

> **■ Note**
> In general, garden apartments are constructed of wood flooring, wood studs, and a wood roof.

Hazards Encountered Aside from the obvious construction hazard of wood adding to the fire loading and potential for early collapse, there are other hazards. These building have balconies with large glass openings. Fire can extend rapidly through these large openings to upper floors and the cockloft. Once the fire has extended into the cockloft, it can spread throughout the building or at least to the area between fire stops.

The interior stairways serve each floor and several apartments on each floor. Although there may be doors on each floor to limit fire and smoke spread, it is not uncommon to find these doors propped open. Stairways may also serve a basement area where often there are other apartments or tenant storage is provided.

Figure 12-22 *A garden apartment building. Balconies can store hazardous materials and be the path for fire spread.*

■ **Note**

Hose lines must be stretched further and there is a high reliance on portable ladders for access to the upper floors, as aerials may be parked too far back from the building.

Garden apartments by virtue of style are set back from the parking lot with landscape around them. This can create access hazards for firefighters (see **Figure 12-23**). Hose lines must be stretched further and there is a high reliance on portable ladders for access to the upper floors as aerials may be parked too far back from the building.

Figure 12-23 *The placement of entrances around garden apartments require long hose layouts, can cause access problems, and may hinder clear communication when giving tactical assignments.*

Strategic Goals and Tactical Objectives   Strategic goals and tactical objectives surrounding garden apartments have already been thoroughly covered in other areas of the chapter and can be applied to these types of buildings; however, the following specific issues should be considered when dealing with a garden apartment fire.

Confinement and extinguishment   Tactical objectives in garden apartment fires, as with other apartment fires focus on confining the fire to the apartment of origin and controlling interior stairways. **Figure 12-24** shows examples of avenues of fire spread in the garden apartment. The first hose line should be stretched by way of the interior stairs to the fire apartment, or in the case of a basement to basement entrance. Backup hose lines should be positioned to support the confinement and extinguishment effort and should cover adjoining apartments and

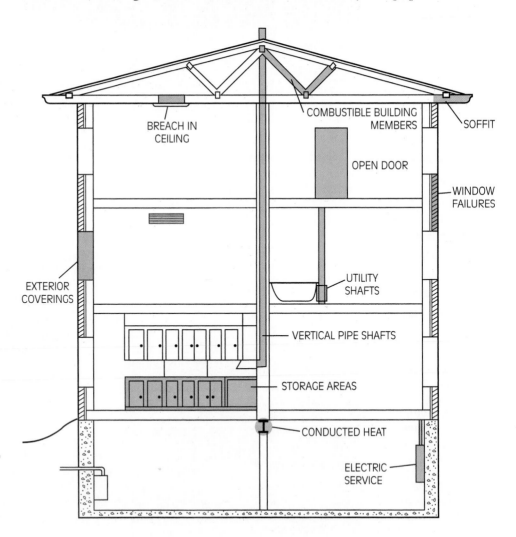

Figure 12-24

Avenues of fire spread in a garden apartment.

apartments above the fire. Hose line deployment to the top floor is also necessary if there is any indication that the fire has extended into the cockloft. Hose lines may have to be placed outside the building as well to protect the large balcony opening, should fire vent from these areas. These exterior streams should not be operated into the fire apartment jeopardizing interior firefighters. Instead, they can be used to prevent spread up the outside of the building.

Ventilation Both horizontal and vertical ventilation is indicated, depending on the location of the fire. To ventilate the fire apartment or hallway, horizontal ventilation can be used. To ventilate the stairwells, vertical ventilation should be used above the stairwell. In the event that the fire enters the cockloft, fire walls should be located and hose lines advanced to support the fire walls, thus limiting fire spread. Trench cuts of the roof may be necessary to further limit the spread of fire through the cockloft.

SUMMARY

Firefighting in multiple-family dwellings, whether they be older apartments, newer apartments, fire-resistive buildings, row frames, brownstones, or garden apartments, will challenge our firefighting expertise.

Construction features in these buildings differ; therefore, the firefighter must understand how these buildings are constructed and where the probable points of extension are. Hazards in these buildings also differ with the type of building and the construction. Fire personnel should be able to anticipate fire and smoke movement and act accordingly. It is equally important no

matter where you work, that members become familiar with buildings, not only in their response districts, but in adjoining districts to which they may respond. Prefire planning and familiarization drills cannot be overemphasized

Application of the fireground strategic goals and tactical objectives is based on the specific construction features and type of building. There is a high hazard to life, both victims' and firefighters', as a result of fires in these structures. Proper hose line placement and ventilation practices are essential to successful outcomes.

REVIEW QUESTIONS

1. Define a multiple-family dwelling.

2. For the purpose of this text, what are six types of multiple dwellings?

3. What are some significant differences between the older and newer apartment houses discussed in this chapter?

4. Where is the first hose line stretched for a cellar fire in an older apartment house?

5. Describe the roof firefighters' general responsibilities at multiple-dwelling fires.

6. True or False? In an older apartment house the interior stairs can always be used to access the roof.

7. List three benefits of horizontal ventilation.

8. True or False? The first line stretched for an offensive attack in a garden apartment

should be positioned to protect interior hall/stairways and to confine the fire to the apartment of origin.

9. Which of the following locations in an apartment building fire should generally be searched first?

A. Fire floor

B. Floor above the fire

C. Top floor

D. The rest of the building

10. How can fire spread at row-frame fires?

A. Building exterior (wood)

B. Common cockloft

C. Common cornices

D. All of the above may be avenues for fire spread.

ACTIVITIES

1. Discuss various types of multiple-family dwellings found in this chapter and compare them to multiple-family dwellings found in your fire district.

2. Review your department's SOGs for multiple-family dwelling fires. How do they compare with those outlined in this chapter?

3. Review a recent postincident analysis of a fire in a multiple dwelling. How did the operations compare to those presented in this text? What improvements could be suggested?

COMMERCIAL BUILDINGS

Learning Objectives

Upon completion of this chapter, you should be able to:

- Identify the common types of construction for commercial buildings.
- Describe the hazards associated with firefighting in these structures.
- List the strategic goals and tactical objectives applicable to fires in these structures.
- List specific tactical objectives applicable to fires in these structures.

CASE STUDY

The silence of a May night was broken by the dispatch of two engines, a ladder, and a battalion chief to a reported structure fire in a combination furniture warehouse/showroom. Upon arrival, the first-due engine found heavy smoke conditions throughout a 100' × 200' metal building. During size-up, it was apparent that the fire was located in the rear portion of the building, which was the warehouse section. Heavy fire was visible throughout this area. A second alarm was requested and on the arrival of the battalion chief, a third alarm was sounded. Companies on the scene secured a water supply and began preparation to attack the fire. The battalion chief reviewed the preincident plan and determined that the showroom was separated from the warehouse by a wall, with two ground-level door openings about 6 feet wide. A crew was sent to the roof to open skylights and hose lines were stretched through the showroom toward the warehouse section. Each 2½-inch line was positioned at one of the door openings to the warehouse. The fire had not extended beyond the warehouse.

With roof operations complete and two 2½-inch hose lines in position confining the fire to the warehouse, the large overhead garage doors were forced open, allowing second- and third-alarm companies to attack the fire in the warehouse with large streams. In a short period of time the fire was extinguished, but overhaul took several hours because of the large amounts of furniture in the warehouse.

With good application of fireground strategic goals and tactical objectives, the result was that the front half of this building, along with the contents, was saved from the fire. Firefighter safety was provided for with a strict accountability process, appointing a safety officer, and sufficient staffing, with the calling of additional alarms. Confinement and extinguishment was carried out according to established guidelines and firefighting principles. A preincident plan was available to help the incident commander in applying fireground strategy. Overhaul and salvage was performed with minimal damage. This was a successful incident for all of these reasons.

■ **Note**
Commercial occupancies may include several different style structures and categories.

■ **Note**
Mercantile properties include supermarkets, department stores, drug stores, and shopping centers. Examples of business properties are a doctor's office, courthouse, and lawyer's office. An industrial occupancy could be a factory or a manufacturing location.

INTRODUCTION

Commercial occupancies may include several different style structures and categories of occupancies (**Figure 13-1**). Included would be mercantile, business, and industrial occupancies. Mercantile properties include supermarkets, department stores, drug stores, and shopping centers. Example occupancies of business properties are a doctor's office, courthouse, and lawyer's office. An industrial occupancy could be a factory or manufacturing location. For the purposes of this

Figure 13-1
Mercantile, business, and industrial occupancies are considered commercial structures.

chapter, we consider a commercial occupancy a general term that includes structures of all of the previously listed types and then some, such as a barber shop, corner grocery store, or restaurant. Certainly, this listing is not exhaustive as there are many different types of commercial occupancies, many of which are often grouped together under one roof such as in office buildings and strip malls.

Fires in these occupancies and structures can be some of the most difficult to control and extinguish, not to mention the increased hazards and dangers to firefighters associated with their various amounts of fire loading and unknown contents, which could be anything from ordinary combustibles to hazardous chemicals.

This chapter presents the common construction types, hazards that may be expected to be encountered, and the strategic goals and tactical objectives that can be employed to fight fires in commercial occupancies.

STRATEGIC GOALS AND TACTICAL OBJECTIVES

Firefighter Safety

Aside from the common firefighter safety concerns discussed thus far, other objectives relating to firefighter safety must be employed in structures of this size and occupancy. Accountability must be at every entry and exit point of the structure. Rapid intervention crews (RICs) must be assigned and located to provide rapid access to the building. If firefighters are using the rear of the building for offensive fire attack, the RIC at the front may be of no use. Consider assignment of more than one RIC when more than one entrance is used (**Figure 13-2**).

■ Note
Fires in these occupancies and structures can be some of the most difficult to control and extinguish, not to mention the increased hazards and dangers to firefighters associated with fighting fires in them with their various amounts of fire loading and unknown contents, which may be ordinary combustibles to hazardous chemicals.

■ Note
Consider assignment of more than one RIC when more than one entrance is used.

Figure 13-2 *Fire in commercial occupancies may span large areas, requiring more than one RIC. In this illustration there are multiple possible entrances, therefore, the need for accountability points at each entrance.*

Fires in these types of structures force firefighters to face multiple different construction types and associated hazards along with multiple different occupancies all under one roof. Not only is the placement of apparatus important but as these structures are becoming increasingly larger in size, they also force firefighters to access deeper and deeper parts of the structure, which creates problems in smoke conditions and low visibility. The height of the ceiling may mask the true intensity of the heat, and a flashover could occur with little warning.

Firefighters working without hose lines during search operations should use a guide rope. These structures, like some apartment buildings, are mazes inside. The added stock that is found in a store adds to the potential to get lost or trapped. Often these occupancies use the rear of the store for storage. This storage often blocks the means of egress, so alternative ways to get out should be identified.

Because some of these structures have exterior walls that are no more than a façade and others utilize the large cast concrete wall sections, they have an increased potential for collapse. Placement of apparatus and personnel should be a main concern from the arrival of the first company.

■ **Note**
The height of the ceiling may mask the true intensity of the heat, and a flashover could occur with little warning.

Search and Rescue

Search and rescue in these structures is time-consuming and staff-intensive, not to mention difficult. However, unlike the residential structures discussed in previous chapters, the life hazard, particularly at night, is reduced.

The search and rescue effort must begin in the areas that are most threatened by the fire and smoke, and work outward from there. Remember, most people that enter any type of building tend to leave or want to leave the same way. For

Figure 13-3
Occupants of buildings generally seek to exit where they came in.

this reason search and rescue efforts should be sure to quickly address these exit areas (**Figure 13-3**) and ensure that these areas stay protected if threatened by the fire spread. The search and rescue effort can be greatly affected and even speeded up if other operations take place at the same time, such as ventilation.

With many of these structures the hours of business may vary. Some may be open every hour of the day. Some of these types of structures not only have the potential for occupants within the commercial occupancies but may also have apartments above the business (**Figure 13-4**). This causes a double concern especially when the products of combustion tend to move vertically rather than in any other direction.

Evacuation

■ **Note**
Evacuation for incidents in commercial structures should begin in the areas of the structure most threatened by the fire and smoke.

Much like search and rescue, evacuation for incidents in commercial structures should begin in the areas of the structure most threatened by the fire and smoke. Typically, in the early stages of a fire in these structures, the adjacent occupancies are evacuated first and work outward from there. For example, if there is a fire in a strip structure that has five separate businesses and the middle business is the one on fire, evacuation would begin with the businesses on both sides of the fire and work outward to the others. Evacuation often needs to be accomplished in these structures, as it is possible for a fire to occur in one part of the building with those in remote areas not even knowing that a fire has occurred. Evacuation in larger commercial structures can be conducted with the use of a public address system, building alarm system, or firefighters, depending on what

Figure 13-4 *Stores may often have apartments on the floors above, significantly increasing the search and rescue need.*

equipment is available. Depending on the severity or potential severity of the situation and the number and condition of evacuees, it may be a better option, or at least an option, to shelter them in place. Complacency and disregard is common in people today and may play an important role in whether a person perceives something as dangerous and leaves when asked or when the building fire alarm sounds.

Exposure Protection

Exposure protection should be applied to both internal and external exposures. In the strip mall-type structures, the exposures are uninvolved stores under the same roof as the fire. In large commercial structures of multiple floors, exposure protection involves protecting from fire spread throughout the building to uninvolved areas. Exposure protection in small commercial buildings in a congested downtown area requires the placement of large hose lines or master streams flowing water on closely exposed buildings. This type of problem is depicted in **Figure 13-5**.

Confinement

Confining fires in these structures can be difficult. In strip stores the fire may extend through both horizontal breaches in the walls and through a common cockloft (**Figure 13-6**).

Figure 13-5 *If a well-involved fire happened in this store, a significant exposure problem would occur to the other two buildings. (Photo courtesy of Amy Mozes.)*

Figure 13-6 *Strip mall with a common cockloft.*

■ **Note**
Do not attack the fire with fewer gallons per minute than are required.

■ **Note**
The hose line stretch may be quite lengthy due to the size of the structure, so the larger the hose, the less friction loss and more water available at the point of attack.

Hose lines should be placed to protect these areas and to support fire walls if they are present. The large open areas make confinement difficult as the fire can spread through convection to uninvolved areas of the building.

In multistory structures, extension of fire through windows is possible and hose lines should be placed to prevent this extension. Vertical openings are also an avenue that must be protected as part of confinement.

Extinguishment

Fires in these structures have the potential of being very large, needing a large force and large amounts of water. Do not attack the fire with less gallons per minute than are required (**Figure 13-7**). The initial lines pulled to be trained on the fire should be those that flow the larger amounts of gallons per minute. If and when multiple lines are working the fire in the same area, be sure that they do not oppose one another. Keep in mind that the hose line stretch may be quite lengthy due to the size of the structure. The larger the hose is, the less friction

Figure 13-7 *Hose lines of sufficient flow must be used to confine and extinguish fires in these structures. (Photo courtesy of Palm Harbor Fire Rescue.)*

loss and more water available at the point of attack. In some large buildings, horizontal standpipes may be available, as in multistory structures. If on arrival the fire is already so large that handlines will be ineffective, use heavy elevated streams, large handlines, and deck guns. Use any and all built-in protection systems such as sprinkler systems, fire doors, and air circulating dampers.

Ventilation

The location of the fire and the type of structure dictate the ventilation needs. Clearly in strip centers when the fire has extended into the cockloft, ventilation is required. Trench cuts may be necessary, but they require sufficient staff, correct tools, and equipment. It may be very difficult to vertically vent these types of structures as they may have a metal roof or a myriad of other construction materials that are more difficult to open up. With the proper tools, equipment, personnel, and training, it can be done relatively quickly. The quickest way to ventilate a strip-type structure is to do it horizontally with the mechanical assistance of fans used in a positive pressure method, but keep in mind that this action may intensify or spread the fire so actions must also be taken to prevent fire spread.

These large structures may typically have built-in roof areas that can be quickly ventilated, such as a skylight or roof vent. The ventilation opening needs to be as close to the main body of fire as possible.

Overhaul

■ Note

These structures also often undergo visual changes to try to keep the buildings looking new and attractive; therefore, voids and hidden spaces created by the renovation must be examined.

Not finding and extinguishing all hidden fire can have the department returning to an even larger fire. Many times these structures have multiple areas of high fire loading or offices that can hide pockets of fire. These structures also often undergo visual changes to try to keep the buildings looking new and attractive; therefore, voids and hidden spaces created by the renovation have to be examined.

The contents of the building may have to be removed in order to ensure complete extinguishment. Although this can be accomplished with relative ease in a residential structure, doing so in some commercial structures with large amounts of stock can be difficult. It may be necessary to request assistance of outside contractors for cranes or other heavy lifting equipment.

Salvage

■ Note

Materials in areas adjacent to the fire should be protected as much as possible from smoke and water damage.

Most of the materials found in these occupancies are items or commodities used in the business or that are for sale. There is potential for high dollar loss of contents. This requires that the incident commander assign the necessary resources to this function. Materials that are not exposed to the fire should be covered or removed. Materials in areas adjacent to the fire should be protected as much as possible from smoke and water damage.

SPECIFIC FIRES

Strip Centers

Construction One common type of commercial occupancy built over the last couple of decades is the strip center and strip malls. In general, these are one-story buildings built with masonry walls and various types of roofs. Depending on the age of the building and the building code at the time, fire walls may be provided between occupancies or when an area reaches a certain square footage. The fronts of these buildings have a lot of glass and may be secured after business hours with roll gates or shutters, as shown in **Figure 13-8**. If the fire involves a store in the middle of the strip, access is limited to the front and rear. The rear of these occupancies has doors, but some are nothing more than walk-through-type regular doors, although depending on the use, they may have a larger garage-type door.

Hazards Encountered During the initial stages, when these types of commercial buildings first began to spring up and gain popularity, in many areas the building codes were not what they should or could have been. This, along with a relatively new type of construction, left some considerable pitfalls that made firefighters' jobs much more hazardous and difficult.

Figure 13-8 *Example of security gates on a business. Extensive use of forcible entry tools may be required.*

■ **Note**

Attic space, though it allows for rapid fire spread, also has another major hazard of which every firefighter must be aware—the roof construction. Almost all of these structures have lightweight wood or metal roof assemblies. These types of roofs continue to plague the fire service due to their potential for rapid failure and collapse under fire conditions.

It is also important to understand that many of these occupancies are transient—what may have been a simple business office yesterday may be a small hardware store today—each of which has distinctly different potential hazards and fire behavior. This is where a good size-up can help operations. Clear communications between the prevention and suppression aspects of the organization are essential to alleviate these occurrences.

The strip centers often have an open attic space that runs horizontally over several businesses, if not the entire strip center. This allows fire to run unchecked horizontally. If the structure has a fire wall that runs up to but not through the roof, it may help to slow the fire spread but will probably not stop it. Do not completely count on this wall as nearly all of these roof decks are metal (**Figure 13-9**). This can allow fire spread through conduction or by the ignition and burning of the tar, which is a major component of the deck or wood deck, which in itself is combustible.

Attic space, though it allows for rapid fire spread, also has another major area of which every firefighter must be aware—the roof construction. Almost all of these structures have lightweight wood or metal roof assemblies. These types of roofs continue to plague the fire service due to their potential for rapid failure and collapse under fire conditions.

The lightweight wood truss roof assembly also significantly increases the fire load as the attic space is filled with match sticks, so to speak. When wood truss roofs, along with the connection points, are subjected to fire conditions, the truss

Figure 13-9 *Typical roof construction for a flat roof.*

Figure 13-10 *Steel trusses weaken and elongate.*

may burn completely through; the metal gusset plates may heat enough to pull away from the wood, or the wood may char enough that the gusset plate is no longer holding the connection point or points together, and the truss can fail.

The lightweight metal roof assembly can still be hazardous, although it in itself does not significantly increase the fire load (storage in the attic space must be considered for potential fire loading). When these roof trusses are subjected to fire conditions, the steel truss not only conducts heat to other areas of the structure, but it absorbs the heat, causing the steel to elongate (**Figure 13-10**) and or increasing the potential for a collapse of the roof or walls.

In either type of these roof assemblies, the big concern is sudden collapse. With these assemblies it is a common occurrence (and it should be expected) that when just one truss or truss section fails, the entire assembly has the potential to catastrophically fail.

The covered walkways that typically run the entire building above the entrances to the various occupancies are an additional hazard associated with this type of structure. Ordinarily these are merely decorative and offer no structural support to the main building itself. These covered walkways can be either cantilevered, as shown in **Figure 13-11**, or they may be supported on the street or parking lot side of the covered walkway by posts or pillars (**Figure 13-12**). Each of these types of walkways has its own inherent hazards.

The **cantilever** type is attached or supported by the main structure on only one end. This cantilevered roof assembly is basically an extension of the main roof and in this case, although decorative, this space is open to the attic and interior of the main structure. A hazard encountered with this type of roof assembly is that it is common practice for businesses to attach heavy signs or other advertising items that may add more weight to the structure than it was originally intended or designed for. This additional weight increases the

cantilever
an overhang supported on only one side

Figure 13-11 *A cantilevered covered walkway.*

Figure 13-12 *A walkway with column support.*

potential for collapse. When the beams are subjected to fire conditions, they can become significantly weakened, if not burned completely through, which can cause a collapse where the roof may fall or tip outward, much like a seesaw. With or without signage present, a collapse potential still exists.

The other type of roof assembly is one that is attached to the wall of the main structure on one end with the other end supported by posts or pillars. This type of roof assembly is merely decorative and is just an attachment to a wall or walls of the main structure. Typically there are no openings between it and the roof or interior of the main structure, but remember, wall breaches and channels or chases may exist due to the placement of business signs or the changing of business signs and other similar building modifications or renovations. Fires in these areas can burn without necessarily affecting the main structure. Actions must still be taken to ensure that the fire does not affect the main structure. These assemblies are typically bolted to a wall of the main structure, so once a fire affects and weakens the connection points, weakens the bolts, or weakens the posts or pillars along with firefighters applying large amounts of water, a collapse is possible and should be considered. This type of assembly is usually constructed of lightweight wood trusses but can also be constructed of steel trusses and again, when one fails, they all can fail.

Another way that these types of roofs are used above storefronts as covered walks is a combination of the two ways just discussed. This assembly looks like a cantilevered roof as it is connected to the structure on one end and the other end extends free outwardly with no posts or pillars used for support. Although it may look this way, it may not be the case. This assembly is also merely just an attachment to the wall of the main structure and is connected in the same manner as described previously. The outside end, which would be supported by a post or pillar, is instead supported by the same wall of the main structure from above. This support from above is typically provided by cables, wood beams, or steel beams.

Another hazard associated with this type of construction directly relates to the various occupancies. Many times, the various businesses within these structures are transient. Because of these frequent occupancy changes, the structure's interior (and sometimes exterior) is continually undergoing some sort of structural modification. For example, if a small office vacates a strip center and the restaurant next door wants to expand its dining area and takes over the additional space, some modifications are going to be made. The wall between the businesses will probably have large openings made to allow for passage from one area to another. Along with obvious structural changes such as this, other less obvious, but equally important changes are often made. These changes might include:

- Breeches of fire walls for duct work
- Poke-through for wiring or piping
- Openings for travel routes
- Pass-through for movement of stock or merchandise
- Openings to accommodate machinery operation

These various openings, along with others, do not occur only during and after building modification. They often occur and exist from the time construction begins.

Strategic Goals and Tactical Objectives

Firefighter safety Firefighter safety specific to strip malls should revolve around the dangerous construction features, including being aware of the potential for roof collapse and the ability for the fire to spread rapidly through the building. The incident commander needs to be proactive in strip malls with regard to firefighter safety. Do not wait for an event to happen and then react to it. Consider whether the firefighters are making significant progress on the fire. If they are not doing so within a reasonable time, then consider pulling crews out of the building before the roof collapses or the covered walk collapses.

Search and rescue Search and rescue for strip malls should focus initially on the unit that is burning. Search should then shift to adjoining units on either side and then the entire building. As the crews move away from the fire they may actually be doing more of an evacuation than a search due to lessened smoke and heat conditions.

A unique aspect of strip malls is that owners tend to modify the interiors. The modification can sometimes include actually building second-floor lofts for offices and storage. These areas may easily be missed during a search in a smoke-filled unit. If stairs are found in a one-story building, the search and rescue group should notify command immediately and then cautiously proceed with the search. Keep in mind that the loft may not have been built to any code and can therefore collapse without warning.

Evacuation   The whole strip mall needs to be evacuated. Occupants will not see the danger if smoke and heat is not impinging directly on their unit and they may be hesitant to leave their stores. All occupants, though, must leave. The potential for fire to spread throughout the structure via cocklofts and common attics is too great.

Exposure protection   When an exposure is threatened, the first efforts in controlling the fire must be to protect the exposure; otherwise, the fire may continue to grow and affect more and more property. With exposure problems in fires involving strip centers, the exposure(s) can be under the same roof, as shown in **Figure 13-13**.

When faced with fires in this type of structure you must consider adjoining occupancies as exposures and take the appropriate actions to protect them:

- Gain access to each adjoining occupancy and have a hose line.
- Open the ceiling areas that are adjacent to the main fire and protect from horizontal spread through the cockloft.
- If fire is already overhead and extinguishing attempts are having no effect, back out and move to the next furthest exposure and repeat the same actions.

Figure 13-13   *Other exposed occupancies may be under the same roof.*

■ **Note**

Gain access to the occupancies on both sides of where the main body of fire is. Pull the ceilings, open walls, check for any kind of opening between occupancies, ventilate, and keep in mind the unnoticed means through which fire can travel, such as through conduction.

Confinement Confining a fire in a strip structure can be very difficult due to some construction features. Gain access to the occupancies on both sides of the main body of fire. Pull the ceilings, open walls, check for any kind of opening between occupancies, ventilate, and keep in mind the unnoticed means through which fire can travel, such as through conduction. Many times a trench cut may be needed to help stop the fire spread. This operation is resource-intensive and can be quite a lengthy effort, so there must be enough resources allocated and the cut must begin far enough in advance of the fire to be effective.

Extinguishment The control of fires in this type of structure requires a rapid offensive interior operation when it is safe to do so. Fire attack and ventilation must take place at around the same time so the attack team can quickly advance while pushing out the fire. The fire attack must be made from the unburned side so as extinguishment takes place, the fire is forced back on itself and can be pushed out of the structure.

Ventilation Ventilation is also a very important operation that must take place during the initial operations. Positive pressure ventilation (PPV) is useful. To accomplish this, first decide which opening will be used to pressurize the structure or portion thereof. Then decide which opening you will use as the exit point of the heat, smoke, and fire gasses. The opening used for the exit point should be approximately one and half the size of the opening used to pressurize the structure. When choosing this form of ventilation, several items must be considered before implementing this action:

- Can it be done safely?
- In which direction is the wind blowing?
- Where is the fire?
- Will this effort spread and worsen the fire?
- Can the structure be pressurized, or are there too many openings?
- Will the exit point force the fire toward any exposures?

One other way to apply PPV in strip centers is to pressurize the adjoining occupancies but not the store that is on fire.

This is also potentially a good type of occupancy for which to conduct a trench cut, not only to ventilate, but to stop or reduce the spread of the fire. Trench cutting is personnel-intensive and somewhat time consuming as compared to other typical methods of ventilation. Firefighters must also keep in mind the building construction and the specific tools that will be needed to accomplish this to ensure the effort will be effective.

Overhaul In strip stores there are many voids and pockets in which a fire can hide. All of these hidden areas need to be sought out and opened up to provide total extinguishment. The attic must be completely searched to find any smoldering spots. This creates a seemingly unending number of void spaces, the potential for having more than one ceiling, numerous openings between occupancies, and areas that used to be openings to the exterior or interior, such as windows, being covered over. These areas need to be opened up and checked in the standard ways, such as opening up the wallboard or wall coverings to check for heat and charring and hidden fire extension.

Salvage Salvage could and should begin in the early stages of an incident and it can begin with something as simple as not breaking out that last window if there is no need. Remember, strip centers house more than one business, so by implementing good salvage operations some of the businesses can quickly reopen. With efforts such as this, the long-term benefit to the community is unmeasurable.

Large Commercial Structures

Large commercial structures (**Figure 13-14**) are also a very common type of building found in most communities. An example of a large commercial structure is a grocery store or supermarket. Every firefighter must have at least a basic understanding of what these structures are built of and how they are assembled.

These structures have been in existence for a long time, and they are becoming larger and larger as well as more common. The owners of these structures are building them so that they can utilize the most space to place as much product and accomodate as many potential customers and buyers in these structures as they can.

Figure 13-14   *An example of a large commercial structure.*

Figure 13-15   *The interior of a large commercial structure.*

Construction   The first, most obvious common construction feature of these structures is the large, clear, open spaces within them (**Figure 13-15**). In older structures of this type, the spans are not as large as those that have been constructed since the 1970s. These structures used the bowstring truss (**Figure 13-16**) or relatively lightweight steel to span the required distances. In the newer structures of the same type, improvements in technology, building methods, and building

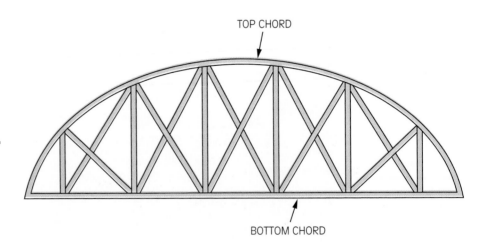

Figure 13-16   *The bowstring truss roof. These roofs are very dangerous under fire conditions.*

TOP CHORD

BOTTOM CHORD

materials have all combined to continually expand and extend the size possibilities. The building material of choice over the last few decades to provide for these large open spaces is steel. Steel can be extruded to just about any length and thickness desired, and multiple steel members can be joined together to extend to the open space desired.

The next common feature is the construction of the exterior load-bearing walls. Most of the older structures typically used concrete block and/or brick as the exterior walls structural material, whereas newer structures, those that have been constructed in the last few decades, have exterior walls made of large concrete slabs. These slabs may be precast off the building site and transported in, they may be precast on the construction site, or the walls may be formed up and then the concrete poured in place (**Figure 13-17**).

Having the structural walls precast, particularly off the construction site, allows the construction to be completed much quicker, thus allowing the construction costs to be lessened and the business to open sooner. Combining these improvements with the improvements made in the production and uses of steel has allowed for many large structures to be built.

Many structures use steel as the main construction material from the walls to the roofs. Basically it serves as the complete skeleton of the building, much like wood does for wood-frame homes. The steel allows for large open spaces with little to no roof support between the bearing walls. Depending on its designed strength, steel can be used in ways never imagined before.

Figure 13-17 *Precast concrete slabs may be used as walls.*

Figure 13-18 *A roof assembly over a large area.*

The roof coverings of these structures are either a metal deck, membrane, or pre- or posttensioned concrete. Rarely anymore is a building with a roof that spans a large area made of wood (**Figure 13-18**).

Although not as common, at least for most places, some areas still have structures of heavy timber or a mill type of constructed buildings made almost completely of wood. The structure may have a brick or other type of façade, but the main structural skeleton is made of and interconnected by heavy pieces of timber.

■ **Note**

Many increases and changes in building and fire codes have been developed to reduce the recurrence of some structural problems and disasters that have occurred involving this type of building.

Hazards Encountered As with the previously discussed strip-type structures, many increases and changes in building and fire codes have been developed to reduce the reoccurrence of some structural problems and disasters that have occurred involving this type of building.

The first hazards discussed are those involving the roof. Whether the structure is old or new, the roofs have certain inherent hazards inherent of which the firefighter must be aware. In older structures of this type, the roof assembly was typically a bowstring truss or constructed with lighter weight steel beams attached to the concrete block bearing walls running the desired span. The bowstring truss roof assembly is a hazard in itself. Like the prefab lightweight trusses of today, the truss is made of wood, and it significantly adds to the fire load and has the potential for rapid collapse. Also with this type of roof assembly, many occupants use the open attic space for storage, thus increasing the fire load and collapse potential. These roofs are typically covered with some type of membrane to keep out the elements. The steel beam roof is typically covered with a metal skin, which is topped with some form of rigid insulation, tar, and gravel. One of the problems with this structural feature is that the roof was originally designed to carry or hold a specific load. Another associated roof hazard is decorative parapets or banding at or above roof level. These are constructed of materials that range from Styrofoam, to rigid insulation, to concrete. Depending on the material of construction, the parapet can contribute to fire spread and toxic smoke exposure, particularly those made of Styrofoam. There is no use in these pieces other than decorative purposes, so they often fail early. It is important to open them up to see what you are dealing with.

As time passed, the ability to provide air conditioning to a large open space was improved. Many of these older buildings did not have air conditioning and were never designed to hold the added roof load of a large heating, ventilation, and air-conditioning (HVAC) unit. However, if these structures added air conditioning, they had no place other than the roof to put these units when they retrofitted their buildings for its use, as shown in **Figure 13-19**.

Obviously, these HVAC units add a significant amount of weight to the roof structure that was not originally intended. When the roof assembly is subjected to fire conditions, the steel absorbs the heat and is more susceptible to collapse due to the introduction of the added roof load. Even without the potential of components that may have been added to the roof to increase the load and cause collapse, the steel can bend and twist when subjected to heat and fire. This action

■ **Note**
Many of these older buildings did not have air conditioning and were never designed to hold the added roof load of a large HVAC unit.

Figure 13-19 *Roof loading is a factor in roof failures.*

could also cause a roof collapse and this often happens relatively unnoticed. The steel also has the potential to elongate by several inches, which could push out the supporting concrete block walls or portions thereof. This may cause not only the wall to collapse, but the roof structure as well.

The roof assemblies of newer structures of this type have basically the same hazards as the older ones. However, some different, new hazards have also developed. Many newer structures use concrete sections as the base of the roof. These concrete roofs must be stressed in some way to prevent them from cracking and falling down. This is done by running steel cables through the concrete and tightening them on ends with turnbuckles, which significantly increases the tensile strength of the concrete. The hazards encountered when faced with a roof like this are numerous. One hazard is the attempt to ventilate a structure that has this type of roof. If firefighters are cutting through the roof to ventilate and one of the steel cables is struck and cut, the results can be disastrous. When a cable is cut, it is under such tension that it will react much like stretching a rubber band until it breaks. The steel cable will rip through the concrete and destroy everything in its path. Obviously, a firefighter or anyone else in the path of this cable will be significantly injured, if not killed. Although this is a serious hazard associated with this type of roof, it is not the only one. The cables, as they are stressed, need to be maintained that way and attached on the outside of the walls of the structure. If fire impinges on these cables or fasteners over time, they will begin to weaken, which in turn provides the potential for failure, thus causing the possibility of roof collapse. Fire can also affect these cables from within the structure where areas of concrete have eroded away or undergone spalling, providing for the same potential effect and for the possibility for the fire to extend to other areas via conduction.

Strategic Goals and Tactical Objectives

Firefighter safety The walls can cause a hazard to firefighters for the same reason that a concrete roof can in the sense of spalling. Spalling of the walls can expose the embedded metal rebar, thus weakening the wall. The concrete is also a hazard to the firefighter, as the spalling concrete may strike and injure them.

The design of these structures can be a safety issue for firefighters. Because of owners' desire to have such wide-open spaces, these buildings have little to no fire-stopping walls within them to help prevent or limit fire spread. Where there are walls in these structures, poke-through, ventilating ducts, and other openings may exist that can allow fire spread, so efforts to locate and control these openings must be made. The size of these structures often requires the need for pillars or posts to help support the roof structure. These columns are often an I beam encased in concrete. If the concrete fails, the I beam can be exposed to heat (**Figure 13-20**). If the posts or pillars are made of concrete, the same spalling concerns exist, and if they are steel, then the concern of elongating, bending, or twisting exists.

spalling
The loss of surface concrete, sometimes forcefully, that occurs when the moisture in concrete begins to expand when the concrete is subjected to fire or exposed to heat

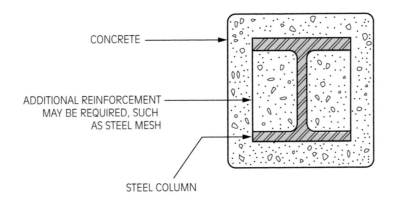

CONCRETE

ADDITIONAL REINFORCEMENT
MAY BE REQUIRED, SUCH
AS STEEL MESH

STEEL COLUMN

Figure 13-20 *Spalled concrete on an encased column will allow exposure to the steel, resulting in possible failure of the column.*

The next safety issue is the covered walkway. This assembly is attached to the wall, typically just the front wall over the main entrance and exit. The walls of these types of structures are typically very high. Thus, the walkway or entry coverings are usually only connected to the wall of the structure on one end and the other end is supported by posts or pillars. These walkways are merely decorative façades and are just an attachment to the wall of the structure. There typically are no openings between the walkway and the interior of the structure, but remember that poke-through and channels or chases may exist due to the placement of business signs and the like. Fires in these areas can burn without necessarily affecting the main structure, although the interior must still be investigated for extension. These assemblies are typically bolted to the wall of the structure, so once a fire affects and weakens the connection points, bolts, or the posts or pillars, a collapse is possible and should be considered. This type of assembly is constructed of lightweight wood trusses but can also be constructed of steel trusses and, again, when one fails, they all can fail.

In the heavy timber type of structure, large spans were impossible to attain. These structures have multiple supporting wood posts throughout the structure. The term *heavy timber* is exactly that; it was not uncommon to have a piece of timber that was 12" × 12" or larger as its outside dimensions. Although the typical definition is considered to be 8" × 8", the larger is common. Occasionally, under certain load requirements such as only carrying a roof, timber may be as small as 6" × 8". These structures withstand the effects of fire on the main structural components for long periods of time, and, with the fire charring or **pyrolyzing** the wood, the fire in itself helps the structural supports withstand being destroyed for longer periods of time. Charring tends to act as an insulator. Of course, the time needed to burn also depends on factors such as the heat of the fire, flammable liquids in the wood, and wind. These structures often have brick as the exterior wall, which is attached to the main structure at multiple places. It is common for these exterior brick walls to quickly fail and come crashing down.

pyrolyzing
a chemical change in wood resulting from the action of heat

■ Note
In the wake of fires
such as Hackensack
Ford and Waldbaum's
Supermarket, it is
inexcusable for an
incident commander
and on-scene
supervisors not to
recognize the truss as
an obvious safety
hazard.

The final firefighter safety issue is that of truss roofs. In the wake of fires such as Hackensack Ford, it is inexcusable for an incident commander and on-scene supervisors not to recognize the truss as an obvious safety hazard. Do not needlessly expose firefighters to this risk to save property. Putting firefighters on or under a truss roof where there is significant impingement on the roof members and no progress is being made on the fire is equivalent to having firefighters fight a car fire on a train track with an approaching train. Sooner or later, someone dies.

Search and rescue Search and rescue is extremely difficult in large commercial structures. A first step should be to make contact with the store owner or manager and determine whether anyone is in the building. If so, try to determine the person's likely location and initially focus the search in that area.

Search groups should search with lights and a rope or hose line. It is very easy to become disoriented in a large structure even if visibility does not appear to be poor. A high ceiling in this type of buildings can mask the typical signs of fire spread and flashover conditions.

Evacuation As with strip malls, the entire building should be evacuated. People generally do not leave a building because a fire alarm has activated until they see smoke or flame. Then panic ensues. Initial engine companies responding to a fire in a large store may be faced with blocked roads and entrances from fleeing customers attempting to evacuate.

Exposure protection These large buildings may have many different internal operations taking place under the same roof, such as offices, production, and storage, as shown in **Figure 13-21**.

These various areas within the structure should also be considered as exposures, at least from the standpoint of the tactical objectives. These structures are

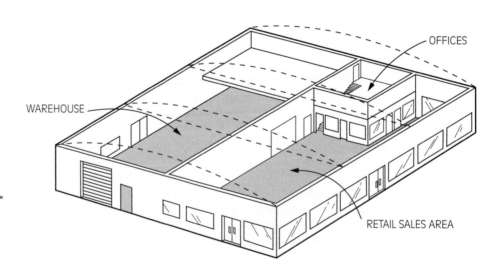

Figure 13-21 *Inside configuration of a large commercial structure.*

OFFICES

WAREHOUSE

RETAIL SALES AREA

commonly isolated from other buildings and may not have a significant external exposure problem except for flying brands. Remember that fires in these structure have the potential to be very large fires and require a very large force with large amounts of extinguishing agent to prevent fire spread to exposures. Large-flow master streams and elevated streams should be placed to protect external exposures on all sides when possible. Proper apparatus placement to meet this objective is critical.

Confinement and extinguishment For interior attacks, hose lines must be positioned to confine the fire to the area of origin, and a planned attack must take place. Often there is a need for numerous hose lines to be operating: Great care must be used to avoid opposing streams. The first line should be placed between the fire and the means of egress, whether that be a hallway or stairway. Backup lines are necessary to ensure adequate flow and to cover internal exposures.

Ventilation When built-in openings are available, they should be quickly utilized if they are located in the area near the fire. Most of the time, vertical ventilation, if possible, is the best option for venting these large area structures. Positive or negative pressure is often ineffective because of shafts, high ceilings, and numerous openings. If it is utilized, it must be done with a plan. With positive pressure ventilation a large opening is needed to place multiple high-volume fans in front of the opening. The exit opening should be high and remote. Additional fans can be used within the structure to keep the air moving in a specified direction, but do not place them in different or multiple directions, as this will only disrupt the process and make it ineffective.

Vertical ventilation can also be performed over skylights and stairway bulkheads to assist in keeping the stairwells free of smoke. Horizontal ventilation also helps increase visibility and survivability for those within the structure, but in structures such as malls, there are few openings to the outside that would be well suited for this specific purpose.

Overhaul Overhaul is difficult due to the characteristically high ceilings and the amount of stock in the buildings. Ladders must be used to reach high areas and to look for extension above hanging ceilings.

The overhaul of the stock will depend on the type of merchandise. For example, furniture stores may have several smoldering couches that must be overhauled.

Salvage If enough personnel are on the scene during the firefighting operation, salvage can begin with placing salvage covers over stock or moving stock away from the fire. Also, firefighters can help to reduce damage from water by creating chutes to channel water away from water-sensitive merchandise. In large commercial buildings, the salvage operations may begin as soon as life safety issues are taken care of and run concurrent with the incident stabilization.

Two- or Three-Story Commercial Structures

Two- or three-story commercial structures (**Figure 13-22**) are really somewhat of a combination of strip malls and large commercial structures. They have many of the same construction features and hazards. Examples of this type of commercial structure are an office building, taxpayer, or mall.

Construction The first, most obvious common construction feature in commercial structures is how they have been designed and engineered. Within the past few decades it has been the architects' and builders' intent to design these buildings to be appealing to shoppers and visitors. Before strict building and fire codes were enacted, many structures were built very poorly.

As construction materials and methods have continually improved, the roof spans that are attainable continue to grow. Usually, the roof has steel as its basic structural material; however, it can be made of wood trusses. Although these structures do not necessarily all have large open spans, many do. Again, in older structures of this type, the bowstring truss or conventional wood-frame truss may be present to span the required distances in portions of the structure such as the roof assembly of a taxpayer or the roof of an anchor store in a mall.

In office-type structures the roof is often supported in many locations. It is important to have an understanding of the materials used to create the support. Steel, wood, and concrete columns are common in structures built in the

Figure 13-22 *A typical three-story commercial structure.*

twentieth century, whereas wood or cast iron could be the roof support in structures built before then. The roofs of these structures are typically flat, made with a metal deck, insulation, tar, and gravel, but in older structures the main component of the roof is wood. The roofs can be pre- or posttensioned concrete. Buildings with a roof that spans a large area made of wood will be designed with trusses, wooden I beams, or wood truss joists.

The next common feature is the construction of the exterior load-bearing walls. Most of the older structures typically used concrete block and/or brick as the exterior walls' structural material, whereas newer structures, those that have been constructed in the last few decades, have exterior walls made of large concrete slabs that are cast on or off the site with steel utilized to tie them together and support them vertically (this type was described in the section on large commercial structures). Because of its many uses and improvements, many structures use steel as the skeleton of the building. This skeleton is then wrapped with a steel skin, glass skin as in **Figure 13-23**, or just about any other construction material in use. It has also become increasingly popular to convert old milltype structures into buildings that house these types of occupancies. Also, as previously described in the section on large commercial structures, these structures are made of heavy timber that is modified or covered up to provide it with a more modern look.

Figure 13-23   *A building with steel framework covered by glass.*

Another common construction feature is the flooring. Probably the most common type of floor is concrete. The floors can be poured in place and supported by some sort of post or pillar from the floor below, or the floor may be pre- or postcast concrete and also supported from below, although fewer supports typically are required. The other type of floor construction is wood. In both the taxpayer and mill type of structure, the main floor structure and surface is made of wood, although it is usually covered with a more visually appealing surface.

The next common feature of building construction is vertical openings. With structures of more than one occupied story, there definitely must be some open vertical chases or paths that run between floors and possibly the entire vertical height. These vertical openings could include:

- Escalators and/or stairwells
- Elevators for occupants and freight
- Ventilation shafts and/or ductwork
- Open chases for piping and electrical service
- Conveyer belts between floors to move stock

These structures may not very often have basements, but they probably have some machinery such as a trash compactor below the surface of the first floor. The taxpayer type of structure has a basement that, much like the attic, is open from one end of the structure to the other.

The ceiling of the basement may be constructed of wood, concrete, or steel or any combination. The floor is supported in many places by posts or pillars that also are constructed of wood, concrete, steel, or cast iron.

Hazards Encountered Whether the structure is old or new, the roof may have certain hazards from poor construction, building modification, poor maintenance and upkeep, or other problems. On the roofs on commercial structures, firefighters must continually be aware of heavy HVAC units, exhaust fans, and the like. As structures continue to be built, the competition for space and the complete utilization of available space mean that this type of building equipment will probably be on the roof. We begin with the roof and attic space on the taxpayer type of structure. As with the previously discussed strip-type structures, a big concern is the open attic spaces that run horizontally over the structure. These types of structures rarely have any type of fire stopping walls, allowing the fire to run unchecked horizontally. Unlike in the strip center-type of attic space, those attic spaces in the taxpayer are often used by occupants for personal storage areas, thus increasing the already high probability of rapid fire spread. If the structure happens to have a fire wall, it will be run up to but not through the roof. Remembering that these roof assemblies and roof decks are mostly made of wood, once the fire burns through the roof, it may easily extend to the other areas of the roof through radiated heat or fire brands. These roofs will be supported by the structure-bearing walls. The roofs of the taxpayer are typically constructed using conventional framing methods. In this type of construction method, the wood sizes used are typically larger than the now normal two-by-four wood sizes, and they were secured using nails and spikes rather than a gusset plate, making them more stable. Modifications to these structures over the years, such as placing air conditioning units on the roof, may increase the load to well above what the structure was designed to carry. Many of the older malls utilized the bowstring truss roof to allow for a more open space on the sales floor for displays and shoppers. Although not all that common in these structures, the bowstring truss adds to the fire loading of the building and has the potential for rapid collapse.

The steel roof is typically covered with a metal skin topped with some form of rigid insulation, tar, and gravel. The hazards associated with this type of roof assembly were described previously in the section on large commercial structures. Those same hazards exist when dealing with these, even without the components that may have been added to the roof to increase the load and collapse potential. When subjected to fire, the steel can bend, twist, and elongate, which could also promote a roof collapse. This may happen relatively unnoticed. The steel, as it has the potential to elongate by several inches, may push out the supporting walls or portions thereof, causing not only the wall to collapse but also the roof structure. Builders are also covering these roofs with a stretched rubber membrane that basically resembles the inner tube of a bicycle tire. These roof membranes have a very spongy feel to them that would normally indicate potential failure, but this is the way this system is normally. This type of roof surface is difficult to cut through, as it gums up saw blades.

Many of the relatively newer structures use concrete sections as the structural base of the roof. The firefighter must be sure to avoid embedded cables to prevent potential failure that might occur if one of the cables is cut during ventilation. Fire can also affect these cables from within the structure where areas of concrete have eroded away or spalled off, allowing for failure and the possibility of the fire extending to other areas via conduction.

Concrete walls can be a hazard for the same reason that a concrete roof can be, due to spalling. Spalling can expose the metal rebar used to provide the wall with more strength, thus weakening the wall. The concrete is also an overhead hazard to the firefighter—as the concrete is spalling, chunks may strike and injure them. The walls of a mill type of structure have heavy timber surrounded by brick as the main structural component. The brick is attached in various areas to the structure with metal strips, which are susceptible to burning away or failing, thus allowing the brick wall to come tumbling down.

Because of building owners' desire to have such wide open spaces, these structures have few or no fire stopping walls within them to help prevent or limit fire spread. Fire can spread via poke-through, ventilating ducts, and any of the other openings or concealed spaces.

Do not forget that structures often require the need for pillars or posts to help support the roof structure. If these are made of concrete, the same spalling concerns exist, and if they are steel, then the concern of elongating, bending, or twisting exists.

If there is a basement, the primary hazard is that there will be limited access and egress. Fires in these open basements quickly extend and advance, deteriorating the main structural components located in them, such as the floor of the first floor and the posts or pillars. Be aware of and check for sagging floor sections by looking at drawers of file cabinets or shifting out of plumb baseboards.

For the most part, these structures do not have the hazard of covered walkways. This does not mean that they do not exist, so the precautions described earlier still must be taken and hazards considered. The most common types of covered walkway are those attached to the wall of the structure on one end and supported by some sort of pillar or post on the other end, or where the roof is attached to the wall on one end with the jutting end being supported from above. Again, these are merely decorative.

Strategic Goals and Tactical Objectives

Firefighter safety When multiple attack teams are utilized in the same area, be sure the hose lines do not oppose one another and never have exterior heavy streams work in the area of interior crews. Be sure to have a backup plan in place if initial operations fail, such as placing lines, personnel, and apparatus ahead of the fire.

Search and rescue The large area of the buildings makes search and rescue demanding. The incident commander must make sure that adequate resources are on the scene to perform effective search and rescue. The task should be broken down into the different areas of the building. For example, there may be three search groups for a three-story building, one on each floor.

Evacuation Evacuation of a two- or three-story commercial building is usually easier than it is for large commercial structures, because most occupants work at the building or frequent it often. They have a better understanding of the layout of the building and evacuate better than hundreds of customers at a grocery store.

Exposure protection The immediate exposures are usually the floor just above the fire or the adjoining rooms or hallways. The exposure in the most immediate danger depends mostly on the construction type.

Confinement The size of these structures and the number of persons in them often require large parking areas that in many cases can cause access problems for firefighters and create situations in which long hose lines stretches are needed. The first lines need to be placed to prevent horizontal spread of the fire. It is imperative to get ahead of the fire and begin the operations of opening walls and ceilings to keep the fire in check. Fire loading and materials, even on upper floors, may require larger lines and or portable deluge guns to keep the fire confined.

Extinguishment Once hose lines are in place on the fire floor, they must be positioned on the floors above, and areas must be examined to determine whether vertical extension has occurred. Remember the top-most floor or the attic space when considering hose line placement. Large and multiple lines should be used in these areas, for if any fire is discovered, it must be quickly contained. The known vertical openings must be quickly accessed and protected, especially the stairwells, which are escape routes for the occupants of the structure and protected egress for companies operating on upper floors. Other than the known obvious vertical openings, hidden openings, such as chases for utilities, must also be located and controlled. Fire can spread vertically both within the structure and on the exterior. The way to prevent autoexposure is to train hose streams on the exterior of the structure above where the fire is venting and attempting to vertically spread. The water will cascade down the structure and significantly decrease or completely stop the fire's exterior vertical spread. Fortunately, fire codes are an important ally to the fire service, because they are continually improving in ways such as requiring smoke-proof and pressurized stairwells, sprinklers, standpipes, and automatic alarms.

Ventilation As with the large commercial structures, vertical ventilation should be used over stairwells to remove smoke and combustion products. Horizontal ventilation can be used on the fire floor and positive pressure ventilation can be of benefit as well. Be careful not to push the products of combustion into other areas of the structure, and ensure that whatever combination of ventilation methods is used, they are working in conjunction with one another rather than in opposition. It is possible to use positive pressure fans to pressurize stairwells to help to keep them clear of smoke and other products of combustion so occupants have a better escape route and firefighters have better operating conditions.

Overhaul The construction type dictates the direction of overhaul. Common to all these buildings are the vertical shafts, which must be meticulously checked.

Salvage Salvage for these buildings uses the same tactics as most other commercial structures.

Stand-Alone Commercial Occupancies

■ **Note**
Examples of this type of occupancy are a convenience store or a corner bar or pub.

As compared to the previous commercial occupancies described in this chapter, the stand-alone commercial occupancy is relatively uncomplicated. Examples of this type of occupancy are a convenience store or a corner bar or pub (**Figure 13-24**). These commercial occupancies are extremely popular and can be found in just about every area of the country. Additionally, manufactured buildings are becoming popular commercial occupancies in some areas, as they are relatively cheap and, more important, very quick to make serviceable as compared to typical construction times. Manufactured occupancies show up typically as offices, such as on a construction site, and can be compared to a fixed single- or double-wide mobile home. Refer to Chapter 10 for more specific

Figure 13-24 *Stand-alone small commercial structures.*

information on manufactured structures. These structures are typically small, but that does not mean they are any less difficult or dangerous.

Construction Stand-alone commercial occupancies are basically of simple construction. These structures were originally built of wood but have progressed to concrete, brick, and steel, although wood is still often used.

Old or new, many of these structures use wood as the main structural component of the roof. The older structures utilized wood that is typically larger than the two-by-four of today. Commonly, the attic of these buildings is open throughout the structure, with no fire wall between the body of the store and the office or stockroom. Most of the time the roofs are flat and, because the building is small, the load-bearing walls are supports for the roof assembly.

The walls of these structures are typically constructed of concrete block but may also be constructed of wood with a brick facade, or they may be made of pre-cast concrete as is becoming more and more popular.

The floors are usually concrete, although they may also be made of wood, especially if there is a basement. In newer structures of this type, the main floor beams may also be made of steel and then the floor surface may be wood or poured concrete.

Many of these structures could have at one time been a home and converted to a business such as an ice cream shop, bar, or restaurant. If this is the case, then it will be constructed with the same materials and in the same manner as a single-family dwelling, as described in Chapter 11.

Hazards Encountered The roofs of these buildings are typically flat, having an open attic throughout the structure. Those constructed of wood can obviously burn through and those constructed of steel can sag, twist, or expand, increasing the potential for collapse. These structures often have heavy equipment such as HVAC units on the roof, which can also contribute to a potential collapse. These structures also often have a large overhang or covered walkway that is cantilevered as described in the section on strip structures. There is also the possibility of a lightweight wood truss roof, especially if the structure has been modified from a dwelling unit. Drop ceilings are also very common in these structures and may conceal fire or other hazards.

The walls, commonly of concrete block, have few openings other than on the front of the building. If the walls are constructed of wood with some type of facade such as brick, then the potential of the facade collapsing must be considered. Walls that use steel as the structural component can expand, twist, or buckle and can promote fire spread through conduction.

The floor may have the same hazards as those associated with the roof if the structure has a basement or is elevated above grade. Unlike the roof, the floor will have several posts or pillars supporting it because of the heavy loads it is intended to hold.

Strategic Goals and Tactical Objectives

Firefighter safety Firefighter safety is compromised if these fires are taken for granted. The small stand-alone building can be a death trap for firefighters. Many convenience stores are overpacked with stock and have tiny offices and storage places that can trap firefighters.

Search and rescue Usually the techniques used for residential fires are sufficient in these buildings. However, the layout and occupant use of the building dictates the search type. Safety ropes or searching off hose lines may be required to ensure that firefighters are not disoriented inside structures that have many shelves, racks, and display cases.

Evacuation Evacuation of the building is necessary but is usually not difficult. Generally, the occupancy loads of these buildings are small and people do not have a tendency to linger in them for long periods of time. Depending on height, proximity, and the wind speed and direction, the evacuation of exposure buildings may be required.

Exposure protection These occupancies are commonly nestled within congested areas so they are, as some of their names imply, convenient. Because of this, when fire occurs, there are often exposure problems. The heavy fire loads common in these structures create a lot of radiated heat. Lines must be positioned between the fire and the exposure, with a continuous effort to limit spread by soaking points of combustibility. Access to the exposed structures must be made to ensure that fire has not extended into them.

Confinement As these structures are often open with few partitions within them, the fire may not be confined to just one room. They may have small rooms within, such as an office or restroom, but they usually are not contained by a fire wall. A fire occurring in one of these rooms can be contained by opening the ceiling around it to stop the fire spread. If the fire is below grade, access can be difficult and hazardous. Be sure to protect stairwells and other vertical openings, such as pipe chases. As most of these structures have flat roofs, there is the possibility of the fire within the structure heating a metal deck roof enough for it to begin and sustain its own combustion. From below, check the bottom of the roof for the signs of this. By directing a stream on these areas, the fire can be contained.

Extinguishment These structures rarely have any built-in protection other than the occasional fire alarm, so the fire can rapidly extend. A quick and aggressive interior attack often controls the situation. Advance lines that are capable of flows of at least 125 gpm. A drop ceiling can hide a fire and contain it in such a way that there is a potential for a backdraft when a portion of the ceiling is opened. Keep aware of the signs of a potential backdraft.

Ventilation Ventilation can be accomplished quickly in these small structures by using the positive pressure method. Be sure this ventilation will not push the fire into uninvolved areas or implement it before attack teams are assembled to quickly extinguish the fire. By implementing this method of ventilation, the structure can be cleared in just a couple of minutes. Often large portions of the front (main entrance side) of these buildings are glass. Removing this glass can also quickly ventilate the structure, but the removal of these large windows negates the option of pressurized ventilation. If vertical ventilation is required, check the roof for stability and use any existing openings such as skylights. Open the roof as close as possible to the base of the fire to help slow its progress.

Overhaul Overhaul can be done by simply removing the affected items from the building. If large areas are involved or if there is the possibility that the fire was caused by an act of arson, then this cannot be accomplished. Overhauling these structures should be done in the same manner as those described previously.

Salvage Salvage is always important—more than just covering or removing product or property, care should be taken to ensure security to the structure, contents, and money that may be within. There may be many refrigerators or other dispensing machines that can be salvaged simply by protecting them from water, whether it be from puddles on the floor to hose streams during overhaul.

SUMMARY

This chapter described a number of different types of commercial occupancies that firefighters may encounter. The descriptions are by no means all-inclusive, but they provide types that can be compared with other structures you may be familiar with. The hazards encountered are some of the most common. The best way to prepare for the building hazards is through preplanning. It is important that firefighters take the time and effort to thoroughly look at all components of the structure and its contents. It is best to begin when the

structure is being constructed. The three incident priorities of life safety, incident stabilization, and property conservation never change, although the way they are attained may combine. For instance, the initial action of attacking a fire rather than beginning a search may be the best option to provide for life safety because of the lack of resources or distance that they are traveling to the scene. Every incident is different and should be treated accordingly, but the actions listed within this chapter have proved effective time and time again.

REVIEW QUESTIONS

1. Describe a small commercial structure.

2. List two considerations for search and rescue at commercial building fires.

3. List two types of occupancies that might be found in a large commercial structure.

4. List four hazards associated with fires in strip-type structures.

5. True or False? A gusset plate is a thin piece of metal stamped to create multiple spike-like protrusions that are used at all of the connecting points of a lightweight wooden truss.

6. A cantilevered beam would be described as

   A. A beam that protrudes horizontally from a structure beyond an exterior wall and is supported on both ends.

   B. A beam that protrudes horizontally from a structure beyond an exterior wall and is supported by cables from above.

   C. A beam that protrudes horizontally from a structure beyond an exterior wall and is unsupported.

   D. None of the above.

7. The incident priority of life safety at a commercial building fire includes

   A. Firefighter safety

   B. Search and rescue

   C. Evacuation

   D. All of the above

8. True or False? Positive pressure ventilation is always more effective than vertical ventilation.

9. True or False? In two- or three-story commercial structures, vertical ventilation should be used over stairwells to remove smoke and heat.

10. True or False? HVAC added to a roof will generally not increase the hazards associated with roof collapse.

ACTIVITIES

1. Visit a structure similar to one of those described in this chapter and conduct a prefire plan of the structure to include the following:

   Type of construction of roof, wall, floor, and interior

   Number of floors

   Location of stairwells, elevators, air shafts, and openings for utilities

   Type and location of built-in fire protection

   Potential number of occupants and variations for time of day

   Potential number of occupants that may be of some special need or help

   Distance to exposures and exposure type

   Water supply location and distance

   Built-in roof openings

   Location of utility and machine rooms

   Location of fire walls

2. Review your departments operational standard operating guidelines (SOGs) in reference to fires in commercial buildings. What changes may be needed? What training may be needed to attain the objectives of the SOGs?

Chapter 14

PLACES OF ASSEMBLY

Learning Objectives

Upon completion of this chapter, you should be able to:

- Define what constitutes a place of public assembly.
- Identify the types of construction for places of assembly.
- Understand the inherent life safety problems in these buildings.
- Know the different hazards associated with firefighting in these structures.
- List the strategic goals and tactical objectives in handling a fire in these structures.

CASE STUDY

In the early morning hours of March 25, 1990, the intentionally set Happy Land fire claimed the lives of eighty-seven persons in a neighborhood club in Bronx, New York. An angry patron and one dollar worth (¾ gallon) of gasoline thrown in the front door turned this two-story nightclub into an inferno. The majority of the dead were found throughout the second floor where they had no warning and no escape. Sixty-nine of the eighty-seven victims were found on the second floor where they had succumbed to the carbon monoxide-laden smoke. There was nothing the fire department could do, as the deaths had occurred before their arrival.

In the late hours of February 20, 2002, the West Warwick, Rhode Island, fire department received multiple calls for a fire at the Station Nightclub. As the last band began their concert that night, they set off a series of stage pyrotechnics that ignited foam sound curtains behind them. At 11:08 P.M., the flames rolled up the walls on stage. Within 30 seconds of the flames erupting, the band stopped playing and the crowd began to evacuate. A building with an occupancy load of 300 had more than 400 persons inside, all trying to escape through exits that were too small and too few. Within 36 seconds after the flames began, three cell phone calls were received reporting the fire. Within 1 minute and 30 seconds, thick, black smoke poured from the windows and appeared to be at floor level. People trying to escape piled up at the doorway; no one else could escape through those doors.

The first West Warwick fire engine with a two-person crew arrived on scene within 5 minutes of the first 911 calls. Three additional engines, one ladder truck, one rescue unit, and a battalion chief, arrived in sequence afterwards. But it was already too late for many patrons. All officers and officer candidates should be required to read the NIST Technical Reports on these fires, as they detail many points that will help an officer in decision making if faced with an incident of this magnitude.

INTRODUCTION

What is a place of assembly? According to the criteria set by the National Fire Protection Association (NFPA), it can be defined as a structure where large groups of people gather, as shown in **Figure 14-1**.

The most important factor to consider is life safety in these structures. Because the concern is the safety and hazards of large numbers of people gathered in one facility, an established minimum number has been set to constitute an assembly occupancy. Although this number can vary, most building codes and NFPA 101, *Life Safety Code,* use a minimum occupancy load of fifty. If the occupancy has the same designated use but less than fifty people, it falls under another occupancy category, such as business.

■ **Note**
The most important factor to consider in a place of assembly is life safety.

Figure 14-1
Assembly occupancies, by design, are made to hold large numbers of people, often in large, open areas.

■ **Note**
The occupants are present voluntarily and are not ordinarily subject to discipline or control. They are generally able-bodied persons whose presence is transient in character and who do not intend to sleep on the premises.

■ **Note**
The major safety problem associated with these structures lies in the large group of people and the population density within these structures.

The definition for *public assembly property* provided in the *National Fire Incident Reporting System* (NFIRS) *Handbook* is "places for the congregation or gathering of people for amusement, recreation, social, religious, patriotic, civic, travel, and similar purposes." Such properties are characterized by the presence or potential presence of crowds, with attendant panic hazard in case of fire or other emergency. They are generally open to the public or may on occasion be open to the public. The occupants are present voluntarily and are not ordinarily subject to discipline or control. They are generally able-bodied persons whose presence is transient in character and who do not intend to sleep on the premises.

The criteria set by the NFPA for an assembly occupancy includes any place where a large number of people congregate for the purpose of entertainment, worship, meet, or await transportation. This definition includes restaurants, nightclubs, concert halls, meeting halls, theaters, convention centers, sports arenas, and transportation centers, to name a few.

The major safety problem associated with these structures is the large group of people and the population density (**Figure 14-2**). According to the NFPA, the population densities can reach 5 square feet or less per person. Anyone who has attended a concert and observed the floor seating can understand how population density can create such significant concerns. In many nightclubs and dance halls, the floors are packed tight with people, sometimes making it nearly impossible to move freely around the facility.

These density factors affect the behavior of the patrons, the capacity of the exits, and ordinary movement on stairs and along hallways and corridors. Think about the last visit you made to a sporting event or theater, and remember the

Figure 14-2 *The population density is a significant life safety concern.*

wait to get out the exit, the crowding, and the bodies pushed together. Now imagine this in an emergency situation with smoke and heat bearing down on the crowd—you can imagine how chaotic conditions can become.

In addition to the crowding conditions, the patrons occupying assembly occupancies are typically not frequent visitors. Therefore, they usually are not familiar with the building layout, the pathways to the exits, the location of additional exits, or other safeguards. They have not taken the time to review the posted fire escape plans, and they have not thought to look for additional exits in case the nearest one is blocked. Most people respond to what is comfortable for them (**Figure 14-3**) and that is to enter and exit by the same route. To leave through any other exit would not feel comfortable and therefore they are reluctant to leave until they reach their exit. In a stadium fire in England, patrons trying to escape a fire would not go over the fence onto the field of play because they had been taught it is not right to enter the field.

For these reasons one of the most difficult issues in public assembly fires is that the damage to human life will most likely have occurred before the fire department arrives. In the tragedy at the Coconut Grove fire in Boston, the fire units were only a block from the structure when they received verbal notification of the fire. A passerby pulled a box alarm for a burning auto. This box alarm brought four engines, two ladders, one heavy rescue, one tower ladder, one division, and one district chief. Upon the arrival of the chiefs, the auto fire

■ **Note**

In addition to the crowding conditions, the patrons occupying assembly occupancies are typically not frequent visitors.

Figure 14-3 *In this picture at least three possible exits are visible. Human nature is that people attempt to leave by the exits through which they entered.*

had been extinguished and the companies were either in the process of taking up or returning to quarters when someone came running down the street yelling that the Coconut Grove was on fire. The deputy chief ran down the street and found heavy smoke coming from the structure. He immediately called for a third-alarm assignment, skipping the first and second. When flashover occurred and flames enveloped the building, he called for a fourth alarm. In only a matter of minutes, most of the fire had been extinguished, but the damage had already been done. There were bodies piled up at the exits and throughout the building.

The Happy Land fire in a Bronx, New York, social club is the second most deadly fire in that city's history. The fire was quickly darkened down by first-arriving crews, but the deadly effects of the fire claimed eighty-seven lives in a very short period of time. The same is true of the Beverly Hills Supper Club, the Station Nightclub, and many other tragic public assembly fires.

HAZARDS

The biggest problem with these and all fires is smoke. When you have hundreds, even thousands, of people gathered in the same structure with smoke spreading throughout the building, the behaviors of these people change quickly. Overriding smoke with a flame front can travel faster than a person can run. These conditions can quickly incapacitate the occupants, causing them to collapse and block the travel of other occupants. In all the case studies of occupied public assembly fires, there were occupants who were caught, trapped, stampeded, and overcome by smoke while trying to escape.

The idea that smoke spread creates widespread panic among occupants is not completely true. In case studies of the Beverly Hills Supper Club, there are reports of people sacrificing themselves to save others. People stayed in the smoke and assisted those who were having difficulty escaping.

Another problem is the building's layout. Patrons may have difficulty in reaching exits or may not use the closest exit. In many public assemblies where there is fixed seating, the occupants must go up or down long sets of stairs to reach their seats. This makes quick escape difficult, as the patrons must move through small aisles and into the flow of people moving up or down the stairs.

Depending on the type of system, the heating, ventilation, air conditioning system (HVAC) can help to either remove smoke or move it into uninvolved areas. The HVAC system can also give a false sense of what you are seeing in the initial size-up. In the MGM Grand Hotel fire, the smoke from the fire was rising out of a superstructure located well away from the actual fire. The smoke was being carried away from the hotel and exhausted out of the air vents away from the actual building.

Travel into the building's driveways and parking areas may be hampered by those trying to leave. There will be people who are determined to get out no matter what chaos or congestion they cause. This will impede your ability to respond (**Figure 14-4**). That is why it is very important to have preset staging areas for the first-responding apparatus.

■ **Note**
A problem to deal with is the number of people in an unfamiliar building who want to get out *now!*

A problem to deal with is the number of people in an unfamiliar building who want to get out *now!*

For the purposes of this chapter, the scope of assembly occupancy is limited to churches, exhibit halls, sports arenas, showplaces, and large nightclubs.

CHURCHES

Construction

The construction features for churches and all places of assembly are many and varied. They can be brick and heavy timber wood frame, such as older churches

Figure 14-4
Apparatus may have a difficult time accessing this building.

and assembly halls, or block and steel trussed, such as many older auditoriums and churches with large attached family centers. With this information in mind, it becomes the obligation and responsibility of the local jurisdiction to visit the various churches in its response area and determine the construction methods and materials used (**Figure 14-5**). Find the inherent characteristics that determine the building's ability to withstand exposure to a fire and how it will limit

Figure 14-5   *Typical church.*

or increase fire growth and smoke spread. In some cases, there is a mix in the construction types. Remodeling and additions to older churches result in mixed building construction styles. Evaluate the building's ability to channel smoke out of the building through skylights, ventilation shafts, and smoke corridors. Evaluate the exits for the quick removal of the occupants. Determine whether the exits could become blocked from a sudden on-rush of occupants trying to escape quickly in an orderly fashion. Where do the exits channel the occupants? Will the discharge of the occupants interfere with fire operations? Will the discharge of the occupants interfere with the placement of arriving apparatus? Will the routes into the facility be compromised by vehicles exiting the parking lots? Is the building totally or only partially sprinklered? Are there large, undivided public areas not equipped with sprinklers where fire can gain considerable headway? Preplan the facility and add the facility to the "target hazard" list that requires more than the average first-alarm assignment.

Hazards

The size and construction type for churches varies greatly. In rural areas we find small, wood-frame churches; in the suburbs there are large, stand-alone churches and those in strip shopping centers; and in the cities we find large gothic churches. Of course, each of these styles could be found anywhere.

■ **Note**

Churches are no longer occupied only during Sunday services.

   A common element among all churches is the life safety concern. Churches are no longer occupied only during Sunday services. There are many reasons for people to be in the churches at any time, any day (**Figure 14-6**). Daycare centers, Bible study groups, choir practice, Boy Scouts, Girl Scouts, and other civic groups hold meetings and occupy churches. This requires a quick size-up of the parking lots and interior of the church.

Figure 14-6   *Church with a daycare center attached. This occupancy has the potential to have occupants every day of the week.*

A second hazard in church fires is the open avenues for quick fire spread. Heat can and will become trapped in the high ceiling area of the sanctuary and build into flashover conditions before it is felt from below. Many churches have basements and lofts. If a fire is to be controlled, the incident commander must determine where the main body of the fire is located and work toward that area. This is difficult with the way many churches are divided into separate classrooms, meeting rooms, nurseries, changing rooms, closets, and other areas.

EXHIBIT HALLS

Construction

The size and construction type for exhibit halls varies greatly. Most are protected by some type of fire alarm or fire suppression system. Since the massive fire that destroyed McCormick Place in Chicago, Illinois, in January 1967, fire and building codes for exhibit halls have changed. The fire started a little after 2 A.M. and was fought by more than 500 firefighters using ninety-four pieces of fire apparatus. The hazards found in McCormick Place included electric extension cords, lack of compartmentation, heavy fire loading of combustible materials, lack of a suppression system, unprotected steel trusses, and no working water supply. The building and contents were a total loss.

Hazards

The first and foremost hazard encountered in exhibit halls is the life safety hazard. These facilities can hold hundreds to thousands of people. Many of these people will know of one way in and one way out. By human nature they will want to leave through the same door they entered. Fire prevention is the key. Fire prevention personnel must ensure that aisle ways are kept open and exits are not blocked. The second hazard is the amount and type of exhibits on display. What is the fire loading? What special hazards do they present? What special needs are possible?

■ **Note**
The second hazard is the amount and type of exhibits on display.

SPORTS ARENAS

Construction

Most modern sports arenas (**Figure 14-7**) can be considered architectural and engineering marvels. The construction types and design features vary as much as

Figure 14-7 *A university sport center.*

the names of the teams playing in them. Generally, though, there are certain designs for specific uses. For example, a hockey or basketball arena is a covered arena with the ice or court in the middle and stadium seating all around. A football stadium may be a bowl, horseshoe, or one-sided stadium. There is no standard for any style sports arena. As a result, occupancy can range from small high school gymnasiums holding a few hundred patrons to college football stadiums, such as Neyland Stadium in Knoxville, Tennessee, which holds more than 96,000 fans.

Other variables include the different types of building materials, the HVAC systems, and built-in fire protection.

It is crucial that fire companies that protect sports arenas have updated preincident plans. Additionally, a good and professional relationship with the people in charge of the arena helps considerably in the event of a fire or other emergency.

Hazards

■ **Note**

People have a tendency to look at large stadiums as outdoor arenas and not as structures.

Other than the obvious life safety hazard, another problem exists in the layout and size of many of these buildings. People have a tendency to look at large stadiums as outdoor arenas and not as structures. This may be the case in high school football stadiums, but professional and semiprofessional sports arenas are actually large high-rise office buildings with attached fields for sporting events. Even after the life safety hazard is addressed, incident stabilization may prove to be taxing and difficult.

Figure 14-8 *At this sports center, a parking garage is attached. If exits are blocked open, smoke from a vehicle fire could affect the arena.*

Another hazard present in sporting events is alcohol consumption. The use of alcohol among patrons can cause problems with strategic goals such as evacuation.

Another hazard associated with sporting arenas is their multiple uses (**Figure 14-8**). Concerts, circuses, and rodeos can present hazards of their own.

NIGHTCLUBS AND SHOWPLACES

Construction

Today's construction methods can be totally noncombustible buildings in which no unprotected structural steel is exposed and all vertical openings are protected by approved doors, such as the new, modern convention and entertainment centers. Nightclubs and showplaces may be constructed in this manner, or they may be made out of any variety of building materials with any level of built-in fire protection. As with all other types of places of assembly, preincident plans are the only true way to determine the construction features.

Hazards

As with some sporting events, a hazard often associated with nightclubs and showplaces is the fact that many of the patrons are drinking alcoholic beverages.

Another hazard comes from the loud music and smoky environment prevalent. This may allow a fire to burn unchecked for a time.

Popular nightclubs can be overcrowded on Friday and Saturday nights, causing a hazard. Additionally, they can be overcrowded during special events, such as holidays and local radio station promotions. When nightclubs are overcrowded, tables and chairs usually are moved and exits inevitably become blocked.

The combination of these problems—intoxicated patrons, late discovery of the fire, overcrowding, and blocked exits—can be disastrous.

■ **Note**
Popular nightclubs can be overcrowded on Friday and Saturday nights, causing a hazard.

STRATEGIC GOALS AND TACTICAL OBJECTIVES

The strategic goals for response to places of assembly change with time of day and activity involved. There are many hours when these structures are unoccupied and empty of any life hazard concern. Then, there are times when there are fifty, hundreds, or tens of thousands of people in these structures. It is very important that the first-arriving companies know the schedule of activities for the facility.

There are tactical objectives that can be easily preset in the preincident planning phase. The response of apparatus along service roads so they do not have to confront exiting traffic can be established. The response of people to specific exit areas for the life safety concerns and control of service systems should be preset. The staging areas and the mandatory meeting points must be preestablished. Do not forget water supplies, sprinkler and standpipe assignments, triage areas, helicopter landing zones, unaffected routes in and out of the complex for medic units, and other concerns.

After establishing these specific preestablished tactical objectives based on specific response criteria, the efforts of the personnel responding should be more effective and efficient. These actions must be rehearsed and learned—not just on paper, but through actual on-site drills to see how long it may take for each objective to be accomplished.

Aside from the preset tactical objectives, remember that the incident size-up dictates additional objectives.

Firefighter Safety

■ **Note**
The biggest firefighter safety issue is accountability.

The biggest firefighter safety issue is that of accountability. Large buildings with odd floor plans and different uses can cause firefighters to become easily disoriented and lost. Patrons have gone to stadiums and entered on the parking lot side, walked inside the building to their seating section, and watched a game. When the game was over, they left by the nearest exit and found that they were on the opposite side of the building from their vehicles. They became disoriented during normal conditions. Add smoke and fire to the equation, and firefighter accountability becomes a critical aspect of firefighter safety.

Another safety issue is the establishment of rapid intervention crews (RICs). Due to the large building and possibility of many victims, the incident commander may see the need to use all personnel in as active roles as possible. An RIC at every entry point should be established, and the incident commander must realize that those personnel are committed.

■ **Note**

An RIC at every entry point for firefighters should be established and the IC must realize that those personnel are committed.

Search and Rescue

Most new places of assembly are designed to enhance evacuation of occupants; however, a thorough search and rescue must be completed. The task of search and rescue will be huge, and the assignments must be done geographically according to the building layout to ensure that no area of the building is missed. Search and rescue assignments should be made so that all personnel involved are aware of their exact assignments. A problem may come, for example, in a large sports arena if the assignment is given to search sections 1 and 2 of the seats. The signs that identify sections 1 and 2 may be obscured by smoke, and the wrong sections could be searched. A better method is to use fixed building components. Instead of sections 1 and 2, the area may be identified as the northeast seating section between exit doors 1 and 2 (see **Figure 14-9**). Any method is acceptable as long as all personnel understand the method and the search is complete.

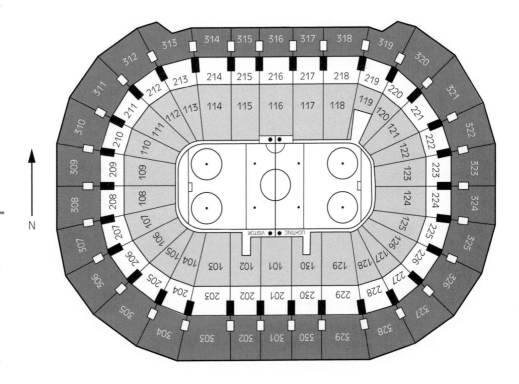

Figure 14-9   *When searching a sports arena, the incident commander should make assignments based on geographic area—north, east, south, and west—instead of section numbers.*

Evacuation

The focus of evacuation is to remove the occupants from the involved areas of the structure as quickly and as safely as possible when there is a significant fire. In large structures we may choose not to evacuate the patrons from the building, but to move them to safe, unexposed areas (see sheltering in place in Chapter 18). The methods of removal of the occupants, safe refuge areas, accountability methods, and other concerns for life safety must be established during preincident planning.

Taking control of the elevator system allows their use to move toward upper or lower floors. Firefighters will be able to move much more quickly and efficiently when they control the elevators. It also prevents patrons from trying to escape by use of the elevators.

The public address system can be used to direct both patrons and firefighting forces. The public address system allow directions to be given to everyone at the same time. But do not assume that everyone will listen or pay attention!

Exposure Protection

Exposure protection for external buildings naturally depends on the proximity of the place of assembly in relation to other buildings. If the place of assembly is heavily involved, the exposure threat may be severe.

Due to the various layouts of places of assembly, exposure protection might also include protecting parts of the structure that are uninvolved, such as attached wings or interior exposures.

Confinement

Locating and confining the fire quickly greatly reduces the hazard of most concern, life safety. Every asset must be considered when attempting to locate the fire. Use of alarm systems that identify detector activation can be helpful. Furthermore, building managers may know the location of the fire. In most cases it is difficult to find the exact location of the fire based on the smoke; however, smoke may lead firefighters in the right direction.

Search and rescue teams should look for fire as well as victims and report to the incident commander as soon as possible. Engine companies should not commit hose lines until the fire is located.

Once the fire is located, confinement to the room or area of origin is the goal. This may be accomplished initially by closing doors to slow the fire's progress until hose lines can be properly placed for extinguishment.

■ Note
Use of built-in fire protection systems is the key to extinguishing fire in places of assembly. This is especially true in large buildings that are longer and wider than the amount of handline hose carried on apparatus.

Extinguishment

Use of built-in fire protection systems is the key to extinguishing fire in places of assembly, especially in large buildings that are longer and wider than the amount

■ **Note**
Use of the standpipe system generally is the most practical means of getting hose lines to the fire.

of handline hose carried on apparatus. If the building is sprinklered, augmentation of the system is a primary goal of engine companies. Use of the standpipe system generally is the most practical means of getting hose lines to the fire. As with all large buildings, the standpipe systems may be divided into various zones, and the fire companies should have prior knowledge of which zones supply which areas.

Ventilation

The control of the HVAC system allows control of the air movement through the building. In some systems, you can remove all interior air and replace it with outside air. In some cases you can direct air flow away from the fire area. The system also allows certain areas to be shut down to prevent the movement of smoke into unaffected areas. Some systems allow for the control of rooftop vents and the opening and closing of certain large overhead doors to facilitate air movement. These activities must only be done by personnel who are quite familiar with the operation. It is important to have the maintenance people meet with command at designated locations.

Overhaul

Overhaul is completely dependent on the type of building construction. Places of assembly can have unique designs such as domes or cathedral-type roofs. Therefore, knowing how fire behaves will direct firefighters to the proper areas for overhaul.

Salvage

Salvage strategies and tactics should be based on the type of occupancy with the focus being on the items most important to building owners. Specific items of concern are presented further in the next section.

SPECIFIC FIRES

The following is a look at some of the concerns when addressing firefighting goals and objectives in specific public assemblies.

Churches

Strategic Goals and Tactical Objectives

Firefighter safety Firefighters are required to make long hoselays into some churches. Depending on the size of the church and the amount of smoke and fire visible, large attack lines may be required. The standard 1¾-inch line may not have the reach or gallons per minute required. Deep penetration into a building could seriously jeopardize the firefighter's ability to safely evacuate. Movement throughout a church can be difficult. Putting firefighters in a smoke-filled church is extremely dangerous due to the potential for them to get lost. Hose lines or lifelines are required.

The other concern is the building construction. What type of trusses are present? How long has the fire been burning on open trusses? Can the structure be ventilated to relieve the heat and lift the smoke conditions? Will a roof collapse cause the side walls to collapse?

Search and rescue If the building is occupied, obtaining input from those who have escaped enhances the incident commander's ability to locate victims. The escapees may know where people were located inside the building before the fire started. They may have knowledge of additional or alternate exits from certain areas. Depending on the size of the church, several search teams may be required. Many times, the church is evacuated with the first sign of smoke.

Evacuation The entire church, rectory, and any attached school or other buildings must be evacuated early in the fire. If the church is located within close proximity to other structures, they also must be evacuated. At the first sign of structural instability, the incident commander must evacuate all firefighters from the church and establish appropriate collapse zones (**Figure 14-10**).

Exposure protection In the case of the downtown area church fire, the exposure problem can be very significant. The structures on each side of the church and to the rear can often become involved. If roof collapse occurs, large firebrands can

Figure 14-10
Collapse zones must be set with considerations for exposures.

become airborne and land on the roofs of buildings close by. In the suburban store-front church, the exposure problem is the same as for any strip mall fire. In rural areas, the incident commander may be faced with wildland fires set by flying brands.

Confinement When dealing with a fire involving a large church, fireground operations must be established quickly to stop the spread of the fire. Depending on the initial observations, crews may have to position their apparatus for an interior attack while leaving themselves capable of employing master streams if the interior attack fails. Tower ladders are extremely effective in assisting in the knockdown of large bodies of fire within the church. This option may be required until an effective interior attack can be prepared. The tower ladders have the versatility to move in and hit the fire through the windows and at the top of the sanctuary where the heat has accumulated. If the fire has progressed to the point where master streams and tower ladders are considered, saving the stained-glass windows is no longer a consideration. Trying to save the stained-glass windows while losing the entire church gets you nowhere. In church fires, large lines are preferred for longer reach and more flow.

Extinguishment The fire in a church can be extinguished if fireground operations are begun quickly and are orderly. Depending on the size of the church and the smoke and fire conditions found, additional help must be summoned early. It is not wrong to have the first-alarm companies setting up for offensive operations while second-alarm companies prepare for a defensive operation. If the interior operations fail, the incident commander is ready for the defensive attack as soon as an accounting of personnel is complete. A large fire in a large church requires many firefighters working from multiple vantage points to bring the situation under control. Because of the size of the sanctuary, the divided interior of the main church, and the rapid fire spread, it may also require a judicious use of master streams and ladder pipe operations to knock down the fire.

Ventilation Churches typically have very steep roofs, making it difficult if not impossible to vertically ventilate (**Figure 14-11**). Many of today's modern-style churches have large roof structures with long angular slopes and metal covering. Rooftop ventilation is not an easy task. In many cases, the building and roof design eliminate the option of using vertical ventilation. The incident commander is faced with the decision of saving the stained-glass windows at the risk of losing the entire building.

Overhaul Overhaul in the church fire is a lengthy process. There are many openings and avenues of fire travel that must be checked. Knowledge of the building construction and inherent features is required. Before any overhaul is considered, a thorough inspection of the structural integrity of the building must be completed.

Figure 14-11 *Typical churches have steep roofs that are difficult to ventilate.*

Salvage Many items hold significant sentimental and spiritual value to the congregation. There are also church records, books, and other cherished items. Before beginning salvage operations, determine a safe location to which to move the items. The less the items are handled, the better.

Exhibit Halls

Strategic Goals and Tactical Objectives

Firefighter safety In the event of a working fire, firefighters may be required to make long hose stretches and work to penetrate deep into the structure. They can easily become lost or entangled among the exhibits. Accountability is always a very important concern. The incident commander must sector the fireground and establish appropriate collapse zones. These actions are designed to enhance firefighter safety.

Search and rescue Initial search and rescue operations are often limited to the removal of victims found in and around exit doors. In many cases, firefighters have not been able to move deep throughout the building to conduct primary searches. We must take for granted that not everyone will evacuate, and therefore a primary search must be completed as soon as possible. If the search team is searching wide areas without a hose line, rope should be used as a lifeline.

Evacuation If a fire occurs when the exhibit hall is crowded, it should be detected early and modern fire alarm systems can be used to direct the patrons

Safety

If the search team is searching wide areas without a hose line, rope should be used as a lifeline.

to leave through the nearest exits. The first-arriving companies will most likely be met with a building that has been evacuated by most patrons. There is realistically no possible way to determine any head count of patrons to determine whether anyone is still in the building. Buildings with adjoining halls, walkways, and common areas may have to be evacuated also. There is a construction practice seen throughout many cities where hotels, shopping malls, and parking garages have building walkways and tunnels from their building into the exhibit halls. This may create more evacuation needs.

Exposure protection The exhibit hall fire can create significant radiant heat if the structure is fully involved. Many exhibit halls have exteriors of concrete and steel, which should help to reduce the radiant heat spread outward toward the sides. Preplanning the exhibit hall in your area will help to determine exposure protection needs.

Confinement In some cases, employees of the exhibit hall are trained in fire response the same as an industrial fire brigade. With the use of interior standpipe and fire hose, these personnel can help to confine a fire in its early stages. Many exhibit halls have a sprinkler system or some other type of fire protection system. If the only response is from the fire department, there will be a delay in confining the fire, depending on where firefighters must enter to locate and fight the fire.

Extinguishment Fire extinguishment can be started with interior standpipe (**Figure 14-12**) hose lines while second-due crews are stretching bigger backup lines. Crews must remember to supply the standpipe and sprinkler systems and standard operating guidelines (SOGs) must require the use of these systems.

Ventilation The ventilation of a large exhibit hall can be a challenge if crews must use vertical ventilation, take out skylights, open rooftop vents, or open the stair shaft at the roof level. With today's HVAC systems, clean air can be brought in from the outside and smoke can be removed from the inside. Before you put all your faith in these systems, remember that wind and other weather conditions can affect their operation. Therefore, get feedback from your interior crews about the effectiveness of the ventilation operation. The idea of cutting large holes in the roof has many limitations and may not be effective.

Overhaul Much of what will burn in the exhibit hall is the exhibits themselves. The buildings have been rendered as noncombustible as possible, so there may not be a significant amount of overhaul of the actual structure. As for overhaul of the contents, utilize the equipment present in the facility, such as forklifts, motorized carts, and other labor-saving devices.

Salvage Depending on the severity of the fire, it might be an option to have the vendors come in and salvage what they can. When they are through, the rest of the materials may be considered trash and can be moved out of the structure to a predetermined location and piled up for removal to a dump.

Figure 14-12   *Many large assembly occupancies have a standpipe system.*

Sports Arenas

Strategic Goals and Tactical Objectives

Firefighter safety Firefighter safety issues specific to sports arenas are much like those for other places of assembly. The size and configuration of the building is the biggest concern, and these can only be made safer with all of the safety issues in place that have been presented thus far in this text.

Large numbers of resources are needed to fight even a small fire in a sports arena, due to the likelihood of panic among patrons and the real possibility of long hose lays, overwhelming search and rescue assignments, and ventilation difficulties (**Figure 14-13**). By calling additional resources early in the incident, the incident commander can ensure enough personnel for proper rehabilitation of crews and thus reduce the potential of stress-related injuries and illnesses.

Search and rescue As occupants of the building are evacuating, search efforts should focus on locating and confining the fire. This action will meet the number-one fireground objective, life safety, by reducing the progress of the fire,

Figure 14-13 *The local football game illustrates the potential life safety problems even in open arenas. A fire, even a small one, under the bleachers can cause a panic.*

Figure 14-14 *Here an ice skating area is enclosed in a mall with other shops. A fire in one of these shops would expose the remaining complex.*

it will give firefighters a starting point to do a primary search for victims that were unable to evacuate.

Evacuation The strategic goals and tactical objectives for evacuation of sports arenas are similar to the methods already presented in this chapter. The biggest concern is the large number of people who will be attempting to evacuate and the possibility for panic and stampeding with resulting injury and death of attendees. Anyone who has been to a professional sporting event knows how long it takes to clear the stadium under the best of circumstances. Under the threat of fire there may be major problems with uncontrolled panic that will be impossible to overcome.

Exposure protection Exposure protection considerations are the same as for all other types of places of assembly (**Figure 14-14**).

Confinement Confining the fire to the room or area of origin is best accomplished through knowledge of the building layout, early response and access to the building, and an action plan that defines the roles of responding units. The best possible way to confine a fire is with an adequate sprinkler system and proper augmentation by the fire department.

Extinguishment Extinguishment of a fire in this type of structure is best accomplished by fire departments who know the building and its features. Firefighters need to know how long hoselays need to be. They need to know the type of sprinkler system, alarm system, and standpipe system that the building has.

Ventilation Ventilation varies drastically depending on the design of the structure. Certainly HVAC systems need to be employed.

Overhaul and salvage Overhaul and salvage are similar to other types of assembly buildings. It varies depending on the type of building and the owner's or occupants' needs.

Nightclubs and Showplaces

Strategic Goals and Tactical Objectives

Firefighter safety The firefighter safety issues, in addition to those present with other assembly occupancies, are concerns about possible contents of the building. Places that might include pyrotechnics as part of a show will have these materials stored. Decorations and other equipment may cause firefighters to become entangled.

Search and rescue Firefighters must respond as quickly, efficiently, and safely as possible to save those who can be saved. In many cases, the devastation to human life will have occurred before their arrival. Rescue is focused on those in the greatest peril, then continues from that point. Primary search may be delayed by the removal of victims in doorways, windows, and other exit points.

Evacuation Evacuation should occur as soon as people realize that a fire has started. Past fires have shown the need for employees to assist in evacuating the patrons. In the Beverly Hills Supper Club, employees made announcements and led people from the building. This same scenario has been played out in other fires. But do not be lulled into a sense of complacency that all employees in these facilities are trained and will react appropriately. All of the first-due resources may need to be committed to search and rescue operations. Prefire plans dictate where and how those first companies must respond and what actions to take.

Exposure protection As with any building fire, exposure protection is a concern. This need will depend on the proximity of other structures.

Confinement Depending on the size of the complex, confinement will be delayed due to long hose line stretches, entering against those trying to escape, and a lack of initial resources. In many of the past fires, the flames were knocked down and confined quickly, but the life loss was still significant. The initial hose line should be placed to confine the fire and protect the means of egress.

Extinguishment The fire may have the chance to spread throughout the building while the first-arriving crews are in the rescue mode. Any presence of smoke or fire within one of these occupancies should result in the staging of at least a second alarm until the situation is further investigated.

Ventilation The crews must vent as they go. Open windows, doors, stairways, shaft doors—whatever it takes to relieve the smoke as the rescues are being made.

Because it is the smoke and heat that kills, the more that is released from the building, the better chance there is of saving people. Positive pressure ventilation is useful once fire location has been determined and hose lines are in place.

Overhaul The type of construction and extent of the fire dictates the amount of overhaul to be completed. The overhaul process is a slow and methodical process, as an investigation will commence immediately after the fire is knocked down.

Salvage Salvage operations are slow, depending on the severity of the fire and the need for investigation.

SPECIAL NOTE

Many large arenas, stadiums, and coliseums will employ fire, emergency medical services, and security personnel to have on site in case of an emergency. The fire personnel generally take charge of the fire alarm control panel and silence the audible alarms and strobes. They then monitor for any type of alarm activation and respond personnel accordingly to ascertain the validity of the alarm. This is done to prevent an alarm activation during a concert or other event. Many concerts and theatrical productions use pyrotechnics that will set off smoke detectors. A false alarm that causes the evacuation of 13,000 people from a concert does not perpetuate good customer relations between the event sponsors, the arena, and the concert patrons. Have you ever seen an indoor football stadium evacuate 60,000 people during a game because of a false alarm? You probably never have or will as it would cause more harm than good.

In some facilities, the fire alarm panel has an in-house intercom system. Control of the fire alarm control panel allows fire personnel to provide messages and directions over the in-house intercom system and to manage elevators, lights, and other systems.

Here are some of the deadliest blazes at clubs and dance halls in the United States:

- *Coconut Grove Club,* Boston, MA: November 28, 1942. 492 dead—unknown cause
- *Rhythm Night Club,* Natchez, MS: April 23, 1940. 198 dead—unknown cause
- *Beverly Hills Supper Club* in Southgate, KY: May 28, 1977. 165 dead—defective wiring
- *The Station Nightclub,* West Warwick, RI: February 20, 2003. 97 dead—stage fireworks
- *Happy Land Social Club,* Bronx, NY: March 25, 1990. 87 dead—arson
- *Dance Hall,* West Plains, MO: April 13, 1928 (explosion). 40 dead—unknown cause

- *Upstairs Bar,* New Orleans, LA: June 24, 1973. 32 dead—arson
- *Puerto Rican Social Club,* Bronx, NY: October 24, 1976. 25 dead—arson
- *Gulliver's Discotheque,* Port Chester, NY: June 30, 1974. 24 dead—arson fire nearby spread to disco

The fire service leaders must continue their pursuit to have sprinkler systems installed in all places of assembly. There are indications the fire protection laws are working. The National Fire Protection Association's (NFPA) database shows a steady, two-decade decline in the frequency and severity of deadly incidents. The number of nightclub fires dropped from 1,369 in 1980 to about 500 in 1999, the most recent year the association has figures. Deaths, injuries, and property damage have also decreased. But as seen in the list, history suggests that some building owners and club operators will cut corners to save money. Patrons' human behavior allows them to forget about the potential danger. Many will never learn to scope out alternative exits when they are in a club, arena, or other place of assembly, and some inspectors will ease up on code enforcement.

Those who have studied club disasters are resigned to the conclusion that no matter how fire codes are written or how buildings are constructed, similar disasters will happen again.

SUMMARY

Some of the worst fires in U.S. history in terms of mass casualties and deaths have occurred in public assembly facilities. These are temporary or permanent structures where large crowds of people come together for purposes as defined in the introduction to this chapter. The people are usually unfamiliar with the layout of the structure and know only the entrance and/or exits they have previously used. Although it has been stated that panic plays a large part in the number of deaths in these emergency scenarios, panic is not usually the main reason, nor is it something the fire department can control.

Most of the damage done to the occupants of the building will have occurred before the arrival of the first-due companies. With adequate prefire planning, companies will know where to go to remove occupants from as many exits as possible. This may involve opening as many doors as possible and literally pulling people from the building. Knowledge of the main exits and how to reach and open them is important.

Quick and effective ventilation of the building is another tactical objective that must be met early in the fire scenario. It is important to use the fastest way to remove smoke from the interior of the structure and away from people. The ventilation methods and location of rooftop vents and HVAC controls must be preplanned.

It is the obligation and responsibility of the local jurisdiction fire department to thoroughly preplan the facilities in their response area. In developing these preplans, remember that fire operations involving hose evolutions may be impeded by the escaping occupants. Therefore, it is important to find other ways to enter the complex. Review your prefire planning methods to ensure that they provide the correct information with regard to the high life hazards found in the public assembly facility.

REVIEW QUESTIONS

1. With regard to life safety, what is the greatest hazard to consider in a fire involving a public assembly building?

2. According to NFPA *Life Safety Code,* what is the minimum occupancy load for considering a structure as being a public assembly?

3. If an occupancy load does not meet the minimum requirements, under what occupancy does it fall?

4. What is the typical building construction method used for public assemblies?
   A. Noncombustible
   B. Frame and brick
   C. Air supported
   D. Varies with type and location

5. Do all the occupants have to be completely evacuated from the public assembly? Defend your answer.

6. What is the most difficult issue to be confronted with public assembly fires?
   A. Smoke movement
   B. Fire involvement

C. Damage to occupants before your arrival
D. Vehicles leaving

7. How can the HVAC system be used to assist in fire operations?

8. What specific issues have a major impact on strategic goals?

9. Choose two main goals to consider in a response to a public assembly.
   A. Life safety and building control
   B. Evacuation and crowd control
   C. Removal of cars from lot and obtaining a water supply
   D. Turning on air conditioning and lights

10. How do preplans affect the response to a public assembly?

ACTIVITIES

1. Visit public assembly facilities within your response district and tour the areas typically unseen by the public to determine optional access points.

2. Review your department's preplan of local public assembly facilities in your jurisdic-

tion. Are they up to date? Is enough information provided?

3. Review the case studies about fires in assembly occupancies available from the United States Fire Administration.

Chapter 15

HIGH-RISE OFFICE BUILDINGS

Learning Objectives

Upon completion of this chapter, you should be able to:

- Define high-rise office buildings.
- Describe various construction methods.
- Describe hazards associated with these structures.
- Describe the building systems present in high-rise buildings that can be of use to firefighters.
- Describe the strategic goals and tactical objectives related to high-rise building fires.

CASE STUDY

Smoke was discovered on the thirty-second floor of One New York Plaza in New York City. The building, a fifty-story high-rise office tower, was preparing for its official grand opening, and finishing touches were being made to the various occupancies. Shortly after several secretaries noticed the smoke, a guard discovered a fire on the thirty-third floor and activated the fire alarm on that floor. Three building employees, two of them security guards, were taking an elevator to an upper floor to notify workers that there was a fire below them. As their elevator reached the thirty-third floor, it stopped and the doors opened, causing smoke, heat, and fire to consume the elevator. The doors warped and they were unable to close them. Two of the workers died in the elevator and the third was critically burned. Another car also stopped on the fire floor, but workers were able to close the door and the elevator took them safely off the fire floor. Workers on other floors became aware of the fire when smoke, traveling through the air-conditioning system, appeared on various floors. Employees started self-evacuation via both elevators and stairways. A high-rise disaster was in the making. It is uncertain exactly when and who reported the fire to the New York City Fire Department, because the fire alarm boxes in the building were not connected to a central station, but once the alarm was received, it was only 3 minutes before the fire department's arrival. At this time fire had control of both the thirty-third and thirty-fourth floors of One New York Plaza. The fire had started in a plenum area on the thirty-third floor below electrical equipment serving the thirty-fourth floor. The first companies to board an elevator to combat the blaze became trapped in their elevator after it failed to operate. Other units were also unable to reach the fire floor due to malfunctioning elevators. Eventually companies arrived on the fire floor but were severely hampered by high heat and smoke conditions. Self-contained breathing apparatus was required to do battle at this fire at a time when their use was not mandatory. Ventilation was both difficult and not performed as aggressively as it would have been at an ordinary building fire. Fear of sending glass shards down to the street below and the inability to determine the effects of horizontal ventilation at high-rise fires were factors. Many windows self-vented at this fire due to severe heat. The fire was eventually extinguished by several handlines, but not before two floors were badly damaged, incurring a loss of several million dollars. Many recommendations were made after this fire, but making changes or retrofitting buildings is a very costly and difficult process that is usually met with stiff resistance from owners, architects, and building groups. Out of this fire, however, and several others came the birth of local law 5 in New York City, which specified rules and regulations as they pertained to new high-rise office construction, a law that remains in effect today.

SPECIAL NOTE

The events of September 11, 2001, which claimed the lives of 343 of our fellow firefighters, have some people questioning the tactics that were used on that fateful day. Although hindsight is 20/20, we in the fire service must learn from yesterday's fires to better prepare ourselves for tomorrow's battles. Though much was learned from this act of terrorism, we cannot lose sight of the general principals of high-rise firefighting. Some suggest establishing a command post away from the building is better than using the lobby as a command post. We disagree. Getting control of the building's systems is best accomplished at the fire command center, located in the lobby. Gaining control of elevators, HVAC, and communications systems must be performed from the lobby. The lobby also serves as a safe location to deploy personnel. All of us must now be aware of the possible signs of terrorism and be able to quickly and effectively adjust our strategies and tactics accordingly. However, we cannot use the events of 9/11 to shy away from our normal SOGs, whether dealing with a high-rise office building fire or a private dwelling fire.

■ **Note**
Twelve people were killed in January 1882 and six were killed in January 1912, both times in high-rises. Although neither of these buildings was very tall, they were beyond the reach of fire department ladders.

■ **Note**
Many jurisdictions have buildings that are five or six stories. If the top stories are beyond the reach of ladders available to local resources, then the structure should be strategically and tactically treated like a high-rise.

INTRODUCTION

High-rise fires have occurred in many areas throughout the country and throughout the world. From tragedies such as São Paolo, Brazil, and the MGM Grand in Las Vegas, to One Meridian Plaza in Philadelphia (**Figure 15-1**) and One New York Plaza in New York, high-rise fires are not uncommon. The first high-rise fires actually occurred well over one hundred years ago. Twelve people were killed in January 1882 and six were killed in January 1912, both times in high-rises. Although neither of these buildings was very tall, they were beyond the reach of fire department ladders. In fact, for the purpose of this text, we define a high-rise building as:

> Any building where the highest floor is beyond the reach of available ground or aerial ladders. These type of buildings will require the application of high-rise tactics.

Many jurisdictions have five- or six-story buildings. If the top stories of these buildings are beyond the reach of ladders available to local resources, then the structure should be strategically and tactically treated like a high-rise. To combat today's high-rise fires, a heavy emphasis is placed on utilizing building systems and resources. From elevators and communications to standpipes and fire pumps, these battles are fought from within, with very little help from ground ladders and outside streams.

Figure 15-1 *One Meridian Plaza in Philadelphia. Photo taken in 1999.*

High-rises require a full commitment of resources from the onset for operations to be successful, and no time should be wasted before requesting them. These fires are the most challenging fires in the fire service.

CONSTRUCTION

The construction features of high-rise office buildings vary, depending on when they were designed and built. In many cases, a good size-up at a high-rise office building can determine whether it contains certain features. In New York City, for example, buildings are described according to the time they were built and this is common throughout the country. There is old, heavy construction and new, lighter-weight construction. Built primarily of steel enclosed in concrete, the older buildings (**Figure 15-2**) are much heavier than today's modern high-rises. Outside walls are of concrete or other masonry construction, and floors are concrete. Significant construction features that affect the way the fire department operates at these buildings include the following:

plenums
part of an HVAC system that provides an avenue for air flow

- Windows that can be opened from the inside
- Lack of central air conditioning or **plenums**
- Compartmentalization

Figure 15-2   *An example of older high-rise construction.*

Newer high-rises (**Figure 15-3**) built from the 1960s on were of much lighter construction than the older, heavier buildings. They are the sleek, steel-and-glass buildings we visualize when we think of high-rise buildings.

There are substantial features in these buildings that were absent in the earlier era. These features include the use of central heating, ventilation, and air-conditioning systems (HVAC) that move air throughout the building, and ceiling plenums that are used to return air to the system. These plenums are also used for the building's electrical and communications equipment. For the most part

Figure 15-3   *Newer high-rise construction.*

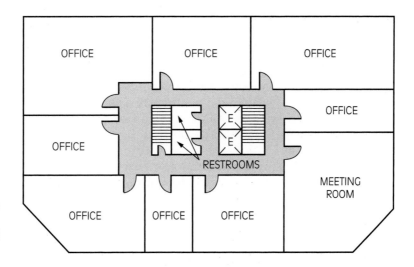

Figure 15-4 *An example of a central core floor plan.*

these buildings contain large, uncompartmented office spaces that allow for expanding fires.

central (or center) core construction
a type of floor layout where components common to every floor, such as elevators and bathrooms, are centered around the core of the building

Central (or center) core construction (**Figure 15-4**) is common in many high-rise buildings, where the elevators, stairs, and support systems are located in the center of the building.

Also in the newer buildings, windows are fixed and inoperable. The only way to vent a window is to break it, usually an undesirable tactic at fire operations in high-rise buildings. In most large cities high-rise office buildings are found by the hundreds, and many of these buildings share common characteristics. A building designed with an HVAC system adds a new dimension to a fire problem and quite a challenge to the incident commander. Not that having a HVAC system is bad: It can be a great feature for the incident commander, if utilized properly, or it could be a nightmare. The same could be said for the use of elevators in high-rises. Fire departments need to control and utilize them, but if used improperly, they can lead to death or serious injuries to both firefighters and civilians, as discussed in the case study.

Building Systems

A key tactical consideration in high-rise fires is that fire departments are totally reliant on the building's systems to assist in extinguishment. Building systems that must be utilized include:

- Standpipes
- Sprinklers
- Heating, ventilation, and air-conditioning systems
- Elevators

- Fire command station
- Communications systems
- Fire pumps

These systems are both varied and complex. Fire departments must master their use in order to bring fire operations to a successful conclusion. A closer look at some of the systems found in high-rise office buildings is needed so these systems can be employed properly.

Standpipes In many areas, standpipe systems are mandated in all high-rise buildings by law or fire code. Some buildings, though not technically considered high-rise buildings, should be defined as such for firefighting purposes and application of strategies and tactics. This definition includes buildings that have stories beyond the reach of available ladders, as discussed earlier. Preincident planning is the only absolute method to determine whether a building has a standpipe system. It can be difficult to stretch hose lines off pumpers and into buildings, although it has been done with success for fires in the lobby or first or second floors of buildings, sometimes in conjunction with handlines off a standpipe.

There may be several risers found in these buildings and it is critical to ensure that fire departments are supplying the system with adequate pressure and with several supply lines. A good rule of thumb is for each pumper responding to a high-rise fire to position the apparatus at a workable fire hydrant located near a standpipe siamese. Firefighters must be aware of falling debris and glass from the fire building. This falling debris may require the engine to be placed away from the building and the hose lines protected. The siamese connection outside these buildings should either be color-coded or have a sign adjacent to it indicating what the siamese is supplying—for example, standpipe, sprinkler, or combination standpipe/sprinkler (**Figure 15-5**).

Depending on the size of the building, consider having a minimum of two pumpers supplying two siamese at the same time during fire operations. Good knowledge of the building prior to the incident will assist in deciding which supply line should be supplied first. If the building has an adequate fire pump supplying the sprinkler system, then the incident commander should consider directing the first supply line to the standpipe siamese before the sprinkler siamese. It is especially important to supply the standpipe system before firefighters begin working off it. Even in wet, pressurized systems, it is not a safe practice to rely solely on the water supply provided through the standpipe system.

In some cities, sprinkler systems are required in high-rise buildings; however, some buildings have been retrofitted and may have sprinkler systems on various floors, as was the case in the Meridian One fire. The fire stopped on the floor that was equipped with sprinklers.

Fire Pumps and Hydraulics All high-rise office buildings should be required to have auxiliary fire pumps, which increase the pressure in the standpipe systems.

! **Safety**
Firefighters must be aware of falling debris and glass from the building. This falling debris may require the engine to be placed away from the building and the hose lines to be protected.

■ **Note**
Depending on the size of the building, consider having a minimum of two pumpers supplying two siamese at the same time during fire operations.

Figure 15-5 *Note the FDC on the right is marked "standpipe," and the one on the left, "automatic sprinkler."*

These pumps can be found singly or in pairs. If found in pairs, both pumps should be activated and the pressures should be set at approximately the same pounds per square inch (psi). In some cases it has been the job of the building engineers to start the fire pumps during a fire, but in many cases they have been assigned other responsibilities or cannot be located upon arrival of the fire department. Many fire pumps start automatically when the alarm system is activated. In either case, a firefighter, preferably one familiar with hydraulics, should be sent to the pump location either to activate the pumps or act as a communications liaison with the incident commander. If there is no building engineer present, the firefighter should consult with the fire safety director or person with knowledge of the pump and make his way to the pump room, usually located in the basement, subbasement, or other utility area. The firefighter should examine the pumps and the diagram of the pump's system that should be posted nearby. Although some systems may be more complex than others, the firefighter should be able to operate them and have a basic knowledge of fire pumps. It is essential that the inlet and discharge gauges be checked and it is critical that the bypass valve is closed so the system can be supplied. Sometimes the bypass valve is left open after the pump testing. Fifty percent of the pump's capacity can be lost with an open bypass valve. Generally, pumps are activated by following certain procedures that should be posted on the control panel. Once the firefighter starts the pumps, he or she should remain at this position to ensure that the required

■ **Note**

Once the firefighter starts the pumps, he or she should remain at this position to ensure that the required pressures are being maintained and that no one shuts the pumps off without permission of the incident commander.

pressures are being maintained and that no one shuts the pumps off without permission of the incident commander. In high-rise buildings that have a common sprinkler/standpipe system, the starting procedure for the pumps is the same. However, one pump is usually automatic, whereas the other is manual. A visual check of the pumps should be made frequently to ensure that the discharge gauges in both pumps are operating at approximately the same pressure. If one pump is substantially lower than the other, the lever should be moved to a higher setting. Reversing the starting order can shut down the pumps. There should be a phone by each fire pump, providing direct communication to the building's command post. Building fire pumps should be able to deliver their rated capacity—at least 50 psi at the highest floor outlet—with a relief valve limiting the pressure to 15 psi above that required to deliver the 50 psi's to the highest outlet.

Heating, Ventilation, and Air-Conditioning Systems (HVAC) One of the most critical aspects of fighting modern high-rise building fires is gaining a handle on air movement throughout the building. It is a foremost concern at high-rise operations that the incident commander use the heating, ventilation, and air-conditioning systems to the advantage of fire operations. With proper use, these systems can limit fire travel and smoke movement throughout the building. In older high-rise buildings, air-conditioning systems usually served only one floor, whereas in modern buildings the air-conditioning system serves multiple floors. Although at first glance the central system may seem complicated, it is quite simple. The HVAC system is broken down into three components: the processing equipment, the supply system, and the return system. The air processing equipment is generally found on the floors containing the mechanical equipment rooms. There may be several of these equipment rooms, which include the air processing equipment, mixing, exhaust, and outside air supply dampers in each building. Dampers are devices that prevent or allow air to enter the HVAC system. Mixing dampers combine return air from the system with outside air. The air flow through the building starts with the processing equipment, which return air from the return shaft, where it is mixed with outside air and then filtered, warmed or cooled, and sent through the supply system.

It is important that the incident commander have some knowledge of the HVAC systems and formulate an operational plan. A good rule to follow on arrival is to verify the fire floor and location while determining the mode of all the HVAC zones. Once the fire floor has been accurately verified, the outside air supply dampers and mixing dampers should be closed, while exhaust dampers should be opened to prevent air from transversing the supply and return shafts. The next step is to reactivate the supply fans to all floors excluding the fire area, supplying fresh air to building occupants, while also pressurizing these zones. Firefighters must monitor all areas of the building, alerting the incident commander of any problems caused by resumption of the supply system. There are many codes governing fire dampers and heat and smoke detectors in high-rise

■ Note
One of the most critical aspects of firefighting in modern high-rise fires is gaining a handle on air movement throughout the building.

■ Note
In older high-rise buildings, air-conditioning systems usually served only one floor, whereas in modern buildings the air-conditioning system serves multiple floors.

dampers
devices to prevent smoke travel throughout the HVAC system; often activated by a fire alarm system

■ Note
It is important that the incident commander have some knowledge of the HVAC systems and formulate an operational plan.

buildings and they may vary from area to area. However, it is essential that all firefighters have a general understanding of how HVAC systems operate and how they can be beneficial at fire operations.

HAZARDS ENCOUNTERED

A clear hazard encountered at high-rise buildings is the sheer size of the building and the number of building occupants. This is a true hazard because it can quickly overcome firefighting forces. The incident commander must get additional resources to the scene as fast as possible. The size of the building also slows the initial hose lines, allowing the fire to grow larger.

Most other hazards come from the building features, such as the lack of opening windows in newer high-rise buildings and the potential for stack effect and stratification of fire gases, as discussed in detail later in this chapter in the section about ventilation.

STRATEGIC GOALS AND TACTICAL OBJECTIVES

Fighting a fire in a high-rise building requires a tremendous commitment of both personnel and supervision. A coordinated effort is made to break up the operation into various sector officers as quickly as possible. A well-planned search effort must be laid out, encompassing the fire floor as well as all floors above the fire, while handlines must be strategically placed and rapidly advanced. The reaction time at a high-rise fire is far greater than for any other type structure. From the time the alarm is received, to arrival time, to getting water on the fire could be upward of 20 minutes or more. Add to this congested traffic and companies responding from long distances, and it is an uphill battle to contain the fire. The initial incident commander should immediately seek out the fire safety director, building maintenance person, or building manager (if not already done by the first-arriving officer) and establish the lobby command post, which will be the center of the operation. It is critical that the correct fire floor is verified and confirmed by firefighters before engine companies board elevators or make their way up stairways. The incident commander should send the ladder company personnel to verify the fire location and guide the engine companies to the best-suited stairway to use as the hookup stairway for handlines. A separate stairway must be used exclusively for evacuation, if necessary. Due to the large floor area to be searched, at least two companies should be assigned to search the fire floor, more if necessary. As additional command officers arrive on the scene, one should be designated to the fire floor to oversee the operations and supervise the handlines advancement and searches on the fire floor and floor above the fire.

A forward staging area should be set up two floors below the fire floor, when possible. From this staging area relief crews are sent to relieve crews in the fire area. The time factor must be considered when firefighters are requesting relief. Firefighters operating on the fire floor must monitor the time left on their self-contained breathing apparatus (SCBA) and allow themselves enough time to get off the fire floor and floors above. This is extremely important for personnel who are not operating with a handline and may become disoriented under heavy smoke conditions. Accountability is very important and, in many cases, ladder personnel should be ordered off the fire floor before additional personnel are sent in. By doing this, verification can be made that all firefighters from a company are accounted for prior to committing additional personnel. Engine companies must hold their positions on their handlines until properly relieved. Separate handlines may be operated from separate stairways, provided these handlines do not oppose one another or compromise evacuation.

Other arriving command officers should be assigned to groups or geographic areas and other incident management system (IMS) positions, including but not limited to search and rescue, logistics, planning, medical, and rehabilitation.

Firefighter Safety

In addition to the safety issues common to all structure fires, there are considerations unique to high-rise buildings.

Depending on the height of the building, it may be impossible for firefighters to haul equipment to upper floors via the stairs and then have the strength to begin the firefight. Before committing to using elevators, firefighters need to understand their operation and whether they are firefighter service-equipped (**Figure 15-6**).

Figure 15-6 *An elevator control panel with firefighter service.*

In New York City local law 5 dealing with high-rise office buildings clearly dictates where and when firefighter service is required. Building fire laws vary from one municipality to another, but most high-rise buildings are equipped with firefighter service features. These features can also be found in smaller buildings. Buildings with firefighter service are a tremendous asset to fire department operations, and all firefighters should be familiar with the proper procedures regarding their use. There are two components of firefighter service: getting control of recalling the elevators and using the elevators safely. Once firefighters have boarded an elevator, they should clear the floor selector panel by hitting the call cancel button. This ensures that there are no surprise stops. The next step would be to hit the next available floor, that is, the second floor, to ensure that the car is responding to your commands. The incident commander and officers must make certain that all firefighters using elevators have full personal protective equipment including SCBA on their backs, with cylinders and pass alarms turned on. Forcible entry tools and radios are also required when using elevators. Remember that firefighters can also look at the "You are here" sign depicting the floor diagram usually found next to the elevator. The location of the stairways should be viewed in relation to the elevator. If a hasty retreat must be made out of the elevator, firefighters may need to know the location of the closest stairway. A basic rule in elevator use is *never* to take an elevator car above the fire if the car also serves the fire floor. Elevators must be used in most high-rise office fires, but they must be used with extreme caution. The effects of heat and fire on elevators can cause serious problems to fire department operations.

A final safety tip on elevators is that swinging car doors open outward, while office doors open inward. When searching an area in heavy smoke, a door that is found to open outward could be an elevator door with no car present. Beware and probe any opening with a tool during a search.

Another safety issue for firefighters in high-rises is that of adequate, effective communication. The incident commander should usually establish his or her communications and command posts where the fire command station is located, most times found inside the building lobby. All types of available communication systems should be sought out and controlled (**Figure 15-7**).

Safety All types of available communication systems should be sought out and controlled.

These interior communication systems include the following:

- Standpipe telephones
- In-house telephones
- Special sound-powered telephones
- Building public address systems
- Fire pump room telephones
- Radios used by building security, if available
- Elevator telephones

These systems must include fire department radios enabling communication with the fire dispatcher and with units on the fireground.

Figure 15-7 *A communications hook-up in a building's stairwell at the standpipe.*

The incident commander must be able to communicate with all the various sector commanders operating throughout the building. Due to numerous radios that will be operating, it is imperative that units be designated specific channels to operate on such as the command channel or primary and secondary tactical channels.

Firefighter fatigue is also an issue in high-rise fires, and quality rehabilitation is necessary. The rehab sector can be set up on the same floor as staging, below the fire floor. The area must be a smoke-free environment and must be an area that will not hamper firefighting operations or interfere with staging. If staging and rehab are located reasonably close to each other, then firefighting forces can easily move from the rehab to staging sectors without using elevators or stairwells.

As with large commercial building fires, accountability is a major concern in high-rise building fires. There may be multiple points of entry into the building. However, one way to maintain some level of control is to have an accountability location in the lobby. This allows the lobby accountability officer to know what companies are on the upper floors of the building. Accountability is also necessary on the upper floors. Depending on the complexity of the incident, each floor may require an accountability location. Company officers must understand when they are not making headway, a withdrawal and regrouping of forces or a defensive attack may be required. It is important to realize that a retreat cannot be made at a nearby window. The only way out will usually be the way you got in, from an interior stair.

Search and Rescue

One of the responding command officers should be placed in command of the search and rescue sector or group. This officer is responsible for the search and safe evacuation of civilians from two floors above the fire through the rest of the height of the building. This sector should have enough personnel to do a floor-by-floor search of the building. Each company or team can search several floors and report its findings to the sector officer, who in turn will notify the incident commander. Additional responsibilities of this officer are to obtain sufficient units to search, prevent unnecessary evacuation, and evacuate only through evacuation stairs. Units operating under this sector or group should bring extra SCBA bottles with them to increase their work time (**Figure 15-8**).

Rapid deployment of ladder companies is crucial at high-rise fires. Upon confirmation of a working fire, it is essential that additional ladder and rescue companies are sent for search and evacuation. The first company has the responsibility to take charge of the buildings elevators, determine the location and extent of fire, and perform search and rescue operations on the fire floor. It must determine the safest way to get to the fire floor and gain as much information about the fire floor as possible, such as floor diagrams (**Figure 15-9**), compartmentation of floor, office areas, communications available, evacuation in progress, and access stairs serving the fire floor.

The first company exits the elevator two or three floors below the reported fire and takes the stairs to the fire floor. A firefighter of the company should

access stairs
stairs that serve a particular floor or area of a building

Figure 15-8
Firefighters should bring extra SCBA bottles into the building with them.

Figure 15-9 *In many cases, the floor below the fire floor has a similar layout. Firefighters may want to walk this floor to be familiar with the layout.*

remain with the elevator and return it to the lobby to pick up additional personnel. This firefighter may remain with the elevator or another firefighter may be designated to take over elevator operations, but the elevator should not be left unattended. Maintaining control of all elevators is imperative. The second company should augment search operations on the fire floor. Office areas are quite large and it may take at least two companies to completely search them. Firefighters must bring additional search ropes to their complement of tools (**Figure 15-10**).

The second company should communicate and coordinate with the first company as to areas searched, location of fire, people trapped, and other items. The third company should take an elevator two or three floors below the fire and be prepared to search the floor above the fire. This company must communicate with the first two companies to get any available information. Upon arrival on the floor above the fire, members should notify the sector officer of what has been found on this floor, such as heat, smoke, fire, and civilians. The sector officer, in turn, notifies the incident commander. Additional companies should be utilized to conduct comprehensive searches of all floors above the fire area and report the results of these searches to their sector officers.

Smaller departments may not have the resources to deploy as aggressively as larger departments. Searches must be made of the fire area and all floors above and below the fire, but they must be made within the limits of safety, and when possible, with the protection of a handline.

Figure 15-10
Firefighters searching floors above the fire without hose lines should use a rope for safety purposes.

Evacuation

Taking control of building evacuation is done during the search and rescue. Initial evacuation should only take place on the fire floor and floor above. Usually companies are assigned to direct the evacuation through a stairwell away from the one used by engine companies. The evacuation stairwell must, therefore, be secured and free from smoke. Civilians have been killed or seriously injured by self-evacuating and becoming overcome in stairways.

■ Note
The evacuation stairwell must be secured and free from smoke.

Exposure Protection

Internal exposure protection involves protecting the unburned areas from extension and this is accomplished during the confinement attempt. The incident commander must also be aware of autoextension, which is when a fire comes out a window on one floor and extends into a window on the floor or floors above the fire floor. Autoextension was introduced with multiple-family dwelling fires and often can occur with high-rise buildings as well. Protecting external exposures in high-rise building fires can be difficult because of the elevation. Heavy streams from the outside are ineffective. Often exposure to adjacent buildings of similar height can be achieved by advancing hoseline into the exposed building and operating them onto the fire building. This also is an advantageous position for fire extinguishment.

Confinement

Confinement of fires in high-rises usually is performed by effective, timely extinguishment. Confining the fire to the area it started in, or was in when you arrived, is the goal. However, this operation can be labor-intensive and time-consuming. Confining is accomplished by advancing interior hose lines to the fire area and attacking from the unburned side. It is also necessary to know the paths of fire extension and place lines in these positions adjacent and above the fire area.

Extinguishment

The specific tasks of the first attack crews is to stretch sufficient hose to reach the seat of the fire (**Figure 15-11**). The first engine should initiate the fire attack, while the second engine, after assisting in getting the line in operation, should stay by the stairway, reserve its air supply, and be ready to relieve the first engine on the handline. The first engine companies are also responsible to supply both the standpipe and sprinkler systems. The pairing of engine companies should be a standard practice at high-rise fires. The first and second engine companies get the first handline in operation, while the third and fourth engines get the second

Figure 15-11 *Extra hose should be laid up the stairs past the fire floor. Then once charged, it is easier to advance.*

handline in operation. The second handline must be stretched and operated at a position to back up and protect the first handline. The second handline should be ready for fire possibly getting behind the nozzle team. All engine companies reporting into work at high-rise fires should bring additional lengths of hose as well as extra SCBA cylinders with them. All handlines operating in high-rise buildings must be at least 1¾-inch in diameter. There is no justification for using less than 1¾-inch; a 2½-inch is preferred, if staffing permits. Remember fire flow requirements: The 2½-inch provides additional flow. Firefighters never use building standpipe hose during fire operations. Some departments that have high-rise buildings in their area will have a difficult time with an aggressive attack if extra resources cannot be moved to the scene quickly. The best strategy for these departments may be to concentrate efforts to confine the fire until assistance arrives.

If a defensive attack must be initiated, firefighters should take a position where they will not be overexposed. Some jurisdictions use a portable master stream device that can be used by one or two firefighters from relative safety. The use of master streams have far more reaching effects than handlines. Their reach, coupled with their cooling ability and large water volume, makes them a great choice for departments with limited resources. If the fire is within reach of outside master streams, such as tower ladders or ladder pipes, they may be a consideration, but there is always a chance of pushing fire into uninvolved areas. Portable master streams have also been used on the fire floors from the interior of the building. This operation is labor-intensive to set up but provides the heavy stream that may be necessary.

■ **Note**
To completely understand ventilation in a high-rise structure, one should be familiar with the various avenues of smoke travel.

stack effect
as heated air rises and then cools, it forms a layer, or ceiling, from which more rising smoke banks

stratification
heated smoke rises until it establishes equilibrium with the surrounding area and then moves horizontally

Ventilation

Ventilation of a fire building is a major priority. Ventilation is the planned and systematic removal of smoke, heat, and gases from a structure. The obvious reasons for ventilation are to remove toxic gases, eliminate hazardous health conditions to occupants and firefighters, locate the fire, and allow smoke to lift. Traditional methods to accomplish ventilation at fires are either by opening or breaking windows, bulkheads, and scuttles, or by cutting the roof. When dealing with high-rise buildings, ventilation is an extremely important aspect of control and extinguishment. Although some methods are basic, others are more complex.

To completely understand ventilation in a high-rise structure, one should be familiar with the various avenues of smoke travel.

When a fire occurs in a high-rise building, the basic fire development and spread is not very different from that in other type structures; however, there is a great magnification. The environment in which it is confined can influence smoke and gas movement. High-rise building fires have shown that two distinct phenomena can occur: **stack effect** and **stratification**. Stratification (**Figure 15-12**) of various gaseous combustion products within a high-rise building means that gases, depending on their relative weights as compared to air, may arrange themselves in distinct layers. Heated air, gases, and smoke rise within a

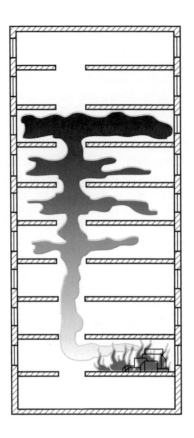

Figure 15-12   *An example of stratification.*

structure via the thermal convection currents formed. After moving away from the heat source, these products of combustion become cooled and their spread, as they travel upward, is reduced. Eventually the heat currents do not affect them. When this happens, stratification starts to occur. Lighter-than-air gases continue to rise, but heavier-than-air gases and possibly smoke particles drift downward. Each part of these gas products then reaches a stratified level (they will stop moving), at which time their concentration in air is increased, and depending on the amount formed, high concentrations of toxic or oxygen-deficient gases collect in remote areas. The effect of this can be disastrous in that gases such as carbon monoxide, hydrogen sulfide, sulfur dioxide, and hydrogen cyanide can accumulate in areas above, below, or adjacent to the fire area. Many of these gases are colorless and odorless. It is possible that areas that have to be searched, examined for extension, or used as sector commands may be contaminated. HVAC systems affect stratification, but at times these systems are shut down during a fire, or the more rapid cooling of gases and smoke caused by these systems may increase the downward movement of some of these products of combustion.

AIR MOVEMENT DUE TO NORMAL AND REVERSE STACK EFFECT

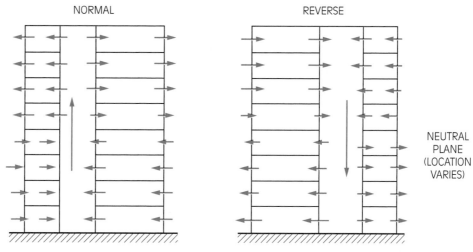

Figure 15-13   *An example of stack effect.*

NOTE: ARROWS INDICATE DIRECTION OF AIR MOVEMENT.

The stack effect (**Figure 15-13**) condition within a high-rise building causes warm air to rise. Pressure differences force air in at the lower levels and out at the upper levels.

A neutral plane, in which air flows in or out of the enclosure at negligible amounts, occurs at or close to the middle of the building. Smoke and gaseous products of combustion can be influenced by this effect. In fires occurring on the lower levels (below the neutral plane), the internal pressures created by the stack effect draw the products of combustion toward any shafts or stairway openings. In fires occurring at higher levels (above the neutral plane), air movement draws products of combustion away from shafts or stairs toward the exterior of the building. In fires occurring within the neutral plane, the stack action has little effect on smoke movement because air movement either in or out is small.

Overhaul

Overhaul of high-rise buildings is labor-intensive. A good preincident plan helps to determine where to direct initial overhaul efforts. The preincident plan can direct firefighting forces to hidden shafts that need to be thoroughly checked. False ceilings should either be pulled or opened so areas above them can be checked for extension. Crews should be assigned to every floor above the fire to check for smoke and possible extension. Additionally, crews should check areas below the fire to verify complete extinguishment.

Salvage

Most high-rise buildings are commercial office buildings, and salvage of specific valuables depends on the owner's or occupants' individual needs. They may request items such as records or computer discs retrieved.

High-rise buildings that people live in, such as large condominiums, need to be salvaged with the occupants' valuables in mind. The occupants may request the fire department to attempt to retrieve items such as pictures, negatives, small safes with important paperwork, and personal items such as clothing.

General salvage operations for high-rise buildings include the redirection of water from upper floors. Sprinkler systems and firefighting operations can dump thousands of gallons of water into a structure that eventually find their way to the ground through the path of least resistance. If firefighters can create that path by use of water chutes, salvage covers, and so forth, then areas of the building that were not damaged by fire can be spared damage by water. Removing toilets can in effect create a drain in the floor.

SUMMARY

The task of high-rise firefighting is a monumental one for the incident commander, fire officers, and firefighters. The incident commander must gather information from the various sectors, formulate strategies, and deploy firefighters. The various sector officers are responsible for their own distinct parts of the operation, which play a tremendous role in a large operation. Company officers must be ever vigilant in operating under mostly adverse and unknown conditions. High-rise fires are not an everyday occurrence and no two are the same. Most firefighters gain their fire experience from fires in smaller structures, such as private dwellings, multiple dwellings, and commercial occupancies. Firefighters must be aware of changing conditions and other variances that the officer or sector commander may not see. All members of the fire service must avail themselves to as much knowledge as possible in the area of high-rise firefighting. Various types of construction features must be understood. Firefighters must understand which features assist and which hamper their operations. The use and knowledge of building systems such as HVAC,

standpipes, elevators, and communications play an integral part of fire operations in these buildings. Fires cannot be fought successfully without them. Fire service personnel must also understand the theories behind smoke movement, such as stratification and stack effect. The idea that colorless, poisonous, odorless gases may develop and kill both firefighters and civilians above and below the fire floor cannot be emphasized enough. Specific operations such as establishing sectors and deploying engine, ladder, squad, and rescue companies must be done in an organized manner. There is no room for freelancing at fires, especially in high-rise buildings. Units must report to the incident commander ready to go to work and then follow specific instructions as to where to operate. The incident commander and all sector officers should maintain logs for their sectors that account for all companies operating and their locations within the building. Units must be available to relieve companies that are operating and be prepared for expanding operations. Additional personnel must be requested immediately, taking into consideration the reflex time involved from recognizing the need for help to actually getting the help. Once help arrives at the scene, it still will take several minutes to get them deployed where needed. One of the most difficult tasks is getting the resources in place. More than sufficient help must be sent if there is any indication at all of a working fire.

Companies can always be returned if they are not needed, but it is very difficult to play catch-up at high-rise fires.

Finally, familiarization drills and prefire planning can assist fire operations. The time to know the layout of these buildings and any unusual conditions is before a fire occurs.

REVIEW QUESTIONS

1. List some major construction differences between older high-rise buildings and newer ones.
2. What is the purpose of building fire pumps?
3. What should be posted near the fire pump?
4. True or False? The bypass valve on the fire pump should be in the closed position.
5. For fire pumps found in pairs how should the pressures be set?
6. Who can authorize the shut down of fire pumps?
7. True or False? HVAC systems serve multiple floors.
8. List some methods of communicating at high-rise fires.
9. Describe stratification and stack effect.
10. What are some methods of ventilation at high-rise fires?

ACTIVITIES

1. Do a review of the tallest building in your department's district. What strategies would you implement for an upper-floor fire?
2. Review your department's standard operating guidelines for high-rise fires. Make a comparison between your department's SOGs and those presented in this chapter.
3. Does your department have the resources to combat a high-rise fire? If not, where would additional resources come from?

Chapter

16

VEHICLE FIRES

Learning Objectives

Upon completion of this chapter, you should be able to:

- Describe the common types of vehicle fires.
- Describe the hazards associated with fires involving passenger cars, vans, and light-duty trucks.
- Describe the hazards associated with fires involving semitrailers and trucks.
- Describe the hazards associated with fires involving recreation vehicles.
- Describe the hazards associated with fires involving heavy construction equipment.
- Describe the hazards associated with fires involving buses.
- Apply strategic goals and tactical objectives to these incidents.

CASE STUDY

A large column of smoke could be seen from the apron of South Trail Fire Department, Station 3 in Fort Myers, Florida. The column appeared to be in the general vicinity of a sizable manufactured home community. As the firefighters began to don their PPE, the dispatcher alerted the station and confirmed the location of the fire. The initial call was for a trailer fire. Dispatch advised that they were receiving multiple calls and the unit was reportedly fully involved. The response to the incident was Engine 22, Ladder 30, Engine 23, Squirt 31, and Battalion Chief 4.

Battalion 4 and Engine 22 were first to arrive on the scene and actually found three recreation vehicles fully involved, with flames impinging on two other vehicles on either side of the fire. The vehicles, large motor homes ranging in size from 30 to 40 feet long, were parked approximately 2 to 3 feet apart in a storage area in the rear of the community.

Engine 22 was directed to pull two preconnected 1¾ lines and attempt to protect the exposures. Ladder 30 and Engine 23 arrived on the scene shortly after the two lines were deployed. As Engine 23 established a water supply via a nearby hydrant, the firefighters on the scene could hear the relief valves from liquid petroleum tanks going off. Engine 23 pulled two more 1¾ preconnected hose lines and attempted to direct the streams in the area of the tanks. The only clue to the location of the burning tanks was a flame shooting horizontally out of the side of the vehicle, which impinged almost immediately on the motor home next to it. The firefighters had to direct the streams toward the base of that flame where they assumed the tank was. This tactical method proved successful and the tanks were cooled. By cooling the tanks, the relief valve closed, allowing the fire to be brought under control.

The crews soon found another problem though. There was a significant lack of access to the interior of the vehicles due to their close proximity to each other. A firefighter wearing typical PPE and self-contained breathing apparatus (SCBA) could not effectively force open doors or advance hose lines through the tight area. Overhaul was ultimately accomplished by laddering the vehicles and entering through the front windshield of the motor homes.

Three of the motor homes were a total loss, two suffered substantial damage, and two others suffered moderate damage. The total financial loss was nearly $250,000. This fire, although classed as a vehicle fire, challenged the fire department to apply solid fireground strategies and tactics.

■ **Note**
The typical car fire can generally be brought under control with one engine company using one hose line and one set of forcible entry tools.

!**Safety**
Plastics, foam rubber, fuel sources, alternative fuel sources such as LP, airbags, drive shafts, air-conditioning systems, bumpers, tires, brakes, and wheels can all present serious hazards to firefighting personnel.

■ **Note**
A company officer must realize that, in the case of vehicle fires, usually he or she remains the incident commander.

INTRODUCTION

Vehicle fires are often treated as routine incidents in the fire service, but they should not be. The typical car fire (**Figure 16-1**) can generally be brought under control with one engine company utilizing one hose line and one set of forcible entry tools. However, car fires can be completely atypical. When cars, minivans, and light-duty pickup trucks burn, they can arguably be considered mobile hazardous material fires. The plastics, foam rubber, fuel sources, alternative fuel sources such as liquid petroleum (LP) and hybrid gas-electric hybrid, airbags, drive shafts, air-conditioning systems, bumpers, tires, brakes, and wheels can all present serious hazards to firefighting personnel. Additionally, cars, vans and pickup trucks are not the only vehicles on the road. Consider the challenges encountered with fighting fires involving semitrailers, recreational vehicles, heavy equipment, and vehicles used for transporting large numbers of people, such as school buses and tour buses.

Each vehicle fire presents its own hazards. Proper strategic goals, tactical objectives, and tactical methods are crucial to a positive outcome and mitigation of the incident. A company officer must realize that, in the case of vehicle fires, usually he or she remains the incident commander. A vehicle fire without complications does not bring a chief officer or even a second engine company to the scene. As discussed in Chapter 4, the company officer must take on the role of all aspects of the incident management team. The company officer usually is the

Figure 16-1 *Vehicle fires are generally handled with one engine company. The firefighters must be aware of potential hazards.*

safety officer, the accountability officer, the liaison, operations, and the investigator. Additionally, many departments do not have the benefit of well-staffed engine companies and therefore the company officers are also involved in extinguishment. If media arrive on the scene, which they sometimes do, especially if the vehicle fire is blocking five o'clock rush hour traffic, then the company officer must also become the public information officer.

The company officer should have experience fighting vehicle fires and be able to recognize the need for additional resources if necessary. Realistically, though, the majority of vehicle fires can be handled with one engine company. Therefore it is crucial to understand the basic concepts of strategy and tactics for vehicles fires.

VEHICLE FIRES IN CARS, VANS, AND LIGHT-DUTY PICKUP TRUCKS

■ Note
Fires in cars, vans, and light-duty pickup trucks usually make up the most majority of vehicle fires to which fire departments respond.

Fires in cars, vans, and light-duty pickup trucks usually make up the majority of vehicle fires to which fire departments respond. These types of fires cannot be taken lightly. Vehicles can be extremely dangerous, even deadly, when approached with a lackadaisical attitude. It is important for firefighting personnel to have a complete understanding of operational knowledge of vehicle fires. Every vehicle is different (**Figure 16-2**). The manner in which doors open and lock, the way hoods and trunk lids open (**Figure 16-3**), the configuration of seats, the location of the battery, the personal protective devices, such as airbags and

Figure 16-2
Passenger vehicles come in all sizes and shapes.

Figure 16-3
Firefighters should be familiar with vehicle components and their locations, and ways to access hoods and trunks.

■ **Note**
Even the most seasoned firefighters regularly come upon vehicles with new hardware or different hazards that they have never encountered.

seat belts, the type and location of the engine, the type of fuel, and so forth, are important components to consider when fighting car fires. It would be impossible to consider every conceivable option or type of vehicle in this text. Additionally, new cars are being designed, engineered, and created almost daily; therefore, even the most seasoned firefighters regularly come upon vehicles with new hardware or different hazards that they have never encountered. Nonetheless, basic principles of firefighting strategies and tactics can be applied generally to all types of cars, vans, and light-duty trucks (including sport utility vehicles). It is still important for firefighters to have a good knowledge of the workings of various vehicles and generalities of various manufacturers. This can be accomplished through reading articles, taking note of how hoods open and doors lock when on the scene of other incidents such as motor vehicle crashes, visiting local car dealers and mechanics, and so forth. It can be considered preincident planning for vehicle fires.

Hazards Encountered

The hazards encountered with passenger vehicles, vans, and light-duty trucks are numerous. These hazards can be broken down into three broad areas:

1. Vehicle components
2. Alternate fuels
3. Hazardous cargo

Vehicle Components Vehicle components are the mechanical and other parts of the vehicle that can be dangerous. Firefighters must keep in mind that the vehicles are not designed to operate correctly when burning. Therefore, one cannot rely on brake systems working or any car part maintaining its integrity under fire conditions. Vehicle components and some potential hazard to firefighters include:

- Drive shafts can explode and throw shrapnel when heated.
- Air-conditioning systems emit deadly phosgene gas when burning.
- Bumpers with built-in shock absorbers can decompress and fly off with explosive force.
- Wheels, motors of some vehicles (such as early-model VW Beetles), and components of the airbag systems are sometimes made from magnesium, which reacts violently with water applied incorrectly.
- A catalytic converter can reach temperatures in excess of 1,500°F when functioning improperly.
- The hazardous by-products of combustion of various plastics and foam rubber.
- Battery acid and exploding car batteries.
- Exploding tires and any compressed component of the engine or drive train.
- Hood and/or hatch pistons.
- Undeployed airbag and safety systems.

Note that airbags, airbag components, and actuators can be virtually anywhere in a vehicle. For example, in late-model BMW sport utility vehicles the side impact airbag actuator is in the "A" post.

Late-model vehicles are typically fitted with driver- and passenger-side front airbags. Other options available now and in the near future include: side impact airbags, side impact air curtains, dashboard knee airbags, rear seat airbags, seatbelt airbags, and foot airbags. Honda has now even made available front airbags for motorcycles. Furthermore, development is underway for front bumper and hood airbags to protect pedestrians in the event of a crash.

Newer model vehicles have passenger-side smart airbag systems that may not deploy in a vehicle crash when there is no passenger or a small passenger; however, this airbag may still deploy under fire conditions.

In fire conditions the firefighters must consider all interior parts of vehicles as potential airbag deployment zones and avoid these areas. Unless there is the need to rescue a victim, it is best to keep all body parts outside the deployment zone.

Most manufacturers utilize bright yellow wires to indicate airbag systems. DO NOT CUT any bright yellow wires in any vehicle.

- Other safety systems. Rollover protection is available in many high-end vehicles, such as Mercedes convertibles.

In such a case the roll bar deploys when the vehicle reaches a predetermined angle (as would occur in a rollover). When this happens, the roll bar activates with explosive force. As with airbags, these areas must be avoided.

Alternate Fuels With the federal government pushing to reduce emissions, car manufacturers are exploring alternate fuel types, such as hybrid gas-electric, hydrogen-powered, biodiesel-powered, and vehicles that operate on liquid petroleum gas (LPG) or compressed natural gas (CNG). Most of these are fleet vehicles, such as those of law enforcement agencies. The fuel burns cleaner but the dangers of having a compressed cylinder of LP in the trunk is obvious. The only way to know a vehicle operates on LP or CNG is by having prior knowledge or by noting a slightly different type of fuel door.

Increases in fuel costs have pushed up sales of hybrid gas-electric vehicles. Many different manufacturers produce hybrid vehicles. In some cases gas-electric is the only option in a specific model, such as in the Toyota Prius. However, in many cases, gas-electric is simply one of the fuel options for the consumer. For example, the Ford Escape and Honda Accord as well as many others have hybrid versions of their standard gasoline-powered vehicles. Identifying these types of vehicles can be challenging, as sometimes the only identifying mark is a differently shaped antenna or a hybrid label on the rear of the vehicle. The other obvious way to identify a hybrid is the bright orange components under the hood. But, as with most vehicle fires, the hood can sometimes be difficult to open. To find out more details about hybrid vehicles it is important to contact automobile dealers in your area. Most manufactures have guidelines readily available for emergency responders.

DO NOT CUT orange wires in a hybrid vehicle.

Other alternative fuel sources are becoming more popular or are on the horizon. For example, hydrogen fuel cell-powered vehicles are under development. One of the obvious issues that may reduce the availability of these vehicles is the lack of a hydrogen fueling infrastructure. Still, because the emission of hydrogen vehicles is essentially water, look for these types of vehicle to be commonplace in the future. Another fuel source on the horizon is in vehicles known as tribrids. These vehicles operate much like hybrids but have a third component utilized for energy that is taken directly from the environment, such as solar power. Other fuels sources that are currently under development include solar-powered and straight electric. However, in most of these cases, the elements will prove standard in that these vehicles have fuel cells or batteries. The type of charging of the fuel cell or battery will vary depending on the fuel type. As new vehicles develop, the hazards encountered by the various fuel sources will be unique and firefighters will have to constantly stay knowledgeable of the changes coming from vehicle manufactures.

Another alternative fuel is E-85. This fuel is a combination of gasoline (15%) and ethanol (85%) that has recently begun to be used in the Midwest, primarily in Illinois and Minnesota. Fires involving E-85 fuels or any gasoline/ethanol mixture containing more than 10% ethanol should be treated the same as

Safety

DO NOT CUT orange wires in a hybrid vehicle.

any polar/water-miscible, flammable liquid. To extinguish these fires, you must use alcohol-resistant foam. E-85 fuels are highly flammable and heavier than air. The fire burns less bright than gasoline, but is still visible in the daylight. An E-85-fueled vehicle does not have any special markings and anyone can buy the conversion kit to retrofit their automobile. There are over six million E-85 vehicles already on the road today in the United States, but not all of them are using the fuel.

Cargo It is quite impossible to know what is in passengers cars, vans, and light-duty trucks. Anything and everything can be in a trunk, back of van, or pickup truck. If the owner is available, it is critical to ascertain whether potentially dangerous items or hazardous materials are in the trunk before firefighting crews approach the vehicle to fight the fire.

Potentially hazardous cargo includes:

- Paint and lacquer thinners
- Gas cans
- LP tanks
- Welding equipment
- Biohazardous waste, such as diabetic needles
- Any unknown item that can be dangerous

> **!Safety**
> It is critical to ascertain whether potentially dangerous items or hazardous materials are in the trunk before firefighting crews approach the vehicle to fight the fire.

Strategic Goals and Tactical Objectives

There are many different types of vehicles; therefore, the strategic goals and tactical objectives apply generally to all passenger vehicles.

Firefighter Safety Firefighter safety applies in the broad sense similar to that of structure fires. As discussed in Chapter 6, proper personal protective equipment, incident command, and accountability apply likewise to vehicle fires. In the detailed sense, firefighter safety when fighting vehicle fires begins with the number-one incident priority, life safety. Protect the firefighter first by properly placing the fire apparatus to block traffic. **Figure 16-4** shows two options that may be used to block traffic. Do not hesitate to block two lanes of traffic. Consider the hose lines stretched in the road and low visibility. If necessary, shut the road down to ensure firefighter safety. When possible, try to position the apparatus uphill and upwind of the vehicle on fire.

The determination of passenger life safety is next and is sometimes easy in a vehicle. If the owner of the vehicle is on the scene and readily available, it is important to find out whether anyone is in the vehicle. If not, then the approach to extinguishment can be very safe, directing hose lines from a distance until the vehicle is cooled enough to make a safe approach. Approach should be made on a 45° angle to the front or rear, not straight on. If no vehicle owner or occupant is in the immediate vicinity or there are suspicious circumstances, then the vehicle must be searched thoroughly. This search requires firefighters to venture much closer to extinguish the blaze, thus placing themselves in a much more dangerous

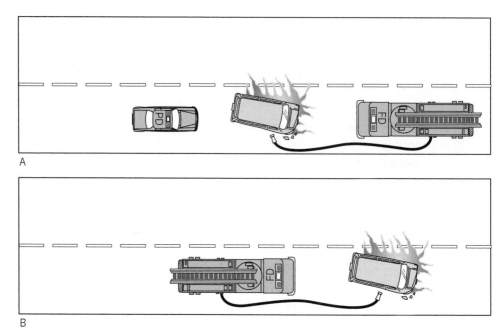

Figure 16-4 *Two traffic control options. A. Law enforcement is blocking traffic. B. The engine company provides the lane blockage. Note that the hose line should come off the curb side of the engine. In some cases both lanes of traffic should be stopped.*

position. It is crucial to understand potential hazards discussed previously in this chapter and approach the vehicle accordingly. Once the conditions permit, the vehicle's wheels should be chocked.

Search and Rescue In vehicles, search and rescue is relatively simple; however, it goes hand in hand with firefighter safety. If firefighters are absolutely sure that no one in is the vehicle, the search can be deferred until a safe approach can be made. However, if there is any question at all, the vehicle must be searched thoroughly. Start with the area most likely to have vehicle occupants, usually the front seat. Also check the floor board area. Check the back seat next, and in passenger vans (**Figure 16-5**), check all the seats.

In pickup trucks, check the bed. This check can be as simple as a visual inspection if the bed is open, or it may take a hands-on search if the bed is covered and there is no or limited visibility. If the scene of the fire is suspicious or the vehicle is abandoned, make sure the trunk is searched. This may be difficult to do expediently due to the difficulties generally encountered when forcing trunk lids. Be aware that the vehicle may be extremely hot, depending on the level of involvement, and visibility inside the vehicle can be at zero. The safest approach to search and rescue is to perform it simultaneously with the extinguishment. The fire should be swept away from the passenger compartments until a search can be made. Additionally, tools may be required to cut the seat belt to retrieve a victim. Be prepared for this operation.

Figure 16-5 *The interior layout of a typical minivan.*

Victims found in burning vehicles can be severely injured. If a victim is a viable patient, consideration must be made to medically treat the patient accordingly. In many cases, sufficient personnel are not on scene to medically attend to the patient and also continue to extinguish the fire. In this scenario, the incident commander must decide whether it is safe to abandon the firefight until adequate personnel arrive on the scene. Considerations to be made in the decision would be dangers involved with the vehicle such as LP tanks, exposures, and other hazards. In cases where only two or three firefighters are on the scene, effective patient care and extinguishment cannot be done concurrently.

Special situations: There may be special situations near areas of the United States where illegal immigration is prevalent. In these areas, particularly near the United States-Mexico border, people may be hidden in cars and trucks. In cars they are usually behind or under seats, although smugglers are consistently becoming more creative. Recently, people were actually found sewn into the seat. In this situation people are placed into the seats after springs are removed and the seat covers are sewn in over them. In trucks, people may also be found in cargo spaces or inside boxes. It is important to listen for noises indicating hidden victims, particularly if there is no driver or there are suspicious circumstances.

Evacuation It is rare to have to evacuate an area as a result of a passenger car fire. However, in cases where there is flame impingement on exposure buildings,

Figure 16-6
Exposures may include other vehicles, structures, vegetation, and hazardous materials storage containers.

the building in danger should be evacuated. Furthermore, if hazardous materials are found in the vehicle and there is a danger to life as a result, evacuation must be considered, depending on the type of material and extent of the release.

Exposure Protection Exposure protection may include protecting buildings or other vehicles near the burning vehicle. Sufficient lines must be placed to keep the fire from extending. Also, if the vehicle is in a parking lot and keys are available to other vehicles that are in danger, they may be moved away from the burning vehicle. Exposure protection may also include protecting dangerous cargo, such as LP tanks (**Figure 16-6**).

Confinement Confinement strategies and tactics should focus on keeping the fire to the area of origin. If the fire is in the engine compartment, the objective should be to keep it from extending into the passenger compartment. Generally, this objective is accomplished through direct extinguishment of the fire. However, a hose line may need to be placed into the passenger area and directed under and over the dashboard to prevent extension.

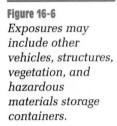

■ **Note**
One handline of 1½-inch or greater is sufficient to extinguish a typical vehicle fire.

▌**Safety**
Avoid the use of booster lines and always be prepared to move to larger and/or more lines.

Extinguishment Water is the extinguishing agent of choice for most passenger cars, vans, and light-duty truck fires. However, the incident commander must be aware of flowing fuel fires and consider the application of foam. For small engine compartment fires, dry chemical extinguishers can be very effective; however, they have no cooling capabilities and reignition must be carefully monitored. Usually one handline of 1½ inch or greater is sufficient to extinguish a typical vehicle fire. Avoid the use of booster lines and always be prepared to move to larger and/or more lines.

Ventilation Ventilation of passenger vehicle fires is not a huge consideration. However, extinguishment is easier if the vehicle's windows are down or the doors are open. In some cases, door windows may need to be broken. If the vehicle is fully involved and the fire is hot enough, the front windshield will usually melt and the side and rear windows will break, thus self-venting.

Careful consideration and size-up must be made to ensure that the burning vehicle is not in a backdraft situation. The indicators are the same for vehicle fires as structure fires. In this case, the only safe way to vent the vehicle is from a safe distance. Usually a straight stream of water can be directed at the side windows. Cool water on the hot glass will cause the glass to break and thus vent the car. Reignition occurs, but firefighters are a safe distance away.

An issue that firefighters will face in the future regarding vehicle ventilation will be the type of glass used for side and rear windows. Vehicle manufacturers have researched for many years materials that will reduce the occurrence of occupants being ejected during vehicle rollovers. One of the solutions to this problem is the installment of impact-resistant safety glass on all the windows of the vehicles, not just the windshield. This feature will obviously create unique challenges when ventilating vehicles.

In passenger vehicle fires inside parking garages, ventilation via fans, smoke ejectors, and positive pressure ventilators may be required to direct the smoke in a desired path. Vehicle fires inside residential garages (**Figure 16-7**) should be

Figure 16-7 *A vehicle fire in a garage that rapidly extended to the structure.*

Figure 16-8 *Be alert during salvage. This is an undeployed airbag.*

treated operationally as a structure fire and ventilation in this scenario is discussed in Chapter 11.

Overhaul Overhauling a passenger car consists of opening all compartments of the vehicle and checking for extension.

Salvage Salvage may consist of moving personal property out of the vehicle, such as the owner's or occupant's items in the trunk or passenger area. If the vehicle registration and proof of insurance can be salvaged from the glove compartment (**Figure 16-8**), paperwork will be easier for the investigating authority and insurance claims easier for the owner.

■ Note
If the vehicle registration and proof of insurance can be salvaged from the glove compartment, paperwork will be easier for the investigating authority and insurance claims easier for the owner.

SEMITRAILER AND TRUCK FIRES

For the purpose of this text, semitrailers (**Figure 16-9**) and trucks can be any type of large commercial truck either pulling a trailer or having a cargo box on the same frame (**Figure 16-10**). The key to the definition is that the vehicles have some type of cargo and they are too large for personal use. The cargo can be anything. This section focuses on fires and not on hazardous materials. However, it is imperative for incident commanders to be cognizant of hazardous materials and apply appropriate strategies and tactics for the incident at hand. For example, there would be different strategies and tactics for a heavily involved truck hauling boxes of cereal than for a fully involved truck hauling pesticides.

Figure 16-9   *The tractor portion of a tractor-trailer.*

Figure 16-10   *A cargo box on a chassis, in this case pulling a small cargo trailer.*

■ Note

Large truck fires are different from passenger vehicle fires and must be treated accordingly.

!Safety

It is critical that company officers and incident commanders be familiar with placarding, container shape, and other indicators that would alert the responders to the possibility of hazardous cargo.

Finding out the contents of the vehicle is a crucial part of the overall strategies and key to safe mitigation of the incident.

Large truck fires are different from passenger vehicle fires and must be treated accordingly. To illustrate, in the field of emergency medical services a pediatric patient is not treated like a small adult. The same is true for large truck fires in comparison to passenger vehicle fires: The basic strategies and tactics can be applied, but critical differences must be understood and implemented.

Hazards Encountered

The hazards encountered with large trucks and semis can be abundant. Potential hazards involving the cargo can be extreme, depending on the cargo. It is critical that company officers and incident commanders are familiar with placarding, container shape, and other indicators that would alert the responders to the possibility of hazardous cargo. See Chapter 18 for further information. Company officers and incident commanders should be educated on the hazardous materials operations, at a minimum.

This chapter focuses on the hazards common to those encountered on all type rigs and not on specific hazards created as a result of particular products being transported.

Figure 16-11   *Tractor cabs can present accessibility problems.*

Figure 16-12   *The air suspension system under a trailer.*

!Safety

When working around or under large trucks, one must be alert to air suspension systems. These systems may fail under fire conditions and cause the apparatus to drop several inches. Therefore, cribbing should be in place before any attempts to work under these vehicles.

A hazard typical to all large trucks is the inaccessibility of the cab area, as shown in **Figure 16-11.** The cab can be several feet above ground and this may cause a safety problem when moving hose lines and tools. One should not rely on the safety or stability of built-in handles or steps. When working around or under large trucks, one must be alert to air suspension systems (**Figure 16-12**). These systems may fail under fire conditions and cause the trailer to drop several inches. Therefore, cribbing should be in place before any attempts to work under these vehicles.

If nonhazardous cargo is burning in a large trailer, access to the trailer can be dangerous. The trailer may have two double doors in the rear and side doors (**Figure 16-13**). All doors should be opened before access is made to allow for better ventilation and provide for better access points. The incident priorities must be implemented in these type fires; therefore, if there is no immediate life safety hazard, then the safety of firefighters should be considered paramount. Incident commanders shall not place firefighters in unreasonably unsafe situations in order to save some type of commodity or product.

Saddle tanks are typical on large trucks. These are the large fuel tanks that are located on the side of the cab, usually under the driver and passenger doors (**Figure 16-14**). These tanks generally carry diesel fuel but can carry gasoline, and they range in size from 25 to 100 gallons each. Saddle tanks are vulnerable and prone to damage from vehicle crashes and road debris. Flowing fuel fires

Figure 16-13 *A roll-up-style cargo door.*

Figure 16-14 *Tractor cab. Note the saddle tank locations.*

must be recognized and dealt with quickly. These flowing fuel fires may be caused by a leak in the saddle tank, and the leak must also be stopped or limited. Newer, large tractor trucks may have ground effects (fiberglass body molding) that obscure the saddle tanks. Prior knowledge of various trucks and styles will be helpful to pick up on this type hazard.

Other hazards that are typical of large trucks and semitrailers are those typical of passenger cars, as discussed earlier in this chapter.

Strategic Goals and Tactical Objectives

The strategic goals for large truck fires vary considerably, depending on the type and size of a vehicle, the exposures that are in danger, and type of cargo. The goals may be to keep the fire from spreading to the trailer or they may be to keep the cargo from igniting. In any event, size-up is extremely important on arrival at a semitrailer fire, especially with involvement in the cab or engine compartment. In addition to typical size-up considerations, it is important to recognize the type of cab encountered.

The various types of semi and truck cabs also dictate the tactical objectives chosen to fight the fire. Engine compartments may be in the front with the hood opening toward the cab or opening away from the cab. Additionally, some older cabs were engineered with the engine under the cab, such as some older Ford Motor Company and International trucks. When the engine is under the cab, special knowledge of the individual truck is needed to open the cab. A good source of this information can come from the truck driver if he or she is available.

A good size-up and strong knowledge base of trucks can also provide information about the type of material of the cab itself. Some materials used in modern cabs are steel, fiberglass, and plastic.

Firefighter safety Firefighter safety on large truck fires is closely related to firefighter safety on passenger vehicle fires. The exception is understanding the possibility of large amounts of hazardous cargo. Furthermore, understanding and overcoming the typical hazards inherent to large vehicles will enhance firefighter safety.

Proper response of personnel also helps to ensure firefighter safety. A large truck fire should have minimum response levels greater than that of passenger vehicles. The exact number is determined by specific incidents and available resources; however, a minimum of two engines and an incident commander is required.

■ **Note**
All trucks and especially semitrailers with cab-over sleeping quarters must be thoroughly searched as soon as reasonably possible.

Search and Rescue All trucks and especially semitrailers with cab-over sleeping quarters must be thoroughly searched as soon as reasonably possible. If the truck driver is not on the scene, the incident commander must consider the possibility that the cab is occupied. A cab-over-style cab should give clues to the likelihood of possible occupants. These style cabs generally contain sleeping areas and other accommodations for the driver. Hose lines should be placed to protect the passenger compartment and sleeper area and a search of these spaces in particular should be conducted as soon as safely possible.

The cargo areas of trucks should also be checked for possible occupants.

Evacuation The level of evacuation in truck fires should be based on the likelihood for movement of the product burning and the level of danger to people in surrounding areas. The evacuation may only be a safe zone around the truck or it may include many square miles if certain types of hazardous materials are involved.

Exposure Protection Exposure protection for large trucks is similar to that of passenger vehicles. Typically the exposures may be buildings, other trucks, or vehicles. A unique aspect of tractor trailers is the possibility of protecting the trailer from a cab fire or a cab from a trailer fire. If the cab is not involved, the latter may be accomplished simply by separating the cab from the trailer. This maneuver requires knowledge in operating and separating tractor trailers. Also, it depends greatly on the location and extent of the fire. Naturally, the safest tactic should be chosen. Properly placed hose lines may be a good option.

Confinement Confinement in a burning truck means confining it to a specific area or areas of the truck. For example, in the case of a truck with the engine compartment heavily involved, the strategic goal may be to keep the fire from spreading to the passenger compartment. This goal can be accomplished by proper extinguishment.

Extinguishment Extinguishment requirements depend largely on the level of involvement. For example, if a large trailer and the material in it is burning, the incident commander should use one of the flow formulas discussed in Chapter 3 to determine the amount of water required for extinguishment. Conversely, an engine compartment fire may only require a single, preconnected $1\frac{3}{4}$-inch hose line. With large trucks, a backup line should also be stretched out for safety and extinguishment.

Alternative extinguishing agents must be considered for different types of fuels and cargo such as dry chemical extinguishers or foam application on gasoline or diesel fires.

Ventilation For the cab and passenger area of trucks, ventilation is usually accomplished through opening doors and/or windows.

Trailers can be ventilated by opening all doors or by cutting ventilation holes in the top or sides of the trailer. It is difficult to vent a trailer if the only opening is the rear doors. Positive pressure ventilation can be utilized in this case if enough room is left at the top of the door to allow the escape of the heat, smoke, and gases. However, positive pressure ventilators should be used in conjunction with properly placed hose streams.

Overhaul All compartments of the vehicles must be checked for extension and to ensure extinguishment. Furthermore, hot metal will steam long after extinguishment. This phenomenon sometimes misleads officers into believing that the smoke they see is actually steam from the hot engine. It is imperative to make sure it is steam and thus prevent a rekindle.

Overhaul in large trucks may also include off-loading the product to make sure the fire is out and no boxes or other materials are smoldering. Heat detectors and infrared cameras are very helpful in both of these overhaul situations. This operation is labor-intensive and requires additional resources.

Salvage In many cases a truck is a person's livelihood, and therefore salvaging as much of the vehicle as possible is prudent. Many trucks have televisions, VCRs, DVD players, and other electronic equipment. Also, there may be personal belongings such as clothes in the cab of the truck.

Salvaging any product from the trailer or box is usually accomplished initially by good strategic goals and tactical objectives, as discussed earlier. Working with the owner/operator helps to determine complete salvage operations. For example, a second rig may have to come to the scene to off-load the material, which may require forklifts or extra personnel that can only be supplied by the owner of the truck.

■ **Note**
Overhaul in large trucks may also include off-loading the product to make sure the fire is out and no boxes or other materials are smoldering. Heat detectors and infrared cameras are very helpful in both of these overhaul situations.

RECREATIONAL VEHICLES

Recreational vehicles (RVs) come in many shapes and sizes and are basically a building with wheels. Some examples include large class A motor homes, class C motor homes, passenger bus conversions, van campers, travel trailers, truck campers, and pop-ups (also known as tent campers).

Class A motor homes (**Figure 16-15**) are large RVs with a modular body on a frame. These are self-contained units that can have air conditioners, refrigerators, bathrooms, and generators.

Class C motor homes (**Figure 16-16**) are much like class A except they have a truck or van chassis and cab with a modular camping unit attached to the frame on the back. This type is attached to the cab and passengers can access the cab from inside the unit.

Passenger bus conversions are large buses that have been converted to provide very comfortable and even luxurious accommodations. These are the type of buses that are characteristic for musicians on a concert tour; however, many people in the general population are purchasing these RVs.

Van campers are usually oversized custom vans that have been converted into campers. They can have amenities such as small bathrooms and cooking areas with refrigerators.

Travel trailers, truck campers, and pop-ups (**Figure 16-17**) are all campers that are not permanently attached to a vehicle. The accommodations and amenities vary greatly.

Figure 16-15 *A class A motor home.*

Figure 16-16   *A class C mobile home.*

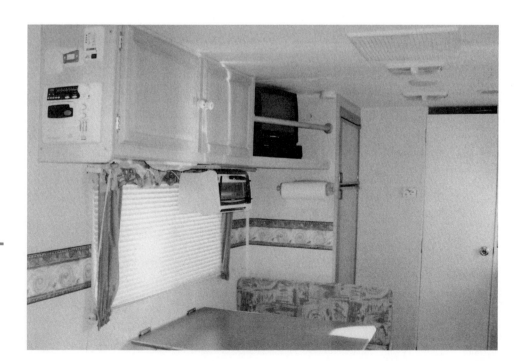

Figure 16-17   *Pop-up (or tent) camper. Many have air conditioning, refrigerators, sinks, stoves, and even bathrooms.*

Hazards Encountered

Hazards specific to RVs include all the hazards thus far discussed in this chapter plus those unique to mobile homes. Hazards unique to RVs include LP tanks (**Figure 16-18**). These tanks can be located anywhere on the vehicle and can vary in size. The only way to know absolutely where LP tanks are located is if they are actually seen or the vehicle owner is available. Further, an LP tank can be a problem should the relief valve open, as in the case study.

Another hazard unique to some RVs is anhydrous ammonia, found in some refrigeration systems on all types of RVs.

Yet another hazard is the vehicles' waste or gray water holding tanks. Gray water is water that is held after use such as washing dishes or brushing teeth. Waste tanks are the tanks that hold human waste from the bathroom on the RV. In either case, if these tanks have waste in them, they could become a bio-hazardous danger if they fail.

Strategic Goals and Tactical Objectives

The strategic goals and tactical objectives for RVs are similar to either passenger vehicles or large trucks, depending on the size and configuration of the RV.

Firefighter Safety The most important aspect of firefighter safety specific to RVs is knowledge and training. Knowledge of various types of RVs and how their

Figure 16-18 *Travel trailer camper. Note the slide out on the side and the location of the LPG tanks.*

Figure 16-19   *Interior of a travel trailer camper. The table converts to a bed. Consider the difficulties of searching such a confined area.*

various systems work is invaluable information. Training on performing search and rescue in very small and confined spaces enhances firefighter safety (**Figure 16-19**). Furthermore, the firefighter must understand fire dynamics and be keenly aware of the speed at which the fire can travel through an RV.

Search and Rescue Search and rescue is crucial in RVs if the RV is suspected to be occupied. Clues are sometimes obvious, such as location. For example, an RV in a campground is likely to be occupied. The rules for one- and two-family dwellings should be applied. Specifically, size-up the situation, considering the time of day, the presence of cars near the RV, and folding chairs and BBQ grills around the RV which give it the lived-in look. Conversely, if the RV is in a storage yard, it may not be occupied. Search in a RV may prove difficult due to the small size of the floor area and unfamiliarity with the layout. In many RVs, tables convert to beds, bunk beds and sleeping areas may be found above the driver's area, and there may be bedrooms in the back of the RV. All the possible sleeping areas must be thoroughly searched.

The search should be completed as soon as reasonably possible with life safety being the number-one priority. This priority may be accomplished through initial extinguishment of the fire with a subsequent or simultaneous search with extinguishment. This may be the case if one engine company is on the scene with three firefighters and the fire is in the engine compartment. In order to save the lives of possible occupants, the most prudent tactic would be to extinguish the fire immediately, containing it to the engine compartment and then searching for victims. This scenario can change with different variables. However, the specific method used depends on the extent of the fire, the type of RV, and the amount of personnel arriving on the scene. The most important factor is to keep life safety foremost on the list of priorities.

If firefighters enter into a burning RV, they are entering into an immediately dangerous to life and health (IDLH) atmosphere and a rapid intervention crew (RIC) should be standing by as is currently required by Occupational Safety and Health Administration (OSHA) and various state agencies.

Evacuation Evacuation of RVs naturally includes the unit itself and any surrounding units that are in danger. Consideration must also be given to greater evacuation if the LP tank is involved or there is an anhydrous ammonia leak.

Exposure Protection Protecting exposures from burning RVs is similar to passenger vehicles. If the RV is a travel trailer, then separation from the tow unit should be considered early in the incident. Stored RVs are usually placed very close together to allow room for as many units as possible, creating an exposure problem. The best protection against exposure is to move the exposed vehicles and RVs away from the fire. This move can only be accomplished if keys can be found. If this is not possible, then hose lines must be placed and streams directed on the exposures to protect them. All RVs have windows, which can make matters worse. The windows allow for ignition of curtains and other combustible materials inside the exposure RV through radiation.

Another option to reduce the fire spread in RV storage areas is to actually create a firebreak by removing one RV on each side of the burning RV and then applying fire streams to the exposures. This too requires a means to move the large RVs.

Confinement Confinement of RV fires is the same as that of passenger cars and large truck fires.

Extinguishment Extinguishment of RV fires depends on the size and extent of the vehicle burning, hence the required fire flow. In dealing with a building with wheels, the fire flow formulas presented in Chapter 3 are applicable. But remember the additional fire load associated with construction materials and contents. The tactical objective for extinguishment is generally the same as that of large truck or passenger vehicle fires. The unique aspect that must be considered in extinguishment is the speed of the burn and the possibility of the hazards addressed earlier in this chapter.

Ventilation Ventilation of RVs can usually be accomplished by opening all windows and roof vents. It is rare to have to cut a hole in the roof of a RV for the sake of ventilation. Additionally, positive pressure ventilation is very helpful in ventilating RVs.

Overhaul Overhaul in RVs is similar to that of passenger vehicles and large trucks. All compartments must be checked for extension.

Salvage It is normal in various parts of the country for people to live in their RVs. Other people spend a considerable amount of time in them. Furthermore, many people have fond, sentimental memories from using their RVs on family vacations and trips. Therefore, salvage on RVs should be as meticulous and careful as with any one- or two-family dwelling. One should salvage an RV as if it were one's own family living in it or using it for vacation. This attitude will ensure a careful, complete job.

HEAVY CONSTRUCTION EQUIPMENT

Heavy construction equipment varies, as is the case with most cars, trucks, and RVs. Heavy construction equipment includes items such as front-end loaders (**Figure 16-20**), backhoes, cranes (**Figure 16-21**), tractors, graders, pavers, and rollers. The hazards encountered and strategic goals and tactical objectives for these type vehicles can fit with those of any one of the other vehicles discussed. However, the following areas introduce specific items that were not mentioned in preceding sections.

Firefighter Safety

Large tires may fail under fire conditions and can cause firefighter injury. Additionally, each piece of equipment has specific safety hazards such as

Figure 16-20 Typical front-end loader at a construction site.

Figure 16-21 *A large crane. These can have a large amount of fuel and lubricating fluids on board.*

hydraulically operated booms and buckets that need to be addressed during size-up and firefighting operations. The hot hydraulic fluid under pressure can present significant safety concerns for firefighters.

Search and Rescue

Search and rescue is usually not an issue for construction equipment fires because most equipment is designed so the operator sits outside. There have been documented cases, though, where operators were stuck in elevated buckets while the vehicle below was burning. Also, large fixed cranes can have the operator's cockpit on top of the crane. This was the case in a nationally known warehouse fire in Atlanta, Georgia. The crane operator was stuck in the crane while the building fire was impinging on the crane. The search was conducted by finding the location of all employees on the scene prior to the fire. The rescue was performed by a firefighter who was harnessed to a helicopter.

Evacuation

Evacuation is usually not an issue with construction equipment fires. However, consideration must be given to evacuating surrounding areas if large equipment becomes unstable and is in danger of falling into occupied buildings or streets.

Exposure Protection, Confinement, Extinguishment, Ventilation, Overhaul, and Salvage

These strategic goals and tactical considerations are addressed in other sections in this chapter and can generally be applied to heavy construction equipment.

BUSES

Strategies and tactics related to bus fires are common to large truck fires with the exception of a high life hazard. As with all vehicles, there is a wide variety of bus types (**Figure 16-22**), such as school buses, public transportation buses, and charter buses. The commonality in all types of buses is that they all have the capability to carry many people. This capability makes search and rescue, confinement, and extinguishment of the fire crucial. All the other strategic goals and tactical objectives can be applied to bus fires from other sections of this chapter.

Search and Rescue

If a fire starts in a bus, most capable people perform self-rescue and exit the bus. Problems arise, though, when the vehicle is involved in a crash and subsequently ignites. People can become disoriented in smoke and may not find their way out of the bus through normal or emergency exits. Search must be conducted as quickly and as safely possible and many fire service and emergency medical service (EMS) personnel must be on the scene to handle a possible mass casualty incident. All possible areas of the bus should be opened, such as emergency escapes, doors, and windows. This must be done in conjunction with proper hose placement so as not to push the fire into unwanted areas.

Figure 16-22 *Buses may have uses other than just transporting people. This bus is a mobile emergency command center.*

Confinement

Confinement must focus on keeping the fire away from passengers and in the area of origin. In most public transportation buses, the engine is in the rear of the bus and doors are in the front. Emergency exits can be located in the rear (on the side). Modern buses designate and design all windows to be emergency exits. School buses typically have the engine in the front with the normal door in the front and emergency exits in the rear and side windows. Regardless of the type or style of bus, protection of the escape routes is imperative.

Extinguishment

Extinguishment of bus fires is characteristically the same as that of all other vehicle fires; however, the unique design of buses and engine access should be considered, as should be the fire flow requirements. The bus driver is indispensable in providing knowledge of the various access points.

SUMMARY

There are literally thousands of different makes, models, years, and types of vehicles on the road every day. No fire service is immune to vehicle fires, and therefore all must have a complete understanding of fighting fires in vehicles.

Universal strategic goals, tactical objectives, and methods can be applied to many vehicle fires because many vehicle components are the same, including general items such as engines, drive trains, tires, wheels, air-conditioning systems, and fuel. Special purpose vehicles require special considerations, such as those created due to their size or the products they haul. Along with universal strategies and tactics, special considerations are presented in this chapter and are applied to vehicles such as semitrailers and trucks, recreation vehicles, heavy construction equipment, and buses.

Semitrailers and trucks are designed to carry some type of load or product. The load is often the critical factor in extinguishment and life safety.

Determining the content of the product being carried can change the strategies and tactics. Thus, it is of utmost importance for fire officers to understand identification of hazardous materials.

Recreation vehicles can be considered mobile, one-family dwellings. Proper search and rescue and fast extinguishment are important aspects of the strategies and tactics.

Heavy construction equipment and buses are unique in the problems they present. Construction vehicles are large and may provide special challenges for search and rescue, such as crane operators. Buses have the possibility for large life safety issues and may create mass casualty incidents requiring many resources.

Every firefighter will fight many vehicle fires in his or her career. Most vehicle fires are considered "routine" by responding personnel. However, there are too many variables and changing designs in vehicles for fires involving them to ever be considered routine.

REVIEW QUESTIONS

1. List three hazards associated with passenger vehicles.

2. Fuel tanks on semitrailers and large trucks that are located on the side of the cab, below the driver and passenger doors, are called what?

3. Describe one type of alternate fuel that may be encountered in passenger and fleet vehicles.

4. What must a firefighter be concerned with if a dry chemical extinguisher is used to extinguish a small engine fire?

5. (A) Cab-over-style truck cabs give clues to the possibility of vehicle occupants. (B) Cab-over-style truck cabs usually contain a sleeping area for the driver.

   A. Statement (A) is correct and statement (B) is false.

   B. Statement (B) is false and statement (A) is correct.

   C. Both statements are false.

   D. Both statements are correct.

6. True or False? Search and rescue is never a concern in construction equipment because the equipment is designed for the operator to sit outside.

7. List three types of recreational vehicles.

8. Class A motor homes can be described as which of the following?

   A. A small van/camper-style RV

   B. A motor home built with an A joist-style frame

   C. A large RV with a modular body on a frame

   D. A tent trailer-style camper

9. A common thread that binds all types of buses is that they all have the capability to do what?

   A. Carry many people

   B. Carry large amounts of luggage under the passengers

   C. Operate on diesel fuel

   D. Have an emergency exit in the rear of the bus

10. In confining a fire on a bus, first consideration must be to protect what?

ACTIVITIES

1. Review your department's standard operating guidelines (SOGs) for response to vehicle fires. Do they include response to large trucks, buses, recreational vehicles, and construction equipment? What changes need to be made to the SOG?

2. Contact the school district or transit authority in your response area and tour a school bus or public transportation bus. Note the various components and systems on the bus. How is the engine accessed? What is the load capacity for the bus? How do the emergency exits work? Where is the fuel tank and what type of fuel is used?

Chapter

17

WILDFIRES

Learning Objectives

Upon completion of this chapter, you should be able to:

- Describe what is considered a wildland/urban interface area.
- Identify the three main types of wildland fires.
- Identify the hazards associated with the various wildland fires.
- List the strategic goals and tactical objectives applicable to wildland fires.

CASE STUDY

Lehigh Acres is a planned community of approximately 50,000 residents. The town is located in southwest Florida, 30 miles east of the Gulf of Mexico. Lehigh Acres is geographically unique in that it encompasses 96 square miles, which are divided into 152,000 residential and commercial lots. Lehigh Acres is only partially populated, which leaves thousands of acres of vacant lots. These lots are overgrown with pine trees, palm trees, scrub, and palmetto pines. A palmetto pine resembles a palm tree with no trunk and burns with volatile force due to the natural oils in the plant. There are many densely populated areas of Lehigh Acres, but new homes are constructed regularly directly on the fringe and in the middle of these wildfire-prone areas. The dry season in southwest Florida is during the spring—the average humidity drops below 50% (very low for Florida), the average wind speed increases above 8 mph, and the average rainfall is less than 2 inches. By the spring of 2006, Lehigh Acres had survived Hurricane Charley (2004) and Hurricane Wilma (2006), leaving downed trees and dead underbrush to add to the fuel load. All that was needed was a spark to set off a wildfire. Such was the setting for a 2,000-acre urban interface fire on April 30, 2006.

The winds on the morning of April 30 blew in excess of 20 mph. At least one of the eventual five separate fires was started by juveniles. The blazes spread so fast and with such intensity that they overwhelmed the Lehigh Acres Fire Control and Rescue District (LAFD) very shortly. Mutual aid was called early in the incident. A total of 240 firefighters from thirty-three separate agencies, including the Florida Division of Forestry, worked to contain the blazes. Two of the smaller fires were contained late in the day, but two of the larger fires had merged into one massive fire that moved in a southwesterly direction, consuming homes, vehicles, sheds, and everything else that was in its path. The hope was to be able stop the fire at a large canal. Up to this point, the fire had jumped roads that were being used as fire breaks so quickly that firefighters could not get ahead of the fire. Flanking the fire for extinguishment did absolutely nothing to prevent its growth and spread. The only means of stopping the fire was to stop the head by getting in front of it and creating a huge fire break with dozers and natural barriers. It was a dangerous strategy and brush trucks were deployed to protect the dozers. The primary function of the engines on the scene was structural protection. Structural firefighters chased the fire from house to house, protecting those they could, evacuating ahead of the blaze, and leaving to burn the structures that were beyond saving. The firefight was heroic. In several cases, firefighters entered buildings that were clear losses, with the roof fully involved, in order to hastily search the structure to ensure no residents were trapped. Too often, firefighters escaped only moments before the roof collapsed. Firefighters saved several people by performing these searches, including a disabled retired science teacher. Fire raged out of control for 10 hours. There were a total of twenty-five structures damaged and fifteen homes were totally lost. Overhaul operations

■ **Note**

The threat of and number of wildfires are ever increasing. More and more people are drawn to and migrating to serene wooded areas throughout the United States.

wildland/urban interface
a wildland area with dwellings or other buildings intermixed, creating a fire problem during wildfires

■ **Note**

The people moving into these areas want to maintain the rustic look that drew them to the area in the first place. In doing do, the homes often have wood siding and/or wood shingle roofs that can easily ignite in a wildfire.

lasted two days. Power poles, lines, and transformers were destroyed, resulting in large areas of the city losing power. The American Red Cross housed sixty people overnight in a shelter and fed the firefighters and other emergency workers 130 pizzas, 600 meals, and 500 bottles of water. Thankfully, no one was killed and there were no major injuries.

This fire demonstrates the critical need for interagency cooperation, strong incident command, solid strategies, and preincident planning.

INTRODUCTION

The threat of, and number of, occurrences of wildfires are ever increasing. More and more people are drawn to and migrating to serene wooded areas throughout the United States (**Figure 17-1**). Consequently, the influx of people in these areas heightens the concern for wildfire. To further compound the problem, a good number of the homes constructed in the termed wildland/urban interface are constructed of easily combustible materials such as wood. The people moving into these areas want to maintain the rustic look that drew them to the area in the first place. In doing so, the homes often have wood siding and/or wood shingle roofs, which can easily ignite in the event of wildfire.

This chapter discusses the wildland/urban interface, preventive measures, hazards encountered, firefighter safety associated with these types of incidents, and the strategic goals and tactical objectives.

Figure 17-1   *Many homes and apartments are built within the interface with little consideration given to the possibility of wildfire.*

THE WILDLAND/URBAN INTERFACE

The term *wildland/urban interface* was coined from the simple fact that people from urban and suburban areas were moving into areas that were previously wildland areas to live and enjoy the peace and tranquility that these areas seem to offer.

Wildfires are very often started by lightning, but as population in remote areas increases, so do the instances of carelessness and arson. Fires in wildland areas often go unnoticed for some time, providing the fire the time it needs to grow. As fires grow, they quickly spread and involve everything in their paths. Once a fire reaches this magnitude, it is a real emergency that may quickly overwhelm local available resources (**Figure 17-2**).

In wildland/urban interface areas the fire service must stress the importance of and need for the help and cooperation of the area's residents. The residents can greatly reduce the effects of wildfire if they take some preventive measures before an incident occurs. The fire service must ask and convince interface residents to follow these simple, yet effective, guidelines:

- Clear dead leaves, twigs, and limbs away from under the structure or porches and for 30 to 50 feet around the structure.
- Keep roofs and gutters clear.
- Remove tree limbs within 15 feet of the ground and thin crowns of trees.

■ **Note**

Residents can greatly reduce the effects of wildfire if they take some preventive measures before an incident occurs.

Figure 17-2   *The summer of 1998 wildfires in the state of Florida required resources from all over the state and the nation. One of the hardest hit areas was Volusia County.*

- Remove dead tree limbs above a structure and remove vegetation within 15 feet of a chimney or stove pipe.
- Remove vegetation from the walls of the structure.
- Mow grass regularly.
- Clear a 10-foot area around liquid propane (LP) tanks.
- Place flammable materials such as oil and gasoline in approved containers away from the base of the structure.
- Stack firewood 100 feet away from a structure.
- Create a defensible space (**Figure 17-3**).

Another factor that may influence the amount of fire or the speed and intensity of the fire is debris that has been left behind or downed from previous tornadoes, hurricanes, or even construction. This debris, if not cleaned up, will significantly increase the available fuel for fire to feed on.

Recall the fire triangle from Chapter 2, with its three sides of fuel, heat, and oxygen. The wildland fire triangle is no different—removing any one of these sides will extinguish the fire. For example, removing the fuel by hand digging, plowing, or back-burning will remove the fuel component. Heat is a problem as it preheats surrounding vegetation prior to the fire actually touching it, causing an ever-evolving and spreading wildfire. Just as plowing or digging a line removes the fuel, it will also reduce the preheating of surrounding vegetation. Of course the typical use of water or a water/foam mixture will also remove the heat. In a wildland fire scenario, removing oxygen is likely impossible; thus, firefighting efforts are focused on the first two sides described.

Figure 17-3
Residents must be made aware of the dangers of wildfires. The proximity of growth to this house would make it difficult for firefighters to defend the property.

This is a good time to discuss "rate of fire spread" versus "fire intensity." These two terms, although interrelated, are distinctly different. The *rate of fire spread* is very dynamic and complex and is affected by types of fuel burning, the slope or aspect at which the fire is burning, and weather conditions. All of these factors contribute significantly to the rate of fire spread or lack thereof. *Fire intensity*, on the other hand, describes just how hot a fire is burning. Although influenced by the same factors as rate of spread, fire intensity is important to understand as it is different. For example, a large debris pile may cause a very intense fire producing a lot of heat, but it is not really spreading anywhere. Each of these areas will influence firefighting efforts in differing ways, from the amount of personnel needed, to the geographic area that must be addressed.

There are basically three types of wildfires, any of which may be in an interface area: a forest fire, a brush fire primarily involving mid-level vegetation, and a grassland fire primarily involving low-level vegetation. Each type is discussed in more depth in this chapter.

■ **Note**
Three types of wildfires may occur in an interface area.

FOREST FIRES

Forest fires (**Figure 17-4**) are the fires that occur in forest areas. Many times we see these fires in the news when the nation's media cover the stories of forest fires in the national parks, such as Yellowstone. Forest fires often originate in

Figure 17-4 *A forest fire.*

remote, heavily wooded areas that are hardly accessible other than by foot or aircraft. Due to their remoteness, at least initially, forest fires often pose little threat to civilians or structures. As these fires continue to grow and develop, the threat and danger to structures and civilians increases, although they may not be of the magnitude of the 1991 Oakland Hills fire because of lower concentrations of people and structures. If a fire continues to grow and overrun natural or artificial boundaries, it may eventually threaten entire towns.

Hazards Encountered

Although a forest fire burns relatively slowly, it burns very hot as compared with the two other types of wildfires. The radiant heat alone from these fires can ignite combustible materials and severely burn firefighters.

Excluding the effects of heat and smoke on the firefighters, probably the most important hazard to consider and constantly be aware of is the weather. It is very important for every firefighter to know at least the following for each day throughout the firefight:

- Wind speed, direction, and any probable or expected changes in its speed or direction
- Air temperature and any probable or expected changes
- Humidity and any probable or expected changes
- Any approaching or expected weather or fronts—for example, the typical Florida summer afternoon thunderstorm, which affects wind speed, temperature, and humidity and places firefighters in more danger from lightning.

All of these affect the fire's direction, growth, and speed of travel, either positively or negatively. Again, every firefighter must be aware of what to expect.

Another hazard of a forest fire is the potential for the ground fire to work its way up various ground and ladder fuels to eventually involve the treetops; this is called a crown fire (**Figure 17-5**). Once a fire is able to work its way into the treetops it can become driven primarily by the winds, and it spreads horizontally very rapidly, overrunning ground crews working toward extinguishment through the construction of firelines and most certainly overrunning the very firelines that were to contain the fire.

Another hazard is the way fire spreads. Understand that fire defies the effects of gravity and quickly spreads up vertical slopes. Therefore, it is imperative that firefighters be ever vigilant in these areas and have a predesignated escape route and area of refuge. It is generally stated that you cannot outrun a fire that is traveling uphill. Wildfire also spreads due to flying brands, material that is burning and can start another fire. Keep watch of not only what the main fire is doing in front of you, but of the area all around you. A brand landing away from the main fire can quickly grow, cutting off the escape route and trapping you. Do not be lulled into a false sense of security that it is a routine fire, because just a slight

Safety The radiant heat alone from forest fires can ignite combustible materials and severely burn firefighters.

Safety A brand landing remote from the main fire can quickly grow, cutting off the escape route and trapping the firefighter.

Figure 17-5 *A crown fire.*

wind shift or different type of terrain or fuel type can turn a slow-moving fire into a whipping firestorm in no time.

The next hazard is associated with the timber. As the fire burns across these areas, the root systems of trees may be damaged to a point that the roots are not holding the tree as they should. When this happens, a tree may fall for no apparent reason and with no help, thus endangering anyone who may be around it. Falling trees are also a concern when hand crews mopping up areas cut down a weakened tree.

Another hazard that exists is from water or retardant drops from aircraft (**Figure 17-6**). Firefighting aircraft play an important and integral role in battling forest fires, but they also create hazards of which firefighters must be aware. The primary hazard is due to the large quantity of extinguishing agent that is dropped. It very often breaks off large tree limbs or even snaps some trees in half. It is the falling tree or parts of the tree that you need to be aware of. Knowing that the extinguishing agent that is dropped creates enough force to snap good-sized trees, consider what it could do to your body. Additionally, when some aircraft pick up water with the various buckets or suck the water into the aircraft, it often picks up debris such as pieces of wood or rocks. This debris heightens the concern for and possibility of injury.

The loss of personnel, even if on paper, must never happen. Forest fires are often very large, potentially covering several hundred or several thousand acres of land. It is very difficult to keep tabs on the typical single-alarm units for a

■ **Note**
Firefighting aircraft play an important and integral role in battling forest fires, but they also create hazards of which firefighters must be aware.

Figure 17-6 *A helicopter water drop.*

structure fire that covers only hundreds of feet of ground space, let alone thousands of acres. A strong incident management system must be in place to properly account for all personnel.

Other hazards concern the firefighters directly. The first is fatigue. Wildland firefighting is extremely physically intensive from the hike in, to the construction of firelines, to the hike out. Forest fires can last days, weeks, and even months. Firefighters must be aware of their own limits and monitor those of the other members of their crew. Further more, exposure to smoke can cause smoke inhalation and the associated health problems.

BRUSH FIRES

Brush fires are wildfires that occur in areas made primarily of mid-level vegetation. For example, the outbreak of wildfires in Florida in the summer of 1998 primarily involved mid-level fuels (**Figure 17-7**). Just as with forest fires, these fires many times occur in somewhat remote areas that are difficult to access. Because these type of fires do not involve heavily wooded areas, it is often possible to access the fire area with wildland firefighting engines, or brush trucks as they are called in many areas such as in the southeastern United States. As compared to forest fires, fires in mid-level fuel grow and travel faster but do not generate temperatures as high as those in forest fires. Although these fires do not burn as

Figure 17-7 *Mid-level fuel fire.*

hot as a forest fire, they still produce amounts of heat higher than the body or combustible structures can withstand. Brush fires tend to be more at the wild-land/urban interface, because building, or more appropriately land clearing, in these areas is easier as it does not involve trying to clear heavily wooded areas on difficult terrain.

Hazards Encountered

Brush fires or mid-level fuel fires have many of the same hazards associated with them as forest fires.

As with forest fires, the primary hazard is the weather. Every firefighter must be aware of and understand how the weather affects the firefighting effort. Every firefighter must know:

- What the winds will be like
- Temperature
- Humidity
- Expected weather changes

The next hazard is that of fire spread. Remember, these fires spread rapidly because of the lighter, more easily combustible fuels. These fires also spread due to flying brands. Be continually aware of your surroundings to keep watch of what the fire is doing. A stray brand landing in an area remote from the main fire can quickly heighten the danger. Be sure to have an escape plan or area of refuge in case the situation quickly worsens, preventing your escape. Keep in mind that there is no such thing as a routine wildland fire.

Falling trees can be a hazard with these fires also. Many of these fires occur in loose or sandy soil that fire can burn down into to eat away at a root system. Although there is not the amount of timber in these mid-level wildfires, the potential still exists and must be taken into account.

Another hazard to be aware of in this type of wildfire is again water or retardant drops from aircraft, but because fires in this type of fuel are generally more accessible to brush and wildland firefighting apparatus, aircraft are not used or needed as much or as often as in forest fires.

Crown fires are not that common during these types of fires, but they can and do happen. The same concerns as those for crown fires in forest fires exist in these areas. A wildland fire that progresses to a crown fire quickly changes the situation and not for the better.

Another hazard is not knowing where the personnel are and what they are doing. Brush fires often cover large land areas, and multiple different fire companies and fire agencies are often involved, thus creating a large number of firefighters to keep tabs on. This is a very important aspect of any firefighting effort, but when multiple agencies are deployed all over the place, it is a must. An accountability system is crucial.

Again, the hazard of the fatigue of the firefighters is important. As with forest fires, brush fires are very physically intensive. Typically, brush fires do not last for the length of time that forest fires do, but they can last several days or weeks. Everyone must keep constant watch on everyone else because fatigue can and will lead to mistakes, unsafe acts, and eventually injury.

GRASSLAND FIRES

Grassland fires (**Figure 17-8**) are the third basic type of wildfire. This type of fire involves low-level vegetation or grassy type of fuels. Fires involving grassland are the fastest growing and moving type of wildfire, but they produce the lowest amount of heat. Grassland fires are typically very accessible by common types of fire apparatus and apparatus specifically designed for wildfires. This helps keep these fires relatively small. Rather than dealing with and trying to control fires in huge numbers of acres, grasslands fires can potentially be only a few hundred square feet. These fires occur most often in urban/suburban areas and pose a real threat of spreading to structures and other property if not quickly controlled.

Figure 17-8 *A grassland fire.*

Hazards Encountered

For the most part, the hazards associated with a grassland fire are the same as those encountered with the two other types of wildfires, but to a lesser degree. Every firefighter must consider:

- Weather conditions
- Fire spread
- Extinguishing methods and agents
- Accountability
- Condition of personnel

With these in mind, a safe and successful operation can take place.

STRATEGIC GOALS

Firefighter Safety

Firefighter safety begins with each firefighter. Firefighter safety is of utmost importance when battling these types of fires because they are vastly different than what the typical urban/suburban firefighter is used to dealing with. Ten firefighting orders that can have a positive impact on firefighter safety at wildfires are given in the following list:

1. Keep informed on fire weather conditions and forecasts.
2. Know what the fire is doing at all times.
3. Base all actions on the current and expected behavior of the fire.
4. Plan escape routes for everyone and make them known.
5. Post a lookout where there is possible danger.
6. Be alert, keep calm, think clearly, act decisively.
7. Maintain communication with your crew, your supervisor, and adjoining forces.
8. Give clear instructions and be sure they are understood.
9. Maintain control of your crew at all times.
10. Fight fire aggressively, but provide for safety first.

A number of hazards associated with these types of fires are not normally hazards associated with structural firefighting, such as air drops of extinguishing agent. The hazards must be known and every firefighter must be aware of them.

These types of fires often require different protective gear than the firefighter is generally used to (**Figure 17-9**). Due to the duration of many of these fires, it is unreasonable to expect firefighters to be outfitted in the typical bunker gear

Figure 17-9
Wildland fire personal protective clothing.

ensemble for they will quickly become overheated, fatigued, and dehydrated. Instead they should be outfitted in wildland firefighting gear, which is much lighter and breathes more easily. Wildland firefighting protective gear typically consists of a good durable pair of work boots with steel toes, a long-sleeved shirt and pants ensemble, or a jumpsuit type of garment made of a fire-retardant material, such as Nomex. In the case of the shirt-and-pants combination, the shirt must remain tucked in the pants. The personal protective clothing is topped off with a helmet or hard hat and some type of eye protection. A fire shelter must also be provided to every firefighter. Although the issuance of all this protective equipment is very important, it is equally important that the firefighter understand how to wear and use all the equipment.

Search and Rescue

For the most part, search and rescue efforts are not a major component of the operations at wildfires. Generally, civilians will have been evacuated, but there are times that search and rescue actions might take place.

■ **Note**
One instance of search and rescue may be with a homeowner who refused to evacuate because he or she wanted to stay and protect the property.

One instance may be with homeowners who refused to evacuate because they wanted to stay and protect their property. These people may end up needing help. It is difficult, at best, to know whether the structure is occupied or whether the residents have evacuated unless you can see people in the house, people around the outside of the structure, or other signs that may indicate whether anyone is in the structure. The occupants may also become fearful for their lives as the fire approaches and may call for help, giving their location as their situation worsens. Another search and rescue instance may be when evacuees become trapped by fire in their vehicles as they attempt to evacuate.

■ **Note**
Another search and rescue instance may be when evacuees become trapped by fire in their vehicles as they attempt to evacuate.

A firefighter that becomes missing or trapped is one search and rescue effort that nobody ever wants to experience. Such an instance is one major reason to have a strong incident management system in place that includes a sector to keep up on accountability.

Evacuation

Evacuation efforts are a huge undertaking. Even small-scale evacuations are time-consuming and personnel intensive. The decision to order an evacuation often involves many more agencies than just the fire department. There are many items to consider regarding evacuation, typically beginning with how the evacuees are to evacuate and what agency or agencies will conduct the evacuation.

Next, you will have to decide on the quickest and safest way to evacuate. Do you want people to drive out, walk out, or is transportation going to be provided? If you want them to drive out or exit on foot, the evacuees must be told in which direction they are to travel. Along with that, assistance from law enforcement agencies may be necessary to ensure the evacuees are able to keep moving and moving in the right direction. In the instances when transportation is provided, you have to first decide on which type of transportation is needed. For instance, is a nursing home that needs ambulances being evacuated or are you evacuating an apartment building and in need of some buses? If you use your imagination, you can come up with many more scenarios that may arise and be relevant to your community. In any case, providing transportation is a giant undertaking and often involves other agencies in the incident.

Who will conduct the evacuation is an important question. Will the fire department go door-to-door advising residents of the evacuation, or will the police department conduct it? No matter who conducts the evacuation, it will be personnel-intensive. Wildfires in interface areas often threaten large areas, so the larger the affected area is, the larger the evacuating force must be. The decision as to who will do the evacuating is often dictated by the fire. If it is a large and fast-moving fire, then the fire department may not have any resources available to do the evacuation. Most of the time, however, the evacuation is conducted as a joint effort among public safety agencies, such as fire and police. When multiple agencies join forces, it expands the number of available personnel and larger areas can be covered, but it is imperative that the evacuation plan be spelled out and all participants are aware of who does what so there is no confusion and duplication of efforts.

Another major consideration when ordering an evacuation is where to send and house the evacuees once the evacuation order is given. Evacuees will need some type of safe shelter, whether the incident is of short or long duration. Often schools are provided as shelters. One big benefit of providing shelter is that it makes accountability of the evacuees possible. This task is also personnel-intensive and time-consuming, but it is far better than trying to track evacuees that have gone in every direction and distance.

Probably the most important issue regarding an evacuation is when to order it. If it is called for too early, evacuees may become impatient and frustrated and try to return to their homes, feeling that the threat was minimal and they left for no reason. If an evacuation is ordered too late, people may not get out in time and may be killed or injured. All you can do is consider the incident's potential and make decisions that err on the side of public safety.

Exposure Protection

The exposures range from the unburned wildland areas all the way to structures and everything in between. Knowing which exposures to protect and how to protect them (**Figure 17-10**) is the key and is described in further detail later in this chapter.

As these fires continually increase in number of occurrences as well as numbers of homes and businesses being damaged and destroyed, the science of fire protection has also advanced. More and more different types of extinguishing agents and fire protection agents are being developed and made available.

■ **Note**

Probably the most important issue regarding an evacuation is when to order it. If it is called for too early, evacuees may become impatient and frustrated and try to return to their homes, feeling that the threat was minimal and they left for no reason. If an evacuation is ordered too late, people may not get out in time and may be killed or injured.

Figure 17-10   *An exposure protection operation using a master stream.*

Class A foam is one type of agent, used in conjunction with water to enable the water to penetrate into ordinary combustibles to either make extinguishment efforts more effective or to apply on unburned materials to help protect them from fire. Compressed air foam systems (CAFs) have been developed that basically inject air and foam into a fire pump to enhance the extinguishing effects of water.

Compressed air foam systems have also proven to be more effective than just water alone. These systems inject air and foam into the discharge side of the pump and allow more efficient and effective use of a hose stream. Like Class A foam, CAFs help the water to penetrate combustible materials, but they also coat the burning material to enhance extinguishment and provide a certain degree of protection of the coated material from fire.

Other protein-based foams and gel-type agents have been developed and continue to be developed that are applied to an exposure to protect it from fire.

Confinement

Confinement of a wildfire is keeping it contained to a few acres or a few thousand acres. Unlike confining a structure fire to the structure of origin, where all your resources are basically working in one place, confinement of a wildfire typically has resources spread throughout the area.

Confinement of a wildfire can be accomplished in many ways. It can be confined by surrounding it with firefighting personnel that may have constructed

Figure 17-11   *One example of a forestry dozer.*

Figure 17-12   *One example of a contract helicopter.*

firelines or stand ready to extinguish the fire as it approaches or both. Other ways to confine the wildfire include:

- Construction of firelines utilizing dozers (**Figure 17-11**)
- Use of firefighting aircraft (**Figure 17-12**)
- Running the fire to a natural boundary, such as a river or lake
- Running the fire to an artificial boundary, such as streets and highways
- Extinguishment of the burning materials
- Backfiring ahead of a moving fire

Extinguishment

The extinguishment of wildfires is dirty, backbreaking work. Many times wildfires occur in areas that have little or no access for wildland firefighting apparatus. Attacking wildfire from the burned area and working outward is the safest method of extinguishment. Unfortunately this method is not always possible or practical. The extinguishment of wildfires should begin by working up the flanks from within the burned area, working toward the head. The head (**Figure 17-13**) of a wildfire is the portion that is the fastest growing and moving part of the fire. Once the flanks are controlled and the head has been knocked down and controlled, then the other sections of the fire can be extinguished and overhauled. Remember, water supply will be a big concern, so utilize your water according to the needs of the incident. For example, lowering the gallon-per-minute flow of the nozzle for adjustable gallonage nozzles or lowering the pounds per square inch of pressure on the handline for automatic nozzles allows the best use of available water.

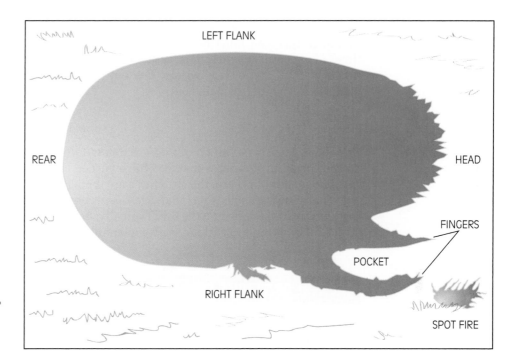

Figure 17-13 *The components of a moving wildfire.*

Overhaul

Overhaul is as important for wildfires as it is for structure fires. Because wildfires occur out in the elements of nature, it is imperative that a good overhaul be conducted, because if the wind were to pick up some hot embers and blow them into an unburned area, another fire could erupt.

When water is available and practical to use by utilizing wildfire firefighting apparatus or short hose lays because the fire is in close proximity to roadways, then the entire perimeter of the fire area should be overhauled. At least 50 feet of the burned areas around the fire should be thoroughly saturated and crews should overturn piles of brush, stumps, and any other areas that may conceal hot spots. The use of backpack-type water bladders or tanks is another way to utilize water in the fire area. Because of the limited amount of water that these backpacks have, they should not be used to put water on any unburned areas, but should be used to directly extinguish any remaining flame and hot spots.

Wildfires often occur in areas that cannot be accessed by firefighting apparatus and overhaul must be conducted by hand. In these instances, hand tools are required to clear swaths of land to prevent the fire from reigniting along its boundaries.

Whether overhaul has to be conducted utilizing fire apparatus, hose, and water, or has to be done with hand tools, it must be done. Wildfires are affected by so many different variables such as wind speed and direction and the relative humidity that you must be very careful to ensure extinguishment. Particular times of concern are when the weather has been and is very dry, which has dried up vegetation, and/or it is windy. The entire fire area must be overhauled and watched to ensure it remains out.

Salvage

Salvage is usually not much of a concern of wildland firefighting. Salvage could be conducted in structures that may have burned as a result of a spreading wildfire, but even in this case, the structures are typically burned down, leaving very few, if any, salvageable items. Salvage operations may also take place in structures that are threatened by the fire. Typically, firefighters would try to protect the structure from the fire and any salvage would have been done by the residents.

TACTICAL OBJECTIVES SPECIFIC TO WILDFIRES

In general, wildfires, whether forest fires, brush fires, or grassland fires, are fought in much the same manner. When your assignment is fire extinguishment, try to gain access to an area that has already burned. This will provide the crew's safety from the fire and an easier way to access the head of the fire. Start extinguishment at the most active flank, working toward the head.

Because of the similarities in the types of wildfires, standard operating guidelines (SOGs) can be established to provide for a safer and organized operation. **Figure 17-14** shows a simple, basic SOG that is specific to wildfires.

The entire scenario and potential actions taken change drastically when combating a large wildfire that involves and/or threatens to involve structures (**Figure 17-15**). There are no hard-and-fast rules as to which structures to protect and which structures to let go, because each situation may have different fuel loads and types, varying intensities of the fire, varying amounts of available resources, and varying available water sources. Keeping all the variables in mind, some conditions or instances should cause you to reassess your decision as to whether to stay and fight a structural fire or to withdraw:

- Your escape route becomes unusable.
- You do not have enough water supply to sustain the threat.
- Spot fires occur and spread quicker than you can control them.
- The roof of the structure is well involved.
- The structure's interior is well involved.

The key in fighting interface fires is to use your resources to accomplish the most good. In other words, if you have to let one structure burn to save ten, then you have accomplished the most good. Each officer and firefighter that is given

■ **Note**
There are no hard-and-fast rules as to which structures to protect and which to let go, because each situation may have different fuel loads and types, varying intensities of the fire, varying amounts of available resources, and varying available water sources.

SAMPLE STANDARD OPERATING GUIDELINE WILDFIRES

SUBJECT: WILDLAND FIRE MANAGEMENT

Wildland firefighting priorities are:

1. Protection of life
    a. Civilian
    b. Firefighter
2. Protection of property
3. Fire containment

Tactical Objectives:

1. First-arriving engine:
    a. Provide size-up (size of fire, type of fuel load, exposures, direction of spread, and any access points for brush units).
    b. Initiate command procedures.
    c. Protect exposures, if any.
    d. Identify water supply resources.
    e. Begin attack, if appropriate.
2. First-arriving brush unit:
    a. Assist with exposure protection.
    b. Gain access to upwind portion of burned-out area, providing apparatus and manpower are not abused or endangered.
    c. Attack fire head from the burned area, working up the most active flank towards the opposite flank.
    d. Extinguish fire and secure fire lines. If division of forestry has provided fire lines, allow fire to burn up to lines.
3. First-arriving water unit:
    a. Assist with exposure protection.
    b. Stage near secure water supply prepared to supply water or assist with extinguishment.

Command Considerations:

1. Command should be established with appropriate geographic and functional sectors.

developed early on. Emphasis should be placed on establishing a rehab sector.

2. In large operations, current weather information should be obtained, referenced, and updated throughout firefighting operations.
3. Keep fire attack unit operations confined to the burned area. Observance for spot fires caused by flying embers or radiant heat should be done by all units. Defensive operations, in front of a head fire and outside burned area, should only be undertaken when life or property is endangered and a secure escape route is available to firefighters.
4. Division of Forestry resources are extremely useful in firefighting operations and save personnel and apparatus wear and tear. In any wildland firefighting operations notify and request assistance from the Division of Forestry, if needed.
5. When multiple wildland firefighting units are utilized, a team concept should be adopted— one unit attacking, with one securing the line and both protecting each other. A reserve wildland firefighting unit should be available in case of mechanical failures or unexpected fire conditions.
6. If a wildland fire does not endanger life or property, the best strategy may be to let it burn to natural or man-made boundaries, where extinguishment will be simplified.
7. Personnel participating in fire attack operations should be provided with PASS devices.

Figure 17-14   *A sample wildfire SOG.*

Figure 17-15 *An example of a successful exposure protection operation.*

an assignment must conduct a structural triage and not just stop at the first threatened structure that they come upon. Scene organization is critical. Unlike a building fire, where the geographic area is small in comparison, the wildfire area may cover miles and miles. Freelancing cannot be accepted.

Initial considerations need to address the safety of the firefighters and their apparatus. In order to accomplish this, several questions must be answered:

- *What and where are the safety zones for the firefighters and apparatus?* If the fire worsens beyond expected conditions, there must be at least one, and better yet two, areas for the firefighters to take refuge. Usually the leeward side of the structure and the interior can be considered safety zones. Firefighters can shield themselves from the initial blast of the fire by using the structure as the shield. Once the fire passes, the firefighters can then take actions to extinguish the structure or the fire impinging on the structure. Look for open spaces that can be used as safe zones. If working along a ridge, position apparatus on the side opposite and below the fire. Keep in mind that an escape route to the safe zone must be maintained and be quickly attainable. Avoid placing apparatus between the approaching fire and the structure you are protecting.

- *Where is the structure located?* If the structure is located in a chimney area or box canyon area, then the concern for firefighter safety may outweigh the defensive actions. If the structure is located on a steep slope or there is a steep slope below the structure, reconsider your decision, as these areas are unsafe. If the structure is near the top of a ridge, then it may be a bit more defendable as it affords the personnel and apparatus to seek safety on the other side of the ridge.

- *From where is the fire approaching the structure?* The answer to this question may be from above, below, or basically across flat ground. A fire advancing from above a structure on a slope is far easier to defend against because of its slower progress (unless winds affect the fire's growth and progress). A fire advancing from below travels much faster and typically has a greater flame length, which in turn preheats the fuels above to further speed the fire spread. Structures that have good clearance or green space around them may still be suspect because of the flame length. Generally speaking, a structure may not be defendable if it is less than twice the flame length from the edge of the clear space. A fire advancing from basically flat ground may also be more defendable because of slower fire spread.

- *What is the exterior finish of the structure?* If the structure has wood siding and/or a wood-shingled roof, it may not be defendable, and one with a better chance to protect should be chosen.

- *What are the fuel conditions?* Fuel type and height will affect your decision. If the fire is in a heavily wooded area as compared with grassland then your decision to protect will vary with the intensity of the fire. Observe the surroundings of the structure to see whether it has enough space to be defendable. If the brush is close to the structure, if the yard is cluttered with combustible debris, if there is a lot of decorative vegetation, or if trees and other vegetation hang above and around the structure, then it may not be defendable as compared to structures that do not have such combustible materials. Heavy fuel loads below a structure may cause you to bypass it for another.

- *What other hazards may exist around a structure?* Look for exposed LP tanks and other hazards such as overhead wires that may burn through.

- *What are the outside conditions and what is expected?* This consideration includes more than just the weather conditions; it also takes into account the time of day. For example, the sun increases the temperature of ground cover the most in the afternoon, which means that the fuels are closer to their ignition temperature. Weather is one huge variable over which we have no control. Winds, humidity, temperature, and so on, must all be considered when deciding to make a defensive stand.

- *What is the available water source?* Having little or no water supply will affect your decision. Be sure to maintain enough water to afford the escape of the crew. Having a water source, whether it be a pool, pond, or hydrant, may allow you to defend areas that you may otherwise decide not to, so long as firefighter safety is not compromised.

All of these considerations need to be continually readdressed throughout the incident. Conditions vary from place to place and incident to incident. A strong prevention/public information campaign can also help by getting the word out to homeowners and building owners as to what they can do to lessen the potential threat to their properties.

SUMMARY

As the number of people that move into and develop wildland areas increases, the incidence of fires in these wildland/urban interface areas also increases. Along with the boom in construction in these areas, residents have had the desire to maintain the country surroundings by constructing and decorating the exterior of structures with the rustic theme, thus having structures that have wood siding and/or wood shingle roofs.

The three types of wildland fires—forest fires, brush fires, and grassland fires—each have hazards specific to them, but they also have many commonalities. The strategic goals for a structure fire still exist, although in varying degrees, when dealing with a wildland fire and must be addressed.

The tactics should be employed with firefighter safety being the primary concern and secondarily to provide for the greatest good. Many times wildfires quickly tax local resources and therefore some tough decisions must be made as to what areas can be defended and what areas cannot. This decision must be made based on predicted fire behavior as much as how difficult or easy it might be to protect various structures, using the information in this chapter.

These fires are often devastating to a community and the fire department must be prepared for such occurrences by preparing through fire prevention and information activities, the preplanning of suspect areas of the community, and education and training of emergency personnel.

REVIEW QUESTIONS

1. Describe how and why wildfires have become such a serious problem.

2. List the hazards associated with fighting brush fires.

3. What is the primary concern when battling a wildfire?

   A. The fire's direction of travel

   B. Firefighter safety

   C. The temperature of the materials burning

   D. Available resources

4. How will evacuation efforts affect the operations of the extinguishment of a wildfire?

5. Describe some extinguishing agents, other than water, that have proved effective in controlling wildfires.

6. What conditions might cause you to reassess your decision to protect an exposed structure?

7. List the ten firefighting orders.

8. Describe a crown fire.

9. Define the head of a wildfire.

10. Describe the components of wildland personal protective equipment.

ACTIVITIES

1. Review your department's SOG for wildfires. How does it compare to the one in this text? What improvements could be made.

2. Note locations within your response area where a wildland/urban interface problem exists. Do you have the area preplanned? What additional resources would be required for a fire of this type?

Chapter 18

SPECIAL FIRES (BASIC INFORMATION)

Learning Objectives

Upon completion of this chapter, you should be able to:

- Describe the various types of special situations that firefighters may encounter.
- Describe the hazards associated with fires involving hazardous materials.
- Describe the hazards associated with fires involving ships.
- Describe the hazards associated with fires involving railcars.
- Describe the hazards associated with fires involving aircraft.
- Describe the hazards associated with fires involving bulk storage facilities.
- Apply strategic goals and tactical objectives to these incidents.

CASE STUDY

A small Cessna approached the runway of a public airport when the pilot began experiencing engine trouble. The pilot was the only person on board the aircraft. The airport was protected by a crash truck with 1,500-gallon foam capability. As the pilot struggled to keep the aircraft flying, he clipped the tops of large pine trees and the plane began to cartwheel through the trees. The plane landed nose down in the ground with the tail section rising against a tree. The body of the plane was severely damaged, but intact. The pilot was alive, but trapped, and a small fire started in the cockpit. The plane crash was approximately 1 mile from the airport in a heavily wooded area off airport property. The first-responding unit to arrive on the scene was an engine company with no experience and little training in aircraft rescue.

Initial crews sized-up the scene and approached it like a vehicle entrapment with large amounts of flammable liquid. A preconnect hose line was pulled to extinguish the small fire and to protect the pilot. Patient assessment began next. The pilot was awake and oriented and only suffering from cuts and bruises.

The first-arriving engine immediately called for a second engine and notified the nearby Crash Fire Rescue agency, who responded with a crash truck. The crash truck arrived on the scene within 5 minutes of the first engine and laid a protective foam blanket on the ground. The plane was approximately 300 feet into heavy timber and thus the crash truck had to stretch booster line to lay down the foam. Meanwhile, the plane was stabilized with rope to keep it from falling over as the engine company crew extricated the victim with hydraulic rescue tools. The patient was transferred to a waiting ambulance and transported to a nearby hospital.

The first-arriving crews were faced with an unfamiliar situation. They took action based on what they knew and applied vehicle extrication techniques. Although the outcome was favorable, it could have been different. Aircraft firefighting and rescue requires specialized training and equipment. Firefighters who routinely respond to specialized incidents such as this should seek training in these respective areas.

INTRODUCTION

■ **Note**
All firefighters, as first responders, have potential to come in contact with special fire situations.

This text thus far has described concepts applicable to all types of fire incidents. Also studies of fires in particular occupancies, vehicles, and wildfires were presented. While these are probably the most common types of incidents that most firefighters will confront, all firefighters, as first responders, have potential to come in contact with the special fire situations presented in this chapter, one of which is shown in **Figure 18-1**.

Figure 18-1   *One special situation that structural firefighters may be called upon to respond to is downed aircraft.*

These special situations, including hazardous materials, ships, railcars, aircrafts, and bulk storage, are each so specialized that they are covered in courses devoted entirely to each. However, because all firefighters have the potential to respond to these incidents, it is essential that incident commanders and firefighters have a basic understanding of what they might expect. All firefighters must be able to size-up the situation, request the appropriate resources, take action if it is safe to do so, and ensure that firefighters are properly equipped.

As shown in the case study, these special situations can and do occur anywhere. The local fire department is usually called on to respond and is expected to know what to do. If, as you study this chapter, you find that your department has a higher-than-average need to respond to the type of incidents discussed—for example, your department is adjacent to an airport, rail yard, or port—then it is recommended that you seek out additional training in those specific areas.

HAZARDOUS MATERIALS

Response to hazardous materials is possible in every community regardless of size or location. When considering a hazardous materials fire, generally the firefighter is dealing with some other type of occupancy or vehicle. A gasoline spill may produce a fire and the fire may only involve the gasoline, but most times the hazardous materials are in a vehicle or building and the fire involves these as well. For example, a fire involving an overturned gasoline tanker is a truck fire with hazardous materials involved. A fire in a warehouse that stores pool chemicals is a warehouse fire with hazardous materials involved. Firefighters must approach with extreme caution, understanding that conventional, or routine, extinguishing agents may be ineffective, and the material may behave differently as exposure to heat increases.

It is necessary for firefighters and incident commanders to approach these incidents with caution and to operate within levels of training and expertise considering the needs of the incident. The Occupational Safety and Health Administration (OSHA) and the Environmental Protection Agency (EPA) have regulations that apply to hazardous materials response. There are also NFPA standards that apply. These regulations and standards must be considered when responding to fire involving hazardous materials. Training levels as recognized in the regulations are summarized in the following list:

- *First Responder Awareness.* Responders at this level are those who are likely to witness or discover a hazardous substance release and who have been trained to initiate an emergency response sequence. At this level of training, these responders would be expected to take no further action other than to initiate the emergency response sequence. Typically, single certified emergency medical service (EMS) responders and law enforcement officers fall into this category.

- *First Responder Operational.* Responders trained to this level are those who respond to hazardous materials incidents as part of the initial response for the purpose of protecting nearby persons, property, and the environment from the release. These responders are trained in defensive tactics without actually trying to stop the release. Their function is to contain the release from a safe distance, keep it from spreading, and protect exposures. First responder operational level requires at least 8 hours of training or demonstrated competency.

- *Hazardous Materials Technician.* Responders trained to this level assume a more aggressive role than a first responder operational level. The hazardous materials technician approaches the point of release in order to plug, patch, or otherwise stop the release of a hazardous substance. Hazardous materials technicians must have received at least 24 hours of training equal to the first responder operational level and demonstrated competency in additional areas.

- *Hazardous Materials Specialist.* Hazardous materials specialists are those individuals who respond to the hazardous materials scene to support technicians. While their duties parallel those of the hazardous material technicians, they require a more direct knowledge of the various substances they need to contain. Hazardous materials specialists may also act as the liaison with federal, state, local, and other governmental authorities in regard to site activities. Hazardous materials specialists must have received at least 24 hours of training equal to the technician level, plus additional competencies as identified in the standard.

- *On-Scene Incident Commander.* Incident commanders who assume control of the incident beyond the first responder awareness level must receive at least 24 hours of training equal to the first responder operational level and, in addition, have competency in command systems, response plans and options, and other dangers associated with hazardous materials.

Hazards Encountered

In order to examine the hazards associated with these materials, it is useful to divide the materials into classes. The Department of Transportation classes hazardous materials into nine numerical classes and two word classes for placarding and labeling purposes (**Figure 18-2**). The categories and a description of the materials and the possible hazards follow.

Class 1—Explosives. The classification has six subdivisions:

    1.1 Materials that have a mass explosion hazard

    1.2 Materials that have distinction through projection hazards

Figure 18-2 *The Department of Transportation placard and marking for hazardous materials can give responders a clue to the hazards.*

1.3 Fire hazard materials

1.4 Materials with a minor explosive hazard

1.5 Materials that are considered insensitive

1.6 Materials that are dangerous but are considered extremely insensitive

Explosive materials might be found at a vehicle accident if a truck transporting the explosives was involved. They may also be a factor in fires involving explosive storage areas. This storage may be at a construction site, in an explosive magazine, or in the home as gunpowder. Through preplanning it is possible to identify storage locations or routes of travel. Whereas some explosives are designed to **detonate**, others **deflagrate**. Those that detonate destroy property and produce a shock wave that may injure or kill. Those that deflagrate may do the same and also add to the fire load and intensity of burning.

Class 2—Gases. The classification is subdivided into four classes:

2.1 Flammable gases

2.2 Nonflammable gases (lower flammable limit of no higher than 13% and a flammable range of greater than 10%)

2.3 Poisonous gases

2.4 Corrosive gases

As with explosives, gases may be found at fixed locations (**Figure 18-3**), in transport, in bulk storage, or in the process of being manufactured. Gases have various chemical properties and are stored in a variety of forms. Gases are either lighter or heavier than air. Firefighters should understand both the properties of the gas involved and its weight in comparison to air, in which air = 1.

---

detonate

to explode with sudden violence

deflagrate

to burn rapidly, which in reality can be considered a low explosive; slower than detonation

---

■ Note

Gases are either lighter or heavier than air. Firefighters should understand both the properties of the gas involved and the weight in comparison to air, in which air = 1.

Figure 18-3 *Fixed locations for liquid petroleum gas (LPG), a flammable gas. Note the poor storage of flammables around the tanks.*

■ **Note**

Class 3 includes gasoline, which is found in any community around the country.

Class 3—Flammable Liquids. This classification is subdivided into three classes:

    3.1 Materials that are flammable with a flash point of less than 0°F

    3.2 Materials that are flammable with a flash point of between 0°F and 72°F

    3.3 Materials that are flammable with a flash point between 73°F and 141°F

The hazards associated with this class involve flammability, clearly a concern for firefighters. The class includes gasoline, which is likely to be found in any community around the country (**Figure 18-4**).

Class 4—Flammable Solids. This classification is subdivided into three classes:

    4.1 Flammable solids

    4.2 Spontaneously combustible

    4.3 Dangerous when wet

!**Safety**

Some materials in class 4, such as calcium carbide, are water reactive. It is necessary to apply specialized extinguishing agents on many of these materials.

Some materials in this class, such as calcium carbide, are water reactive. It is necessary to apply specialized extinguishing agents on many of these materials.

Class 5—Oxidizing Substances. This classification is subdivided into two classes:

    5.1 Oxidizers

    5.2 Organic peroxides

Oxidizing substances supply their own oxygen to the burning process. The peroxides are very unstable when heated or improperly stored.

Class 6—Poisonous Substances. This classification is subdivided into two classes:

    6.1 Poisons

    6.2 Infectious materials

Figure 18-4
Aboveground storage for gasoline and diesel fuel.

These substances are dangerous to health or life. They may be found in solid, liquid, or gas form.

Class 7—Radioactive Materials. This class is divided into three levels depending on the amount of radiation released. Radioactive materials are used in a number of processes. Be particularly alert around medical facilities that often use radioactive materials for diagnostic testing.

Class 8—Corrosive Material. Corrosives are the materials that are either acids or bases. The pH of the material is either higher than or lower than that of human tissue; therefore when exposed, tissue is destroyed.

Class 9—Miscellaneous Materials. This classification is subdivided into three classes:

9.1 Miscellaneous dangerous materials

9.2 Environment hazardous

9.3 Dangerous waste

These materials are hazardous, but based on their properties, they do not fall into one of the first eight categories.

ORM-D—Consumer Commodity. This classification is subdivided into five classes, A, B, C, D, and E, and is used for other regulated materials.

Strategic Goals and Tactical Objectives

Hazardous material incidents present a specialized set of problems and therefore require specialized strategic goals and tactics objectives. However, for the first responder, or firefighter, some application of tactical objectives can be applied to the eight strategic goals presented in this text. Remember that any actions taken must be in compliance with regulations governing hazardous waste operations. The firefighters must be trained to the level of the operation and have the proper equipment and resources on hand to safely and effectively take said action.

Firefighter Safety Firefighter safety can be accomplished at hazardous materials incidents often by doing nothing (**Figure 18-5**) or taking defensive actions only, keeping the firefighter away from the effects of the chemical. However, to ensure the highest level of firefighter safety, proper identification of the product is necessary and then notification of the hazardous materials team. Furthermore, a decontamination process should be set up in order to quickly decontaminate any firefighter who might be exposed to the chemical involved. Proper personal protective equipment (PPE) as well as an accountability system is necessary, even if only defensive actions are taken by the firefighters. Hazardous materials incidents are often of long duration. Rehabilitation should be a consideration early in the incident.

■ **Note**
Firefighters must be trained to the level of the operation and have the proper equipment and resources on hand to safely and effectively take action.

!Safety
To ensure the highest level of firefighter safety, proper identification of the product is necessary and then notification of the hazardous materials team.

Figure 18-5
Sometimes in order to provide for firefighter safety, responders must not take offensive action until the product is identified and appropriate resources are in place.

Search and Rescue Search and rescue tactical objectives at a hazardous materials incident are complex. It is assumed that the area requiring search and rescue is an area impacted by a chemical. Although common search and rescue techniques could be employed, the fact that the environment is impacted by hazardous materials requires thought and deliberate decision making.

The first step in the decision process is to determine the level of risk present if the firefighters were to attempt a rescue. Three levels of risk associated with rescue at a hazardous materials incident are defined in the following list:

- *Low risk.* The low-risk category is based on four assumptions:
    1. The hazards are known and are not expected to increase.
    2. The PPE available is appropriate to the task.
    3. Responders are properly trained for the task.
    4. There is a high probability of a safe and successful outcome.
- *Calculated risk.* A calculated risk involves the same assumption as does the low-risk model, with one exception. That exception is the degree of certainty that the hazards are not expected to increase, assumption number 1. In other words, although unlikely, the hazards may increase, and the action plan takes this into consideration and has a contingency plan.

Sheltering in Place

Sheltering in place is used when, for whatever reasons, the victims must be protected, but it is impossible to move them or it would expose them to more harm if moved. In this case an accepted option is to shelter in place. To shelter in place, the victims are protected by the barriers created by the shelter. Ventilating systems that bring in air from the outside should be shut down and any opening to the outside closed or sealed. When considering this tactical objective, the incident commander should consider:

- The duration of the release. In many cases the worst is over when firefighters arrive.
- The number of potential evacuates. Where will they go?
- The available resources.
- Can the people be adequately protected where they are?

Also for consideration is that studies of actual incidents have shown that generally sheltering in place is safer than evacuating citizens. The use of evacuation may result in panic, injuries, and in some cases deaths unrelated to the event itself. Even with considerable notice, there have been many occasions of problems associated with an evacuation.

- *Unacceptable risk.* An unacceptable risk has no chance for success. Assumptions 2 and 3 may be valid, but 1 and 4 are not, which makes a significant difference. In these cases firefighters should not be committed to locations inside of the product. The risks associated with search and rescue far outweigh the potential gains.

Should the incident commander decide that the risk is low and the rescue should be attempted, the common tactical methods would be employed to meet the objectives. Proper PPE should be worn and the same search procedures employed as for other types of incidents, with the exception that those rescued who were in contact with the product must be decontaminated. Furthermore, those who have the best chance for survival should be rescued first.

■ Note
Those rescued who were in contact with the product must be decontaminated.

Evacuation For the goal of evacuation, two objectives must be considered. The first objective is to notify those who must and can be moved. This can be accomplished through public address (PA) systems, the media, local emergency broadcast networks, and door-to-door canvassing if the product does not prohibit. The second objective deals with those who need protection or evacuation but cannot be moved for whatever reason (see box, "Sheltering in Place"). The reason might be time, or numbers, or condition of the people. In this case, the objective is to shelter in place.

Exposure Protection When the hazardous material is involved in fire, exposure protection may be necessary. Tactical objectives are to place hose streams between the fire and the exposure property. The property might be buildings, dwellings, vehicles, and so on. Exposure protection may also be implemented when there is no fire. For example, an unignited leak of a flammable gas, such as liquified petroleum gas (LPG), which is heavier than air, may be directed away from exposures or ignition sources with hose streams. It cannot be overemphasized that these tactical objectives should only be employed if it is safe to do so. Furthermore, recall that some hazardous materials react with water. Therefore, if water is used as an exposure control, the runoff must not be allowed to come in contact with the product.

Confinement There are some confinement objectives that first responder operational level training personnel can employ. In terms of hazardous material response, these are often referred to as *defensive actions*. The tactical objectives are designed to limit the area of impact or stop the forward spread of the material. The goal is confinement. The tactical objectives include damming, diking, and retention (**Figure 18-6**). Application of these objectives should be evaluated in terms of safety and not putting personnel in the product.

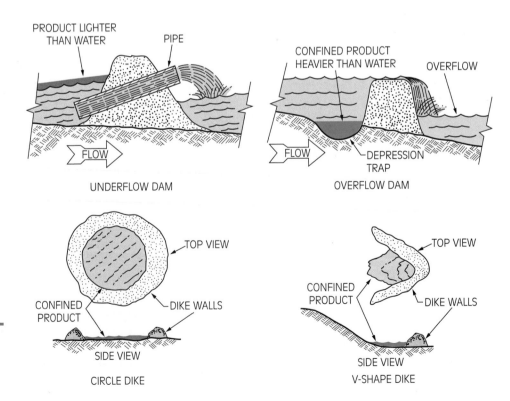

Figure 18-6 *An example of damming, diking, and retention.*

Extinguishment Provided that the material involved is not water reactive and it is safe to extinguish the fire, then common methods can be performed. The tactical objectives are to extinguish the fire by delivering water at the point of burning in a sufficient flow to cool and extinguish the fire. Keep in mind that hazardous materials often burn hotter than common combustibles and therefore may require higher water flows. Another tactical objective that results in extinguishment is to remove the fuel. This is particularly applicable to flammable gas fires. If the firefighters can be protected or if the fuel shutoff is remote from the actual burning, the fuel can be shut off. Never should a gas fire be extinguished if the gas supply cannot be stopped immediately after the fire is out.

Ventilation Tactical objectives for ventilation may include the ventilation of the building to prevent a buildup of a flammable gas. Remember, if the gas is lighter than air, the building should be ventilated in the upper portions. Heavier-than-air gas requires ventilation at lower levels.

Overhaul There is limited need for overhaul at a hazardous materials incident from a first responder perspective. By definition, overhaul requires entry into the product and should be handled by the hazardous materials team.

■ **Note**
By definition, overhaul requires entry into the product and should be handled by the hazardous materials team.

Salvage Tactical objectives related to salvage involve limiting the loss of property as a result of exposure to the material. Although like overhaul, salvage may require personnel to enter the product and therefore should be left to the hazardous materials team, there are some cases when the firefighters can perform salvage operations. When possible, tactical objectives can include covering or moving materials away from the product. Or it may be part of confinement by diverting the product away from valuable property.

SHIPBOARD FIRES

■ **Note**
Very few ships are exactly alike, but they present similar hazards.

Shipboard fires also present special hazards and require special training and resources. Ship fires are sometimes fought at sea and must be handled with the resources on board. However, they can be at port (**Figure 18-7**) and the local fire department may be called on to respond or assist, so firefighters should have some basic operational knowledge. Ships can be small or very large, often larger than most buildings in a jurisdiction, and situations are often complicated by limited access. Very few ships are exactly alike, but they present similar hazards.

Hazards Encountered

Ship fires present many hazards to the firefighters. The cargo can be anything from large numbers of passengers with limited cargo, to few crew members with

Figure 18-7 *A cargo ship in port.*

hazardous materials as the cargo. Cargo ships may carry up to twenty passengers, as well, adding to the life safety hazard. Fire on ships presents hazards in terms of unfamiliar surroundings and very punishing conditions, as firefighters often must enter from the top of the fire and go down into narrow confined spaces to seek the seat of the fire. Land-based fire crews will have considerable difficulty in moving around on a ship, as the ship is compartmentalized and has very narrow hallways and narrow, steep stairwells (**Figure 18-8**). Compartmentalization creates a maze that would be difficult to an unfamiliar person under the best conditions but is worsened by smoke and limited visibility. Shipboard firefighting operations require lifelines, PPE, SCBA, and strong command procedures and accountability. Another concern occurs when water is applied to a fire. As opposed

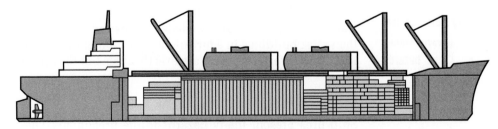

Figure 18-8 *Cargo ship interior layout.*

GENERAL CARGO VESSEL

fire control plan
a required plan on a ship depicting fire control procedures and resources specific to the ship

to a building fire where water freely runs off, in a ship the water stays on board. This can produce a problem with the ship's stability. All ships, including passenger ships, carry a considerable amount of fuel, usually hundreds of thousands of gallons, which can present serious issues.

One help to the firefighter is a ship's prefire plan, called the fire control plan (FCP). Firefighters responding to a ship fire will probably find the FCP the most valuable resource available because it has most of the information an incident commander needs to fight a fire on the ship. Included will be these:

- General layout of the ship
- Dimensions
- Firefighting systems, which can include:
  - Hose and hose reels to deliver water
  - Cartwheeled foam, dry chemical, halon, or carbon dioxide systems
  - Fire mains
  - Foam systems
  - Main standpipe and the international ship-to-shore connection
- Communication systems
- Plans for each deck of the ship

The incident commander should obtain the FCP as soon as possible after arrival at the incident. The FCP is required to be stowed in a marked location outside the deckhouse. There are required signs to direct firefighters to the location of the FCP.

Strategic Goals and Tactical Objectives

Application of strategic goals and tactical objectives in the case of a ship fire assumes that the firefighters are properly equipped and trained to safely undertake the operation.

Firefighter Safety Firefighter safety objectives should start with asking the question: Are the crews trained and equipped to handle the situation? Ship fires require firefighters to go into confined areas with limited means of access and egress. The buddy system must be used, with backup crews available. Lifelines should be used to provide exit paths. Accountability systems should provide for a rapid accounting of personnel both on the ship and in hazard areas. Rehabilitation may be needed more frequently because of the construction of the ship and the need to go long distances to reach the seat of the fire.

! Safety
Lifelines should be used to provide exit paths.

Search and Rescue On ships with passengers, search and rescue operations may be necessary. As in an apartment building fire, the tactical objectives should focus on sectoring the ship and having crews search assigned areas. Search of areas

Table 18-1   *Comparison of building component terms to ship terms.*

Structural Term (Building)	Ship
Walls	Bulkheads
Floors	Platforms or decks
Ceiling	Overheads
Windows	Portholes
Doors	Hatches

where potential victims are most exposed should be done first. Terminology is important to search and rescue teams. **Table 18-1** lists a comparison of ship terms to those with which firefighters are more familiar.

Reference should be made to the FCP before crews are assigned areas for search and rescue.

Evacuation   If the fire is at sea, the only option for evacuation is using lifeboats. However, in port, evacuation can be performed as it would be for other structures with the exception of egress routes, which can be limited by the capacity of the number of ramps leading off the boat. Another tactical objective for evacuation may be to evacuate the area or the port around the vessel. This would be applicable when a cargo ship is on fire and the burning cargo or smoke from the cargo would be hazardous to health or if the potential for an explosion were present.

Exposure Protection   Most often the exposure problem presented will be to buildings at the port or to ships in adjacent slips. Hose lines of sufficient flow should be placed on these exposures to prevent extension. In some cases, exposed boats may be moved away from the danger area.

Confinement   Confinement objectives focus on keeping the fire in the area of origin. This can be helped by closing bulkhead doors to try to slow the spread of fire and smoke. Furthermore, where possible, hose lines can be placed in adjacent areas to protect avenues of vertical and horizontal fire spread.

Extinguishment   The extinguishing objective depends on the material that is burning. If the material is hazardous, specialized agents might be necessary. Total flooding of confined areas may also be effective for some materials. The use of the built-in protection systems should be part of the tactical objectives plan. Use the FCP as a tool.

Ventilation Ventilation is difficult due to the ship's configuration. Tactical objectives involve removing or opening hatch covers and turning ventilators away from the wind. Positive pressure ventilation may also be employed once the fire is confined to a particular area.

Overhaul Overhaul is necessary in both the fire area and adjacent areas. Objectives are to open voids and search along common vertical and horizontal travel routes. As ships are constructed with several levels, the area above the fire should be examined for extension.

Salvage Efforts should be made to protect, inasmuch as possible, cargo and property on the ship. This may involve moving cargo out of the hazard area, using salvage covers, and routing water away from high-value property.

RAILCAR FIRES

■ **Note**
Trains can carry large amounts of people or cargo. The cargo is often hazardous.

Railroads pass through many jurisdictions. Any department that covers an area where there are train tracks has the potential to respond to a rail incident. Trains, like ships, can carry large amounts of people, or large amounts of cargo. The cargo is often hazardous.

Hazards Encountered

The hazards encountered with railcars depend on the type of railcar. Railcars can be classed into eight basic types, and the hazards associated with each are given in the following discussion.

Locomotives Locomotives are the workhorse of the train. They are large and have several hazards associated with them. Aside from their size and weight, diesel locomotives carry large amounts of diesel fuel and lubricating oil, and they produce high amounts of electricity with the generators. Basically a diesel electric locomotive is a generator on rails, and the biggest hazard is associated with the electricity being generated by the diesel engine.

Electric locomotives run on electricity from overhead wires that can carry current between 25,000 and 50,000 volts.

Boxcars The boxcars gets its name from its appearance, that of a box with wheels. Boxcars carry a variety of commodities that may or may not be hazardous and have varied levels of combustibility. The boxcars are designed to protect the cargo from the elements and avoid contamination. Often they are made of wood, which adds fuel to the fire. Some boxcars may contain a refrigeration system. The compressor is electric and is powered by a diesel generator attached to the

Figure 18-9 *A cargo box on a flat car.*

boxcar, which has tanks for the fuel supply (usually number 1 or number 2 diesel fuel oil). **Figure 18-9** shows a cargo box on a flat car.

Flatcars Flat cars do not have wall enclosures or tops, as shown in **Figure 18-10**. They provide no protection from the elements. The size of the flatcar and its capacity can vary depending on the load. Aside from wood used in construction and possibly the cargo, they do not present any special fire problems.

Intermodal Equipment Intermodal cars are basically flatcars with intermodal containers placed on them. Intermodal containers (**Figure 18-11**) allow the cargo to be transferred rapidly between different modes of transportation. For example, an intermodal cryogenic container could be shipped by sea to a port, transferred by crane to a truck, trucked to a rail yard, loaded onto a flatcar, transported to a rail yard across the nation, and loaded back onto a truck for delivery to the point of use, without ever having to offload the product. The hazards associated with

Figure 18-10 *A typical flatcar.*

Figure 18-11 *An intermodal container can be shipped by sea and then land without having to offload the cargo.*

intermodal equipment is basically the hazard of the cargo. The cargo can be flammable gases, liquids, cryogenics, and so on.

Gondola Cars Gondola cars have flat bottoms and walls and may or may not be covered. Gondola cars may be constructed of wood but are most often steel. The hazard of the cargo may contribute to the fire situation.

Hopper Cars Hopper cars have fixed sides and ends. The floor is sloped to one or more hinged doors. They are used to transport dry bulk materials, such as fertilizers, chemicals, salt, flours, and grains. Depending on the cargo, they may or may not be covered. The hazard of the cargo may present significant problems.

Passenger Cars Passenger cars are those railcars designed to carry people or perform functions for passengers, such as luggage storage or dining. Hazards associated with passenger cars begin with the number of persons on board. Unlike freight trains, which have small numbers of train crews, passenger trains may carry several hundred people. Passenger cars have combustible interiors. They also are equipped with cooking systems, air-conditioning systems, and electrical systems, all of which can contribute to the fire problem and are of tactical concern for the firefighter.

Figure 18-12 *One type of rail tank car.*

Tank Cars Tank cars (**Figure 18-12**) vary greatly. Although tank cars can carry products that are as safe as milk, when a firefighter thinks of a train derailment and a major emergency, a burning tank car comes to mind. The capacity of tank cars can range from just a few hundred gallons to 45,000 gallons of product. Tank cars may have one or more compartments. They may be nonpressurized and carry a commodity such as gasoline, or pressurized and carry liquified petroleum gas (LPG). There may be tank cars that are interconnected by pipes, forming a tank train. Furthermore, several 1-ton containers of chorine are commonly transported on a flat car, called a *multiunit tank car*. The product and the type of tank car play an important role in determining how to apply strategic goals and tactical objectives. For example, if flame is impinging on a pressurized railcar containing LPG and sufficient water supply cannot be established and applied to the point of impingement in a sufficient quantity, evacuation might be the only valid strategic goal.

Strategic Goals and Tactical Objectives

Railcar incidents produce some of the most notable incidents to which a firefighter can respond. If the train is a passenger train, then the potential exists for large numbers of casualties. If the accident involves a freight train carrying hazardous materials, the magnitude of the incident may affect hundreds or thousands of people in addition to the responders.

Firefighter Safety Firefighter safety objectives should include all those common for all incidents. It is essential that all other rail traffic be stopped. This can be accomplished by contacting the dispatcher for the railroad. Fire department dispatchers should have the contact number for rail dispatchers if a railroad is in the department's response area.

Because these incidents can be spread over a large geographic area, access to the scene should be controlled as part of the accountability system. Identification of the cargo of the train must be undertaken and a process put in place to limit firefighters' exposure to vapors or products of combustion. The potential for explosion and BLEVEs (boiling liquid-expanding vapor explosion) during tank car fires must also be evaluated. If the tank car shell at the point of flame impingement cannot be cooled, firefighters should withdraw and evacuate other persons in the hazard area.

If the rail locomotive is powered by electricity, then either overhead wires or a third rail will be present. In either case, the fire department should have the railroad tracks preplanned and a response plan in place.

Search and Rescue Nonpassenger trains travel with a minimum of people, usually a crew of two to four. Should an accident or fire occur, firefighters should account for the crew, which should be in the lead locomotive if the train is in transit. The crew, if not injured, should be able to provide valuable information on the cargo and hazard of the cargo on the train. Much information on the cargo and position of the railcars can be found on the consist or wheel report. The train crew should have both of these documents. Tactical objectives related to search and rescue should be employed in a passenger train incident. Priority areas should be developed depending on the time of the incident and the proximity to the fire. At night, most passengers are probably in the sleeping areas, whereas during the day, they may be in the observation or dining cars. The train should be divided into sectors and a complete search, inasmuch as conditions permit, should be conducted.

Evacuation Should the products on the train be hazardous, or if the products of combustion are hazardous, then evacuation must be ordered. As with any hazardous materials situation, the physical and chemical properties of the materials must be evaluated and the evacuation distance must be appropriate. Some of the largest evacuations in history have been a result of rail incidents.

Exposure Protection If the firefighter is able, any exposures should be protected with hose streams. Often railroad tracks run close to buildings, through rail yards, and through neighborhoods. Exposure protection objectives using some of the defensive methods discussed previously can be applicable for protecting exposures from runoff of product. If fires involve adjacent passenger cars in a train accident, priority should be given to prevent extension to the passenger compartments.

Confinement Confinement objectives, when possible, involve the placement of hose streams in order to prevent further progress of the fire. If the incident commander chooses the extinguishment option of allowing the product to burn out (see Extinguishment goal), then streams to confine the controlled burn will be necessary. Confinement of fires in passenger compartments of trains can be handled

BLEVEs
boiling liquid-expanding vapor explosion; describes the rupture of a container when a confined liquids boils and creates vapor pressure that exceeds the container's ability to hold it

■ **Note**
Nonpassenger trains travel with a minimum of people, usually a crew of two to four.

similarly to those in buildings or apartments. Hose lines should be placed in order to prevent fire spread into paths of egress and to support search and rescue operations.

Extinguishment If extinguishment is chosen as an option, the objective when common combustibles are burning is to apply water in a sufficient quantity and form to put the fire out. This involves the placement of hose lines and ensuring a continuous water supply. Because train incidents have occurred and will occur in areas without fire hydrants, water supply can be a problem. In these cases, the use of tanker shuttles is necessary. Water supply is sometimes a factor in the decision to let the product burn out. One extinguishment method might be to remove the fuel side of the fire triangle. In this case, the fire is allowed to burn the materials until they are gone.

Ventilation Access for ventilation may be difficult if the train is derailed. Once the location of the fire is determined, positive pressure ventilation can be used in passenger cars to improve search and rescue operations and fire control, if there are not too many openings. Remember that passenger train cars are interconnected to allow a passenger to walk car-to-car on the train. It is important during ventilation operations to ensure that smoke and gases are vented at each car entrance.

Overhaul Overhaul is necessary to search for extension and make sure complete extinguishment has taken place. Interior partitions should be inspected for signs of horizontal fire spread. Overstuffed seats and furnishings should be removed or opened up.

Salvage Where safe, salvage of undamaged product or personal property should be performed. Generally this is removing what can be removed from the train or hazard area. If a railcar catches fire but has not derailed, rail crews may be able to assist by removing uninvolved cars and property from the immediate area.

AIRCRAFT FIRES

Aircraft incidents may occur just about anywhere. If your jurisdiction provides fire protection to an airport or adjacent to an airport, then the chance of an aircraft incident is increased. Aircraft incidents may involve small private aircraft with few persons aboard, airliners with several hundred persons aboard, cargo aircraft that may carry some hazardous materials, or military aircraft. First responders may be able to take limited actions as part of an initial response.

Hazards Encountered

Regardless of whether they are designed for cargo or passengers, fixed-wing aircraft all have several similarities in terms of hazards. These include a large amount of fuel, hydraulic fluid, and, in some cases, oxygen equipment. Aircraft engines can be of three different types: reciprocating, jet propulsion, and turbo-prop. The fuel for the aircraft depends on the engine type. The reciprocating engine uses high-octane aviation gasoline, or avgas. Avgas is very flammable. Jet engines and turboprop engines use a commercial high-grade kerosene fuel known as Jet A, which has a higher flash point than avgas, but a lower ignition temperature. Military aircraft with jet engines use a blend of kerosene and gasoline. Each of the fuels adds to the fire problem.

Aside from the fuel hazards, there are hazards associated with the aircraft itself. The aircraft, at least larger aircraft, have a hydraulic system for the controls. The hydraulic lines may carry pressures of up to 5,000 psi, and there is the hazard of the hydraulic oil itself. Larger aircraft have some type of onboard oxygen system. Aircraft also have auxiliary power units (APU) to supply power when the plane's engines are shut down. The APU usually operates off of the plane's fuel supply. On military aircraft, explosive devices are used to dislodge the canopy or ejection seat, and there is the possibility of ammunition. On crop-dusting aircraft, there may be residual poisons or insecticides. Radar units on the front of the aircraft may be operating after a crash and emitting radiation unless shut down.

Aircraft tubing may be identified by words, symbols, or color markings. The color-coding system for the tubing is given in the following list:

Fuel	Red
Lubrication	Yellow
Hydraulic	Blue and Yellow
Compressed Gas	Orange
Breathing Air	Green
Fire Protection	Brown

Rotary-wing aircraft may carry anywhere from one to fifty people in flight. They are compact and all systems are close together, creating a hazard for the passengers. Fuel carried on helicopters can range from 20 to 700 gallons. Fuel cells are usually carried in the center of the fuselage or in auxiliary tanks mounted on pods. The main rotor blades on large helicopters can have a span as much as 60 feet. Main rotor blades may dip as much as 4 feet from the ground during start-up and shutdown. Further, the tail rotor spins faster than the main rotor and may not be seen when spinning. Extreme caution must be used when approaching rotary-wing aircraft.

Strategic Goals and Tactical Objectives

Aircraft incidents that occur on field at larger, commercial airports are most often handled by an on-site aircraft crash and firefighting department (**Figure 18-13**). However, those that occur off field or at many small general aviation

airports around the country are served by the local fire department. The strategic goals and tactical objectives presented in the text and their applicability to aircraft incidents are described in the following section.

Firefighter Safety Aircraft incidents by nature of size and fuel load can present problems from a safety standpoint. The cargo should be identified and hazardous materials researched. Radioactive materials are sometimes carried on aircraft and may be present because of the plane's radar. Precautions should be taken against this exposure. Aircraft incidents are usually of long duration, so rehabilitation and crew relief must be planned for. As with an incident when large numbers of persons are injured or killed, critical incident stress management is indicated. Entry to the crash site should be limited. Firefighters must also be protected during operations from high-pressure hydraulic lines, oxygen lines, and fuel lines. See **Figure 18-14.**

Search and Rescue Search and rescue begins with the area around the crash site. Injured persons who can walk should be directed into safe areas. Search and rescue in the plane's passenger compartment follows basic principles. The search and rescue operation must be supported by hose streams. Access may be gained by cutting through the aircraft fuselage or windows. However, the easiest access is through the plane's doors. Firefighters must use caution in opening the doors as this may cause the emergency chute to deploy, potentially injuring the firefighters.

Evacuation Evacuation of the plane's passengers will probably occur before the fire department arrives. However, because of the hazards associated with the incident, the evacuation of nearby buildings or other transportation routes may be necessary.

⚠Safety
Firefighters must use caution in opening the doors as this may cause the emergency chute to deploy, potentially injuring the firefighters.

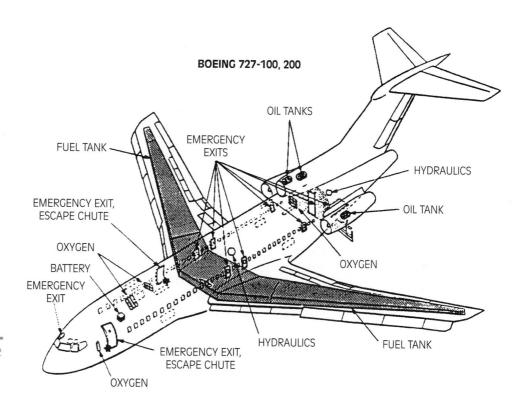

BOEING 727-100, 200

OIL TANKS

FUEL TANK

EMERGENCY
EXITS

HYDRAULICS

OIL TANK

EMERGENCY EXIT,
ESCAPE CHUTE

OXYGEN

BATTERY

EMERGENCY
EXIT

OXYGEN

HYDRAULICS

FUEL TANK

EMERGENCY EXIT,
ESCAPE CHUTE

OXYGEN

Figure 18-14 *Aircraft components on a Boeing 727.*

Exposure Protection The fire associated with the aircraft may be large and hot because of the high volume of fuel. Exposure protection is necessary and can be accomplished with high-volume hose streams.

Confinement Confinement can be accomplished with the use of hose lines between the fire and exposures. Foam, necessary on the burning fuel, can also be helpful in confinement, as unignited fuel can be blanketed and ignition prevented.

Extinguishment Extinguishment of fuels or other fluids requires the use of foam. Enough foam should be on hand to completely extinguish the burning fuel. Extinguishment of other combustibles, such as the interior of the aircraft, can be accomplished with water streams.

Ventilation On fires involving the interior of the aircraft, ventilation is required. As cutting through the shell of the aircraft may cause the firefighter to cut through fuel or hydraulic lines, this should be done only when absolutely necessary. Positive pressure ventilation can be used through doors and window exits, provided there are not too many openings.

Overhaul Once life safety has been taken care of and the fire extinguished, the aircraft should be left intact, inasmuch as possible, for the federal investigators. While some overhaul may be necessary to ensure complete extinguishment, only as much as needed should be performed.

Salvage As with overhaul, investigators will want to find the plane in the condition it was before the incident. Therefore, little if any salvage should occur.

BULK STORAGE FACILITIES

Bulk storage facilities hold a large volume of products in a variety of containers, sizes, and shapes (see **Figure 18-15**). The good thing about bulk storage facilities is that unlike with trains, vehicles, and aircraft, you either have them or you do not. A bulk storage facility does not move, and therefore affords firefighters a chance to preplan the facility and be familiar with the product or products stored. Preplanning should cover all the routine preplanning subjects and, in addition, list the volumes of product, the physical and chemical hazards, the technical assistance available, and the level of evacuation that might be required.

■ **Note**
A bulk storage facility does not move; therefore, it affords firefighters the chance to preplan the facility and be familiar with the product or products stored.

Figure 18-15 *A bulk storage facility for liquids.*

Hazards Encountered

The hazards associated with bulk storage focus on the product, the type of tank, construction of the tank, and the physical layout of the tanks. Although products may vary greatly, some generalities can be made. Products may include flammable liquids, flammable gases, liquid corrosives, and heavy fuel oils, to name a few. However the bulk storage may include flour and other equally hazardous commodities because of dust explosions. Because the firefighters have the chance to preplan, the opportunity to be very familiar with the product should be embraced. Types of tanks vary between low-pressure, or atmospheric, tanks, to those that are under pressure. The tanks may be vertical or horizontal, above or below ground. Low-pressure tanks have different types of roofs, including the cone roof, floating roof, floating roof with geodesic dome, and covered floating roof. Pressurized tanks hold gases that are liquified or stored in bulk under pressure. Bulk tanks are generally steel-constructed and may be laid out close to one another in a tank farm arrangement. However, they are required to have built-in safety features. For example, they may be required to employ deluge water or foam systems. The tank must be diked so that any overflow or product release fills the dike and does not spread to other areas.

Fires in tank farms or around bulk storage tanks present a unique challenge. The product may spill and burn while surrounding the tank. Adjacent tanks may fail as a result of extreme heat. Boilover can occur.

Strategic Goals and Tactical Objectives

Firefighter Safety Bulk storage facilities are very dangerous for firefighters responding to an incident. Access to diked areas should be limited to only those cases when it is absolutely necessary to save a life. Within the diked area are aboveground pipes, valves, and numerous other trip-and-fall hazards. In the dark or after extinguishment efforts have begun and water fills the dike area, these obstructions may be impossible to see. Furthermore, should a boilover occur, firefighters would be trapped in the dike. Crew should not be placed in line with the ends of tanks containing pressurized gases. Other common safety procedures should be in place and full protective clothing, appropriate for the materials, worn. As with many of the other incidents described in this chapter, these can be long-duration incidents and require strong accountability and rehabilitation procedures.

Search and Rescue Search and rescue may be performed if firefighters can safely enter the area. Because most bulk facilities are not public places, staff at the facility should be able to give the incident commander a good account of the number of people missing and an approximate location of where they were last working. These areas should be searched first.

■ **Note**

Fires in tank farms or around bulk storage tanks present a unique challenge. The product may spill and burn while surrounding the tank. Adjacent tanks may fail as a result of extreme heat. Boilover can occur.

! **Safety**

Access to diked areas should be limited to only those cases when it is absolutely necessary to save a life.

Evacuation Aside from evacuation of the facility, very often at these events surrounding homes, buildings, and transportation routes must be evacuated because of the fire or the large amount of hazardous smoke and gases. Access to the area should be denied to anyone except those responding to the incident. Some bulk facilities may have sirens or other warning devices to alert residents in the local area of an emergency at the facility. These alarms can be used if the situation warrants an evacuation of area residents.

Exposure Protection Protection of exposed tanks can be paramount in these types of incidents. Large streams should be placed on exposed tanks. Preplanning allows for the knowledge of the presence and the operation of fixed systems that may be at the site. Water should be applied in large quantities to the vapor space of pressurized containers, and to the sides below the liquid line of large liquid or non-pressure tanks.

Confinement The dikes that should be in place will help with the confinement effort. Through information obtained by preincident planning and advice from the plant employees, valves and piping may permit some of the burning product to be pumped off into another tank. Pipes exposed to the fire should be shut down to prevent addition of fuel if the pipeline would rupture.

Extinguishment The extinguishing method depends on the product involved. Agents might include foam or other specialized agent. Application of the agent is also a tactical decision based on the fixed protection that is in place, such as prepiped deluge systems for foam or piping for subsurface injection.

Ventilation Ventilation of the bulk tank involved will probably not be an issue. There may be concerns for other nearby buildings or enclosed structures. Depending on the extent of the incident, these areas should be evacuated so ventilation may still not be a concern.

Overhaul Like many of the hazardous materials-related incidents, overhaul should only involve what is necessary to ensure that the fire is out. Further overhaul and cleanup should be left to specialized teams.

Salvage Aside from valuable property that you might be able to remove safely from the hazard area, salvage is not an issue.

ELECTRICAL SUBSTATIONS

Electrical substation fires occur for numerous reasons—mechanical failure, lightning, overheating from energy demands, and so on. The substation fire is usually a class B fire after the electric has been shut down. In many cases, the

Figure 18-16
Electrical substations contain a number of hazards for responding firefighters.

fire department will have to stand by until they have been assured by the electric company that the power has been shut down and the lines de-energized (see **Figure 18-16**).

In a substation fire in Tennessee, a transformer at the substation exploded and spilled onto the ground more than 4,000 gallons of mineral oil that caught fire and burned. It took the electric company crews more than 45 minutes to isolate and shut down the power so firefighters could begin firefighting operations. Until they could begin fighting the fire, the crews evacuated the surrounding homes and businesses and protected exposures. While waiting, a foam bank was established, and once the power was off, the crews moved in with foam lines and quickly extinguished the fire.

In a substation fire in Texas, the fire crews were confronted with a massive fire when a transformer exploded, releasing more than 10,000 gallons of mineral oil. Due to the high heat and intensity of the fire causing damage to the overhead high-voltage lines and powerline tower, the fire crews had to stay far back from the scene until the power could be shut down. The fire began at 10 A.M. and was controlled with high water flows from a tower ladder by 1 P.M.

As with both of these incidents, a command officer must understand that evacuation and exposure protection are the only activities to undertake until the power has been shut down. Once the power is off, the fire crews will be confronted with a flammable liquid fire.

Another situation involving electrical transformers is the underground electric vaults found in many downtown areas. When these fires erupt, the firefighting considerations are the same as for a substation. The more important concern here is that explosions from these underground vaults can catapult manhole covers more than 100 feet into the air. Keep your people out of the area until the power company shuts the power down. This will usually cause the fire to begin burning with less intensity and fire crews can begin their attack. Identify the other manhole covers in the area and do not place people or equipment near those manhole covers. Once the power is off, foam can be flowed into the manhole from a distance and allowed to extinguish the fire.

SUMMARY

Firefighters may be called on as first responders to incidents involving hazardous materials, ships, railcars, aircraft, and bulk storage facilities. Each of these types of emergencies is complex and requires a certain level of expertise. This chapter presented some guidelines as basic knowledge of these emergencies necessary for all firefighters.

Chemicals are categorized into nine classes based on their hazards. These chemicals can be found at incidents involving the production, transportation, and storage of the chemical, or at its place of final usage. Chemicals may be found just about anywhere, from the large industrial complex to a single-family home with a pool or barbecue. Firefighters must be alert to their potential presence.

Numerous forms of transportation can produce significant incidents and hazards to responders. Ships may be a passenger ship with many people aboard, a cargo ship with automobiles, or a tank ship with liquified natural gas. Railcars may carry passengers, freight, or tankers with flammable liquids or gases. Even passenger trains present a high level of hazardous materials, because the locomotive carries diesel fuel and there are refrigerants for the air-conditioning. Aircraft, while mostly thought of as carrying passengers, also present hazardous material challenges because of fuel, hydraulic fluid, and oxygen.

Bulk storage facilities store very large quantities of liquid or gaseous materials that are often flammable or have other hazardous properties. These facilities probably have a number of tanks close to one another presenting extreme exposure problems. Because the hazard is recognized, several private fire protection systems are often in place. These systems are of help to the firefighters. Electrical substation fires will also present special hazards to firefighters, including the fact that the substation may still be charged with electricity.

The applications of strategic goals and tactical objectives are not as clear for these incidents as they are for other types of incidents. However, some application can be made and these have been presented in this chapter. The complexity of incidents discussed in this chapter cannot be overemphasized. Firefighters and incident commanders must have additional knowledge and training in these areas if they are to operate safely and effectively. Many courses are available that cover these specialized areas.

REVIEW QUESTIONS

1. A flammable gas would fall into what hazardous materials class?

   A. Class 1
   B. Class 2
   C. Class 3
   D. Class 4

2. A firefighter that responds to a hazardous materials incident as part of the initial response for the purpose of protecting nearby persons must be trained to what level?

   A. First Responder Operational
   B. First Responder Awareness
   C. Hazardous Materials Technician
   D. Hazardous Materials Specialist

3. A material such as caslcium carbide, which is dangerous when wet, is considered in what hazardous materials group?

   A. 3.2
   B. 2.1
   C. 1.3
   D. 4.3

4. List four ways that the public could be notified of an evacuation after a hazardous materials release.

5. List five subjects that a ship's fire control plan should cover.

6. True or False? Cutting through an aircraft's fuselage is the desired method of gaining access to the interior passenger compartment.

7. Fuel used for military jets that is a blend of kerosene and gasoline is called _____.

   A. Avgas
   B. Jet A
   C. JP-4
   D. Jet B

8. List three hazardous materials that might be found at a passenger train derailment.

9. True or False? A ship's bulkhead doors may be closed to help firefighters confine the fire.

10. True or False? Exposures are of little concern at bulk storage fires because of the protection afforded by the dike system.

ACTIVITIES

1. Review your department's standard operating guidelines (SOGs) for response to aircraft, ship, train, or bulk storage facility incidents. What changes might be needed? Are training of the personnel and the availability of specialized equipment consistent with the requirements of the SOGs?

2. Review your department's preplans for target hazards. How many contain hazardous materials? Are specialized extinguishing agents required? If so, are they available?

Chapter

19

PUTTING IT ALL TOGETHER

Learning Objectives

Upon completion of this chapter, you should be able to:

- Apply fireground management concepts, the strategic goals, and tactical objectives to a simulated fire in a single-family dwelling.
- Apply fireground management concepts, the strategic goals, and tactical objectives to a simulated fire in a multiple-family dwelling.
- Apply fireground management concepts, the strategic goals, and tactical objectives to a simulated fire in a commercial building.

CASE STUDY

In late afternoon fire crews are dispatched to a reported structure fire with multiple phone calls reporting fire and smoke coming from a commercial building. Three engines, one ladder, a heavy rescue and a battalion chief respond. From a block away the first arriving engine sees smoke rising into the air and reports to the dispatcher that a working fire is in progress. Upon arrival and without taking notice of the fire location, the first engine crew begins an interior attack to find the fire.

The battalion chief arrives next on the scene and observes heavy black smoke pouring from the roof of the single story commercial building. The BC runs into the business owner and is told that everyone is out of the building. The battalion chief pushes the owner aside and follows the hoseline to the front door and stands watching as the firefighters are working to get their line stretched to the back of the building through the smoke. The battalion chief fails in his responsibilities to get a view of at least three sides of the building. He overlooks the bowstring truss roof set on the concrete block building and the fire that is rolling through the attic of the building.

Upon the arrival of the ladder company, he orders them to the roof to begin ventilation operations. The battalion chief then follows the ladder crew as they move their ladder truck to the side of the building and begin to raise the ladder to the roof. He watches as the men take their axes and saw and begin their climb to the roof. The battalion chief then walks around to the front of the building to see what the other crews have chosen to do.

Instead of setting a command post and providing orders to the incoming apparatus, the battalion chief wanders around the fire scene watching as a spectator might have done. The crews commence to work on whatever duties they decide needs to be completed. The battalion chief is seen again walking around the fire scene as the other firefighters and officers enter the structure. Within 30 minutes of the firefighters arrival the bowstring roof begins to collapse pushing the side wall out toward the ladder truck. The firefighters on the roof scramble to safety while the firefighters inside become trapped.

In today's fire service, the Incident commander should use pre-plans to begin a mental size-up of the building before arriving on scene. Then begin obtaining information from first arriving companies that will lead to the Strategic Goals and Tactical Objectives needed to fight the fire and accomplish the goal of firefighter safety at the incident. In a logical thought process, the IC will consider the incident priorities while completing an accurate size-up of the incident. The officer's knowledge of building construction would have led to understanding that a bowstring roof sitting on concrete block walls is a dangerous building. The heavy smoke and fire conditions in the attic would have led to the strategic decision to keep firefighters off the roof and out from under bowstring trusses. There was not a life safety concern, except for the firefighters on scene.

The Incident Commander will establish a command post, assign a safety and accountability officer and conduct a well managed and disciplined fireground.

INTRODUCTION

This book has presented concepts for the effective management of fires, including the study of fire behavior, fire extinguishment, incident management systems, strategic goals, safety, company operations, private fire protection systems, post-fire considerations, and the management of fires in specific occupancies and of nonstructural fires.

The goal of this chapter is to introduce fires in three of these situations and give examples of how the concepts can be applied to these specific fires. In each case, the study involves a particular type of fire. A graphic showing the building and a graphic of the floor plan are provided along with a description of the fire situation. If appropriate, a preincident plan is provided. Also the size-up information is given as well as suggested strategic goals and tactical objectives based on the information presented in this text.

Hypothetical Fire Department

For the purpose of these scenarios we use a four-station fire department in Anytown, USA, which provides a full range of fire suppression, prevention, hazardous materials response services, and emergency medical services (EMS). The community is a mix of single- and multifamily residences, light commercial buildings, and two 10-story high-rises. Part of Anytown was built in the early 1900s; however, in the past 10 to 15 years there has been a growth spurt mostly of garden apartments and townhouses. **Table 19-1** outlines the resource capabilities of the department. Anytown Fire Department uses an IMS that refers to the structure using the terms side 1, 2, 3, and 4, with side 1 being the address side.

Special Note: For the purpose of the examples in this text, the staffing levels of engines and ladders is set at three. This number is less than optimum staffing and is used as an illustration only. Staffing will vary based on local conditions. What must be realized is that seventeen people are used on the first and each additional alarm in these scenarios. This is the number of firefighters needed for a safe and effective fire attack. Engine and ladder company effectiveness increases as staffing increases.

The department enjoys a very close relationship with similar surrounding departments and has common standard operating guidelines (SOGs) as well as an automatic aid agreement that allows for closest unit responses and predetermined assignments for up to four alarms. The department has several staff officers who also respond routinely to working fires. Included with staff is a fire chief, an operations chief, an EMS chief, and a training/safety officer. Response times for these staff members vary at night, but during normal working hours they can be expected to respond rapidly.

Table 19-1   *Anytown, USA, fire and rescue resources.*

Apparatus	Type	Staffing per Shift	Notes
Engine 1	1250 gpm class A pumper 750-gallon water tank	1 company officer 1 FF/driver-operator 1 firefighter	
Engine 2	1250 gpm class A pumper 750-gallon water tank	1 company officer 1 FF/driver-operator 1 firefighter	
Engine 3	1250 gpm class A pumper 750-gallon water tank	1 company officer 1 FF/driver-operator 1 firefighter	
Engine 4	1250 gpm class A pumper 750-gallon water tank	1 company officer 1 FF/driver-operator 1 firefighter	
Ladder 1	75-foot aerial ladder truck	1 company officer 1 FF/driver-operator 1 firefighter	
Squad 1	Heavy rescue squad	1 FF/driver-operator 1 firefighter	Squad has truck company equipment on it.
Ambulance 1	Advanced Life Support, transport-capable ambulance	2 FF/paramedics	Paramedics are cross trained and provide fire suppression functions.
Ambulance 2	Advanced Life Support, transport-capable ambulance	2 FF/paramedics	Paramedics are cross trained and provide fire suppression functions.
Battalion 1	Command vehicle	1 Battalion chief	

Note: FF means firefighter.

SINGLE-FAMILY DWELLING SCENARIO

The first scenario involves a one-story single-family dwelling as shown in **Figure 19-1. Table 19-2** lists the size-up factors.

Figure 19-2 shows a floor plan of the structure. Note the fire's location and possible extension probabilities. **Figure 19-3** shows the street and adjacent properties.

Strategic Goals and Tactical Objectives

The following is the application of the strategic goals as presented in this text to this specific incident.

Figure 19-1 *Single-family dwelling scenario structure.*

Table 19-2 *Size-up factors for single-family dwelling fire.*

Factors	Description
Environment	
Time	3:00 A.M.
Building construction features	Single-family dwelling of ordinary construction. Houses in this area were built about 15 years ago.
Weather	70°F. Wind not a factor.
Height	One story
Area	2,300 square feet
Occupancy	Residential
Access	Road and driveway access on Side 1. Sides 2, 3, and 4 are yards and do not provide access for fire vehicles.
Terrain	Level
Resources	
Personnel responding initial alarm	Engine 1 with 3 personnel Engine 2 with 3 personnel Engine 3 with 3 personnel Ladder 1 with 3 personnel

(continued)

Table 19-2 *(Continued)*

Factors	Description
Resources	
	Squad 1 with 2 personnel
	Ambulance 1 with 2 personnel
	Battalion 1 with 1 person
	17 total personnel on initial alarm
Personnel needed	Varies depending on the assignments
Apparatus responding	3 Engines
	1 Ladder
	1 Squad
	1 Ambulance
	1 Battalion chief
Apparatus needed	Varies depending on assignment
Water supply	Hydrants are throughout the area. The closest is 200 feet away and has a flow of 1,000 gpm.
Built-in private fire protection	None
Need for special extinguishing agents	None
Response time	The first engine and ladder will be on scene within 4 minutes. The balance of the first-alarm assignment will be on the scene within 8 minutes.
Conditions/Situation	
Fire location	The fire is in the family room.
Fire extension probability	Internal exposures are present. External exposures not likely. Houses are 50 feet from each other.
Type of fuel	Common household
Fire condition on arrival	Heavy fire visible out rear patio door; heavy smoke throughout the house.
Life hazard	High based on
	Time of day
	Cars in driveway
Property conservation concerns	High
Possible incident duration	1–2 hours

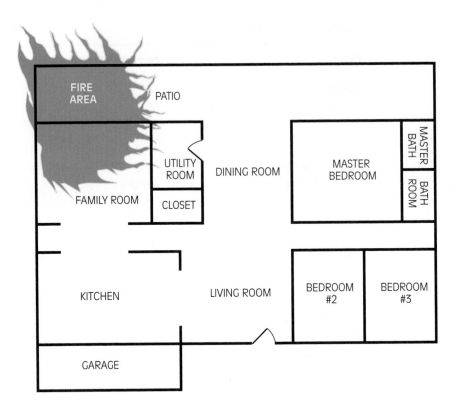

Figure 19-2 *Floor plan of the single-family dwelling.*

Firefighter Safety To accomplish the goal of firefighter safety at this incident, several tactical objectives must be accomplished. It is presumed that department SOGs provide for the wearing of appropriate personal protective equipment (PPE), the implementation of an incident management system (IMS), and accountability procedures. At the house fire, the first-arriving engine's driver serves as the accountability officer. This driver will be operating the fire pump and other arriving units that are to be assigned in the hazard zone will give this operator their accountability tags. The IMS provides that the first-arriving engine officer will pass command in favor of a fast attack. The battalion chief will assume command on arrival.

The tactical objectives to be employed for firefighter safety and the units assigned are noted in the following list:

- IMS in place, Battalion 1 as the incident commander. A safety officer should be assigned as soon as possible.

- Offensive fire attack will be the initial mode of operation.

- A rapid intervention crew (RIC) will have to be in place prior to interior fire attack unless the incident commander feels there is an imminent life safety risk prior to the assignment of an RIC.

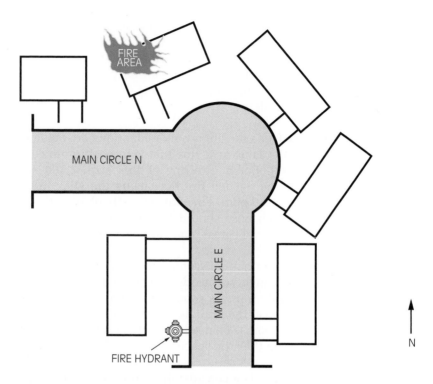

Figure 19-3
*Surrounding
properties.*

- Attack and backup lines of appropriate size for this fire (1¾-inch in this case)
- Rehabilitation will be set up after the initial fire attack is complete. As this type fire should be of relatively short duration and the environment factors are mild, rehabilitation may just be brief medical evaluation and fluid replacement.

Search and Rescue Unless the occupants of the home are accounted for, an interior search must be conducted in this situation. Considering the time of day, the bedrooms and paths of travel to exit and behind these areas should be searched first, as should behind doors and under windows.

Evacuation Evacuation of the fire structure is covered during search and rescue. However, there may be evacuation needs of surrounding areas if the fire was not controlled and the operation changed to a defensive mode.

Exposure Protection Interior exposures will be protected during fire confinement and extinguishment phases. External exposures will become a factor if a defensive operation occurs.

Confinement The confinement of this fire is critical to prevent extension to the rest of the dwelling and to support search and rescue activities. In this situation, the first hose line should be stretched through the front door and positioned to hold the fire in the family room. This not only supports the confinement effort, but also supports and protects the means of egress for search and rescue purposes and firefighter safety.

Extinguishment Once the fire is confined, efforts are made toward extinguishment. In this case, the same hose line that is in position to confine the fire is used for extinguishment. For a dwelling of this size, 2,300 square feet with a 20% involvement, the needed fire flow using the NFA formula is estimated to be approximately 150 gpm. This flow is easily obtained using a 1¾-inch hose line. Another 1¾-inch hose should be positioned to back up the first line, further protect the means of egress, and support the search and rescue operation.

Ventilation The fire has self-vented through the patio doors. Further ventilation should initially involve the opening of windows to support search and rescue by horizontal ventilation. As soon as possible, positive pressure ventilation (PPV) should be set up at the front door with exhaust control accomplished by closing interior room doors. The need for vertical roof ventilation is unlikely unless the fire extends into the attic space.

Overhaul Overhaul is a consideration to ensure that the fire is completely extinguished. Once the visible fire is extinguished, crews should open the walls and ceiling in the family room. Enough area should be exposed until unburned material is apparent. The attic area should also be checked either through the ceiling openings or by the attic access. The eaves outside of the patio door should also be checked for extension.

Salvage Once the fire suppression efforts are taken care of, emphasis must be placed on salvage. In this case, covers and floor runners may be used to cover carpets and furniture in unburned areas. The removal of the smoke in the bedroom areas may help in the effort to salvage clothes and other personal belongings.

The Company Assignments

In order to accomplishment the strategic goals and objectives just discussed, company assignments must be made. **Table 19-3** depicts these typical assignments both initially and throughout the incident. The units are listed in order of arrival.

After the fire, a postincident analysis (PIA) should be conducted. As discussed in other chapters, the PIA should focus on the incident and point out strengths and weaknesses of the operation.

Table 19-3 *Typical company assignments.*

Unit	Initial Assignment	Other Assignments
Engine 1 (E1)	Advance a 1¾-inch hose line through the front door to the family room for fire containment and extinguishment. This line will be supplied from tank water until a water supply can be established. The driver of E1 will be the accountability point.	Once the fire has been extinguished and this crew has been through rehab, they will assist in salvage.
Ladder 1	Since there is no need to operate the aerial ladder, all three personnel from Ladder 1 are available for assignment. The company officer and one FF will assist E1 by forcing entry. Once inside, they will begin a search of the priority areas. Once the hose line is in place, the driver of Ladder 1 will begin to horizontally ventilate the windows from the outside while checking for utility issues.	Once the primary search is complete, this crew will begin overhaul operations and salvage.
Squad 1	The initial two persons out team.	Once relieved by E3 being assigned as RIC, the squad will set up for PPV and assist with overhaul and salvage.
Battalion Chief 1	Incident command	
Ambulance 1	Will assist Ladder 1 with the search.	Once the primary search is completed, the Ambulance 1 crew will set up a rehab sector. Should a victim be found or other need for EMS occur, the Ambulance 1 crew will provide initial care and Ambulance 2 will be called to transport if the need arises.
Engine 2 (E2)	Will reverse lay a supply line from E1 to the closest hydrant. The E2 driver will make the hydrant connection. The company officer and FF from E2 will advance a second 1¾-inch hose line from E1 into the structure for backup and support purposes.	Once the fire is extinguished, the E2 crew will perform the secondary search.
Engine 3 (E3)	Upon arrival, the entire E3 crew will become the RIC.	Once the need for the RIC is over, this crew will assist with salvage and overhaul.
Staff Officers	Safety will be assigned to the first arriving staff officer trained in this function. Other staff officers will be assigned to • Public information • Resident's needs • Assumption of command if the incident escalates	

Note: FF means firefighter.

This fire scenario was an example of a basic room and contents fire that fire departments across the country respond to frequently. The fire was handled appropriately with the first-alarm assignment of units and personnel. All of the necessary strategic goals and tactical objectives were accomplished while minimizing the property loss and having a high concern for firefighter safety.

MULTIFAMILY DWELLING SCENARIO

This simulation involves a four-story, multifamily dwelling, as shown in **Figure 19-4**. The building was built in 1930 and is of ordinary construction. Since this simulation requires additional resources, there will be additional alarms. For the purpose of this simulation each alarm after the first will get the same amount of personnel and equipment. **Table 19-4** describes the size-up factors.

Figure 19-5 shows a floor plan of the structure. Note the fire's location and possible extension probabilities. **Figure 19-6** shows the street and adjacent properties.

Figure 19-4
Multifamily dwelling scenario structure.

Table 19-4   *Size-up factors for multifamily dwelling fire.*

Factors	Description
Environment	
Time	11:00 A.M.
Building construction features	Multifamily dwelling of ordinary construction built in the 1930s.
Weather	65°F. Wind not a factor
Height	Four stories
Area	5,000 square feet per floor. Each apartment is about 400 square feet in area.
Occupancy	Residential
Access	Road access on Sides 1 and 4. Side 1 is a street, Side 2 is a similar apartment house, Side 3 is an alley, and Side 4 is a street.
Terrain	Level
Resources	
Personnel Responding	Engine 1 with 3 personnel Engine 2 with 3 personnel Engine 3 with 3 personnel Ladder 1 with 3 personnel Squad 1 with 2 personnel Ambulance 1 with 2 personnel Battalion 1 with 1 person 17 total on initial alarm and subsequent alarms.
Personnel Needed	Varies depending on assignments.
Apparatus responding	3 Engines 1 Ladder 1 Squad 1 Ambulance 1 Battalion Chief
Apparatus needed	Varies depending on assignments.
Water supply	Hydrants are throughout the area. The closest is 50 feet away and has a flow of 1,200 gpm.
Built-in private fire protection	None
Need for special extinguishing agents	None
Response time	The first engine and ladder will be on scene within 4 minutes. The balance of the first-alarm assignment will be on the scene within 8 minutes. The balance of each additional alarm will be on the scene within 8 minutes of request.

Table 19-4 *(Continued)*

Factors	Description
Conditions/Situation	
Fire location	The fire is in apartment 3B on the third floor.
Fire extension probability	Internal exposures present, specifically the fourth floor and cockloft. External exposures likely on Side 2.
Type of fuel	Common household
Fire condition on arrival.	Heavy fire visible out two third-floor windows of apartment 3B. Heavy smoke in the third-floor hallway and in apartment 4B. Medium smoke in adjacent apartments and the fourth-floor hallway.
Life hazard	Medium because of time of day
Property conservation concerns	High
Possible incident duration	2–3 hours

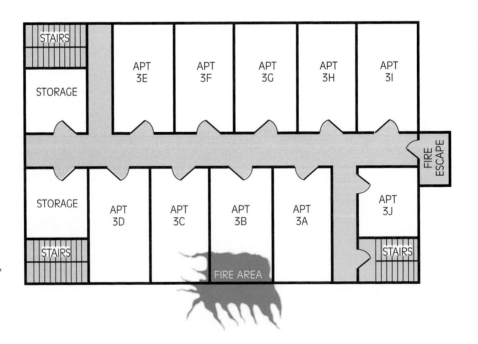

Figure 19-5 *Floor plan of the multifamily dwelling.*

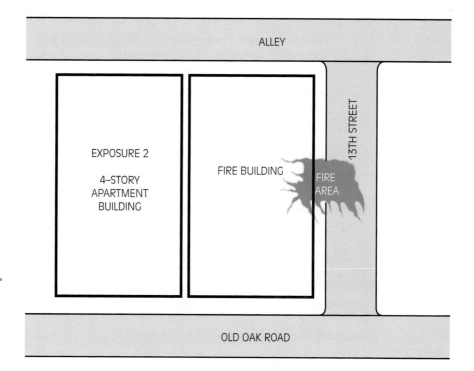

Figure 19-6 *Street and property adjacent to the multifamily dwelling.*

Strategic Goals and Tactical Objectives

The following is the application of the strategic goals as presented in this text to this specific incident.

Firefighter Safety To accomplish the goal of firefighter safety at this incident several tactical objectives will have to be accomplished. Again it is presumed that department SOGs provide for the wearing of appropriate PPE, the implementation of an IMS, and accountability procedures. At this fire there are several entrance points into the fire building. The first-arriving companies at each of these potential entrance points will be the accountability point for that entrance. However, due to the complexity of the incident and the building, an accountability officer should be assigned as soon as possible. The IMS provides that the first-arriving engine officer will pass command in favor of a fast attack. The battalion chief will assume command upon arrival.

The tactical objectives to be employed for firefighter safety and the units assigned are as follows:

- IMS in place, Battalion 1 as the incident commander. A safety officer should be assigned as soon as possible, as should an accountability officer.
- Offensive fire attack will be the initial mode of operation.

- An RIC will have to be in place prior to interior fire attack unless the incident commander feels there is an imminent life safety risk prior to the assignment of an RIC. In this incident, a second RIC should be considered since the building is large in area with multiple entry points.
- Attack and backup lines of appropriate size for this fire (at least 1¾-inch in this case)
- Rehabilitation will be set up after the initial fire attack is complete. As this type fire should be of medium duration and the environment factors are mild, rehabilitation may just be a brief medical evaluation and fluid replacement.

Search and Rescue Unless the occupants of the entire building are accounted for, an interior search must be conducted. Unless there are other obvious needs, the search should progress as follows. The apartments on the fire floor should be searched first, starting with the fire apartment and working out into adjoining apartments. The floor above should be searched next or preferably simultaneously, if staffing permits. On the fourth floor, the search should start with the apartment above the fire and work outward.

Evacuation Evacuation needed for of the occupants. While the products of the fire have not yet affected first- and second-floor occupants, they should be evacuated. Evacuation of exposure buildings will depend on fire extension and, if necessary, a switch to a defensive operation.

Exposure Protection Interior exposures must be protected during fire confinement and extinguishment phases. External exposures will become a factor if a defensive operation occurs.

Confinement The confinement of this fire is critical to prevent extension to the rest of the building and to support search and rescue activities. In particular, initial hose lines must be placed to keep the fire from entering the common hallway. In this situation the first hose line should be stretched through the front door up the east stairwell and positioned at the entrance to apartment 3B to hold the fire in that apartment. This supports both the confinement effort and protects the means of egress for search and rescue purposes, evacuation, and firefighter safety. Another hose line should be positioned in the east stairwell to the fourth-floor apartment 4B to support confinement by preventing vertical extension.

Extinguishment Once the fire is confined, efforts are made towards extinguishment. In this case, the same hose line that is in position to confine the fire will be used for extinguishment and a second hose line will be positioned to assist. For this apartment of 400 square feet with a 100% involvement, the needed fire flow using the NFA formula is estimated to be approximately 130 gpm. This flow is easily obtained using a 1¾-inch hose line. The backup 1¾-inch hose should be positioned to back up the first line, to protect the attack line crew, and to extinguish hot spots as the attack crew moves throughout the fire apartment.

Ventilation The fire has self-vented through the windows. Further ventilation should initially involve opening of roof openings over the stairwells in order to clear the stairs of smoke and heat to support search, rescue, and evacuation. Windows in adjacent apartments and those above should be vented as needed. Use caution and coordination when opening windows above the fire apartment so as not to increase the fire spread. As soon as possible, PPV should be set up at the first-floor entrance to the stairwell to further ventilate the stairs. Other PPV fans can be used once the stairs are clear at each floor stairwell opening. This allows ventilation of each apartment, using the apartment doors as exhaust controls. Vertical roof ventilation other than stairwell openings will not likely be needed unless the fire extends into the attic space.

Overhaul Overhaul is particularly important at this fire. Once the visible fire is extinguished, crews should open the walls and ceilings in the fire apartment. Again, enough area should be exposed until unburned material is apparent. The adjacent third- and fourth-floor apartments must also be checked, paying close attention to vertical shafts such as those for utilities. The apartments below the fire apartment also require investigation. The attic area should also be checked, either through the ceiling openings or by the attic access. The eaves outside of the roof above the fire apartment should also be checked for extension.

Salvage In multifamily dwellings, the need for salvage increases with each apartment. Efforts should be focused on the entire building. This will include steps to prevent further smoke and water damage to the adjunct apartments. Water damage is very likely in the second- and first-floor apartments directly under the fire apartment.

The Company Assignments

In order to accomplish the strategic goals and tactical objectives just discussed, company assignments must be made. **Table 19-5** depicts these typical assignments, both initially and throughout the incident. In this situation a second alarm should be called for on arrival of the first engine. The units are listed in order of arrival.

After the fire a PIA should be conducted. The PIA should focus on the incident and point out strengths and weaknesses of the operation.

This fire scenario was a simple example of a fire in an apartment. It presents a greater challenge than the previous fire as it was above the ground and in a much larger building. The fire was handled appropriately with the resources available from the response of two alarms. The strategic goals were met through proper application of tactical objectives.

Table 19-5   *Typical company assignments.*

Unit	Initial Assignment	Other Assignments
Engine 1 (E1)	Advance 250 feet of 1¾-inch hose line through the front door and up the east stairs to the third floor. Enter the hallway and place the hose line at the fire apartment door. This line will be supplied from tank water until a water supply can be established. The driver of E1 will be the accountability point for this entrance	Once the fire has been extinguished and this crew has been through rehab, they will assist in salvage.
Ladder 1	Position the apparatus at the corner of Sides 1 and 4 and place the ladder to the roof. This ladder will provide a secondary means of egress for the roof team. The company officer and FF will access the roof via the interior of the Side 2 exposure building. The operator will remain with the apparatus and become the accountability point for firefighters going to the roof. The assignment will be for roof ventilation.	Once the roof operations are complete, this crew will assist search and rescue on the fourth floor.
Squad 1	The two FFs will assist E1 by forcing entry. Once inside, they will begin a search of the priority areas.	
Battalion Chief 1	Incident command	
Ambulance 1	Will be initial two out	
Engine 2 (E2)	Will forward lay a supply line to E1 from the closest hydrant. The company officer, driver, and FF from E2 will advance a 3-inch hose line to the third-floor stairwell landing in the east stairs, which will have a gated wye on it from E1. The E2 crew will bring a 100-foot 1¾-inch standpipe pack, connect to the gated wye, and advance to apartment 4B on the fourth floor, checking for extension and conducting primary search.	Once the fire is extinguished, the E2 crew will perform the secondary search.
Engine 3 (E3)	E3 entire crew will take a 100-foot standpipe pac up the east stairs to the third floor, connect to the wye, and advance to apartment 3B to back up the attack crew.	
Staff Officers	Safety will be assigned to the first-arriving staff officer trained in this function. The second-arriving staff officer will be assigned accountability. Other staff officers will be assigned to • Public information • Resident needs • Assumption of command as incident escalates • Direct operations of fire floor	

(continued)

Table 19-5 *(Continued)*

Unit	Initial Assignment	Other Assignments
Second-Alarm Units		
Engine 4	Become RIC 1	
Engine 21 (Mutual Aid)	Assist in search/evacuation of floors one and two.	
Engine 22 (Mutual Aid)	Secure a secondary water supply for Ladder 1, should a defensive attack be ordered. Relief of E1 Crew.	
Ladder 21 (Mutual Aid)	Begin overhaul on the third floor.	
Ambulance 2	Set up treatment and rehab area.	
Squad 21 (Mutual Aid)	Assist in search/evacuation of floors one and two. Identify salvage concerns.	
Deputy Chief	Assume command. Assign Battalion One to operations.	

Note: FF means firefighter.

COMMERCIAL BUILDING SCENARIO

This scenario involves a one-story strip mall, as shown in **Figure 19-7**. The building was built in 1965 and has concrete block walls with a metal roof. The building houses eight stores and has a common cockloft throughout. This scenario requires additional resources; therefore, there will be additional alarms. Each alarm after the first will get the same amount of personnel and equipment. **Table 19-6** describes the size-up factors.

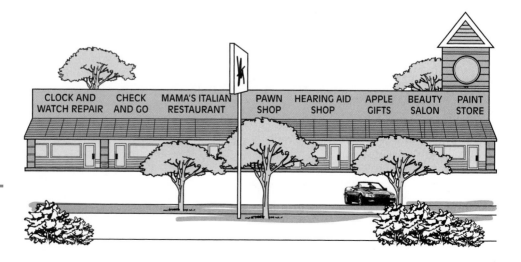

Figure 19-7
Commercial building, strip center, structure.

Table 19-6   *Size-up factors for commercial building.*

Factors	Description
Environment	
Time	2:00 A.M.
Building construction features	Strip stores of ordinary construction built in 1965.
Weather	55°F. Wind not a factor
Height	One story
Area	400 × 50 feet
Occupancy	There are eight stores in the strip. From north to south: 1. Clock and Watch Repair 2. Check and Go 3. Mama's Italian Restaurant 4. Pawn Shop 5. Hearing Aid Shop 6. Apple Gifts 7. Beauty Salon 8. Paint Store
Access	Side 1 is a parking lot. Side 2 is a street, Side 3 is a narrow alley, and Side 4 is a street. Access is available from all four sides.
Terrain	Level
Resources	
Personnel responding	Engine 1 with 3 personnel Engine 2 with 3 personnel Engine 3 with 3 personnel Ladder 1 with 3 personnel Squad 1 with 2 personnel Ambulance 1 with 2 personnel Battalion 1 with 1 person 17 total on initial alarm and subsequent alarms.
Personnel needed	Varies depending on assignment
Apparatus responding	3 Engines 1 Ladder 1 Squad 1 Ambulance 1 Battalion Chief
Apparatus needed	Varies depending on assignment
Water supply	Hydrants are throughout the area. The closest is 100 feet away and has a flow of 1,500 gpm.
Built-in private fire protection	The paint store is protected by an automatic sprinkler system.
Need for special extinguishing agents	Possible foam required for thinners and other flammable liquids in the paint store.

(continued)

Table 19-6   *(Continued)*

Factors	Description
Resources	
Length of time for resources to be on scene	The first engine and ladder will be on scene within 4 minutes. The balance of the first-alarm assignment will be on the scene in 8 minutes. The balance of each additional alarm will be on the scene within 8 minutes of request.
Conditions/Situation	
Fire location	The fire is located in the Italian restaurant.
Fire extension probability	Good—internal exposures, specifically the cockloft and immediate adjacent stores. External exposures not likely.
Type of fuel	Class A combustibles in fire building. Possible natural gas-fed fire in restaurant. Various fuels possible if the fire extends into exposure stores.
Fire condition on arrival	Heavy fire visible out the front windows of the restaurant. Heavy smoke in the Check and Go and the Pawn Shop. Very light haze noted in all other stores.
Life hazard	Low due to time of day
Property conservation concerns	High
Possible incident duration	3–4 hours

Figure 19-8 shows a floor plan of the structure. Note the fire's location and possible extension probabilities. **Figure 19-9** shows the street and adjacent properties. As this is a target hazard in Anytown, **Figure 19-10** provides the prefire plan.

Strategic Goals and Tactical Objectives

The following is the application of the strategic goals as presented in this text to this specific incident.

Firefighter Safety Department SOGs provide for the wearing of appropriate PPE, the implementation of an IMS, and accountability procedures. Like the multiple-family fire, at this fire there will be several entrance points into the fire building and hazard area. The first-arriving companies at each of these potential entrance points will be the accountability point for that entrance. However, due to the complexity of the incident and the building, an accountability officer should be assigned as soon as possible. The IMS provides that the first-arriving engine officer will pass command in favor of a fast attack. The battalion chief will assume command on arrival.

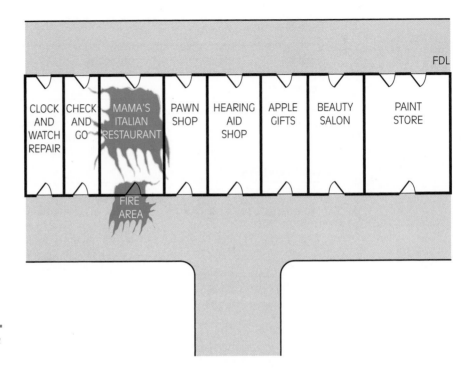

Figure 19-8   *Stores in the strip center.*

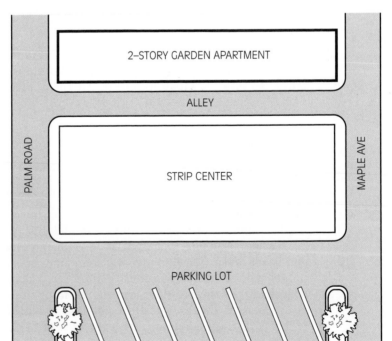

Figure 19-9   *Plot plan and surrounding properties.*

ANY TOWN FIRE DEPARTMENT
PREINCIDENT PLAN

Building Address: 2468 Arch Way	**Emergency Contact Name:** John Arch
Business Name: Shoppes of Anytown	**Phone Numbers:** 111-555-2222

Building Description: Multioccupancy Strip Center

Construction Type: Block Exterior Walls. Wood Interior Walls Covered with Sheetrock

Roof Construction: Metal Decking with Metal Truss. Common Cockloft

Floor Construction: Concrete Slab

Occupancy Type: Multioccupancy	**Hazards to Personnel:** Ordinary combustibles. Flammable liquids in paint store

Water Supply Location #1: East 100′ **Available Flow:** 1500 **GPM**
Water Supply Location #2: West 200′ **Available Flow:** 1500 **GPM**
Water Supply Location #3: Rear/East 150′ **Available Flow:** 1500 **GPM**

Estimated Needed Fire Flow
Based on a Single Story Involvement

Level of Involvement	25%	50%	75%	100%
Est. Needed Fire Flow	104	208	312	416

Fire Behavior Predictions:
Fire may spread horizontally throughout cockloft. Heavy fire loads in store. Lack of private detection, may result in late notification.

Anticipated Problems: Heavy Fire Conditions. Rapid Spread. Possible hazmat from paint store.
If paint store is involved, watch run-off and smoke travel in area.
Rear access is tight alley. Rear access to store is through a walk-through door only

Private Fire Protection:

Sprinklers: ■ Standpipes: ☐ Fire Alarm/Detection: ☐ Other: ☐ (Note Below)
Sprinklers in paint store only.

Completed by: Engine 1	**Date:** 1/15/20xx	**Last Update:** 1/15/20xx

Figure 19-10 *Preincident plan—commercial building scenario.*

The tactical objectives to be employed for firefighter safety and the units assigned are noted in the following list:

- IMS in place, Battalion 1 as the incident commander. A safety officer should be assigned as soon as possible, as should an accountability officer.
- Offensive fire attack will be the initial mode of operation.
- An RIC will have to be in place prior to interior fire attack unless the incident commander feels that there is an imminent life safety risk prior to the assignment of an RIC. In this incident, a second RIC should be considered since the building covers a large area.
- Attack and backup lines of appropriate size for this fire ($1\frac{3}{4}$-inch in this case) possibly $2\frac{1}{2}$ if staffing permits.
- Rehabilitation will be set up after the initial fire attack is complete. As this type fire should be of medium duration and the environment factors are mild, rehabilitation may consist of a brief medical and fluid replacement.

Search and Rescue Because of the early morning hour of this incident, the life safety concern for occupants should be low. A primary search will not be possible initially because the restaurant is fully involved. Efforts should be made to search adjoining occupancies that have a smoke condition. This can be done in conjunction with the deployment of hose lines.

Evacuation Should there be any occupants in any of the other stores, they should be evacuated. Should the fire extend into other stores that might have hazardous products, such as the paint store, consideration should be given to evacuation of surrounding residential properties.

Exposure Protection Interior exposures must be protected during fire confinement and extinguishment phases. External exposures will become a factor if a defensive operation occurs.

Confinement The confinement of this fire is critical to prevent extension to the rest of the building. Initial hose lines must be placed to keep the fire from entering the common cockloft. As the fire is fully involved and no special hazards exist in the adjacent stores, the first hose line should go into the Pawn Shop because of the number of stores on the south side of the fire. Ceilings must be opened immediately to check for extension and to confine the fire to the restaurant. The next hose line for fire confinement should be positioned in the Check and Go store. Again, get the ceilings opened to confine the fire to the restaurant.

Extinguishment Once the hose lines are placed and the fire is confined, efforts can be made toward extinguishment. In this case, the engine companies assigned in the rear should advance a $2\frac{1}{2}$-inch hose line through the rear door and push the fire toward the front of the restaurant, which windows have already self-vented from the fire. Once a second engine arrives in the rear, a second $2\frac{1}{2}$-inch hose line should be advanced to back up the first. For this store of 1,250 square feet with a 100% involvement, a needed fire flow using the NFA formula is estimated to be approximately 416 gpm. This flow is obtained using the two 2½-inch hose lines each flowing 200 gpm.

Ventilation The fire has self-vented through the front window. Further ventilation should initially involve opening of roof openings over the restaurant, if it is safe to do so. Vertical ventilation using skylights should be performed over Check and Go and the Pawn Shop. PPV may also be used in these stores. A trench cut may be considered if staffing is available and the fire cannot be stopped from below.

Overhaul Once the visible fire is extinguished, crews should open the walls and ceiling in the restaurant. Again, enough area should be exposed until unburned wood is apparent. The adjacent stores should be examined as necessary to ensure complete extinguishment.

Salvage Efforts must be made to protect the property from fire, smoke, and water damage in the adjacent stores. The Pawn Shop may have valuable jewelry or other items of value. The Check and Go store may have a lot of paper records essential to the business. Salvage efforts should be extended to all properties; however, if the fire is confined to the restaurant, the ventilation of the light smoke may be all that is necessary for salvage in the stores further from the fire building.

The Company Assignments

In order to accomplish the strategic goals and objectives just discussed, company assignments must be made. **Table 19-7** depicts these typical assignments both initially and throughout the incident. In this situation a second alarm should be called for upon arrival of the first engine. The units are listed in order of arrival.

Further ventilation of adjacent uninvolved stores will be performed by crews on the scene once initial fire containment is complete. Salvage operations will be conducted in conjunction with overhaul by crews when their initial assignments are complete.

Should the initial action not confine and extinguish the fire, a third alarm should be ordered immediately. If initial crews are overly fatigued, a third alarm may also be considered after the initial operations to complete overhaul and salvage.

After this fire too, a PIA should be conducted. The PIA should focus on the incident and point out strengths and weaknesses of the operation.

This fire scenario was an example of a fire in a strip store-type commercial building. These type fires present challenges unique to them. For example, multiple occupancies with multiple types of material in each and a common cockloft. The fire was handled appropriately with the resources available from the response of two alarms; however, it is possible that a third alarm may be needed. The strategic goals were met through proper application of tactical objectives.

Table 19-7   *Typical company assignments.*

Unit	Initial Assignment	Other Assignments
Engine 1 (E1)	Advance a 150-foot 1¾-inch or 2½-inch hose line through the front door of the pawn shop. This line will be supplied from tank water until a water supply can be established. The driver of E1 will be the accountability point for this entrance.	
Ladder 1	Position the apparatus in front of the pawn shop. Crew will enter the pawn shop with E1 to assist in opening ceilings and confining the fire. Ladder 1 operator to place aerial ladder to roof.	
Squad 1	Enter the Check and Go with E3. Assist with confinement efforts.	
Battalion Chief 1	Incident command.	
Ambulance 1	Will be initial two-out crew.	
Engine 2 (E2)	Will forward lay a supply line to the rear of the strip center and prepare for fire attack with a 2½-inch handline.	Once the fire is extinguished, the E2 crew will perform the secondary search.
Engine 3 (E3)	E3 entire crew will take a 200-foot 1¾-inch handline off of E1 into the Check and Go with the squad after forward laying 5-inch supply line to E1 and E3.	
Staff Officers	Safety will be assigned to the first-arriving staff officer trained in this function. The second-arriving staff officer will be assigned accountability.	
Second Alarm Units		
Engine 4 (E4)	E4 will respond to the rear of the building and take the second 2½-inch handline into the rear for fire attack.	
Engine 21 (Mutual Aid)	RIC 1 in front of building.	
Engine 22 (Mutual Aid)	Secure a water supply for Ladder 1, should a defensive attack be ordered.	
Ladder 21 (Mutual Aid)	Go to the roof to perform ventilation as needed.	
Ambulance 2	Set up treatment and rehab area.	
Squad 21 (Mutual Aid)	Become RIC 2 in the rear.	
Deputy Chief	Assume command assign Battalion 1 to operations.	

SUMMARY

This text has presented an integrated set of strategic goals and tactical objectives designed to meet fire incident priorities. These are used in conjunction with a knowledge of fire dynamics, extinguishing methods, safety, incident management, and common company operations. Together these all can be applied to almost any fire situation to mitigate the situation in a safe and effective manner.

This chapter put these concepts into play on three different fire scenarios. While there may be other ways to accomplish the same result, these are accepted practices. It is clear that a certain amount of resources are needed to handle these incidents. Resources, including both personnel and equipment, must be called for early in the incident.

After the incident, considerations must also be given to proper demobilization and postincident analysis.

REVIEW QUESTIONS

1. In the multifamily scenario, if two of the same size apartments were involved, what would the needed fire flow be? (Use the NFA formula.)

2. In the commercial building scenario, what would the needed fire flow be if three stores were involved? (Use the NFA formula.)

3. True or False? A PIA should be conducted on all three of these incidents.

4. True or False? In the single-family dwelling scenario, the bedroom and exit paths should be searched first because of time of day.

5. True or False? In the multifamily dwelling scenario, the top floor should be searched first.

ACTIVITIES

1. Using the information in the single-family dwelling scenario, apply your initial response, in terms of personnel and equipment, to the incident to accomplish the strategic goals.

2. Using the information in the multifamily dwelling scenario, apply your initial and additional alarm response, in terms of personnel and equipment, to the incident to accomplish the strategic goals.

3. Using the information in the commercial building scenario, apply your initial and additional alarm response, in terms of personnel and equipment, to the incident to accomplish the strategic goals.

ACRONYMS

AFFF	Aqueous film-forming foam
APU	Auxiliary power unit
ATC	Alcohol-type concentrate
BLEVE	Boiling liquid-expanding vapor explosion
CAFS	Compressed air foam system
CFR	Code of Federal Regulations
CISD	Critical incident stress debriefing
CISM	Critical incident stress management
CNG	Compressed natural gas
EMS	Emergency medical services
FCP	Fire control plan
FDC	Fire department connection
FEMA	Federal Emergency Management Agency
FFFP	Film-forming fluoroprotein foam
FGC	Fireground command
FIRESCOPE	Firefighter Resources of California Organized for Potential Emergencies
HVAC	Heating, ventilation, and air-conditioning
IAFF	International Association of Firefighters
IC	Incident commander
IDLH	Immediately dangerous to life and health
IMS	Incident management system
ISO	Insurance Services Office
LP	Liquid petroleum
LPG	Liquid petroleum gas
NFA	National Fire Academy
NFIRS	National Fire Incident Reporting System

NFPA	National Fire Protection Association
NIIMS	National Interagency Incident Management System
NIOSH	National Institute for Occupational Safety and Health
OS&Y	Outside screw and yoke
OSHA	Occupational Safety and Health Administration
PA	Public address
PAR	Personnel accountability report
PASS	Personal alert safety system
PDP	Pump discharge pressure
PIA	Postincident analysis
PIV	Post indicator valve
PPE	Personal protective equipment
PPV	Positive pressure ventilation
PVC	Polyvinyl chloride
RECEOVS	Rescue, exposure, confine, extinguishment, overhaul, ventilation, salvage
REVAS	Rescue, evacuation, ventilation, attack, salvage
RIC	Rapid intervention crew
RPD	Recognition-primed decision making
RV	Recreational vehicle
SCBA	Self-contained breathing apparatus
SOG	Standard operating guideline
TDR	Tender delivery rate
USFA	United States Fire Administration
VES	Vent-enter-search
WPIV	Wall post indicator valve

GLOSSARY

Accelerator Device designed to speed the operation of a dry pipe valve.

Access stairs Stairs that serve a particular floor or area of a building.

America Burning: The Report of the National Commission on Fire Prevention and Control A 1973 report to the President of the United States describing the nation's fire problem.

Autoextension When a fire goes out a window or door and extends up the exterior of a building to the floors above or the cockloft.

Balloon frame construction Style of wood-frame construction in which studs are continuous the full height of the building.

BLEVEs Boiling liquid-expanding vapor explosion; describes the rupture of a container when a confined liquids boils and creates vapor pressure that exceeds the container's ability to hold it.

Booster tank The onboard water tank for an engine.

Bowstring The most dangerous truss is the bowstring. It is easy to identify by its curved top chord. These trusses are common in bowling alleys, skating rinks, and other large buildings requiring a long, uninterrupted span.

Branches Used in the IMS to establish and maintain a manageable span of control over a number of divisions, sectors, and groups.

Bulkhead/scuttle The opening from a stairway of ceiling to the roof.

Cantilever An overhang supported on only one side.

Capacity hookup A hookup to a fire hydrant designed to supply the full volume of the pump.

Central (or center) core construction A type of floor layout where components common to every floor, such as elevators and bathrooms, are centered around the core of the building.

Chord and web A type of truss construction in which the chord is the main member either on the top or bottom and the web is the vertical member.

Chords Top and bottom members of the truss.

Code of Federal Regulations (CFR) The document that contains all of the federally promulgated regulations for all federal agencies.

Collapse zone The safety zone set up around a fire building where the potential for a collapse exists; should be the full height of the highest wall.

Command The highest level of responsibility and authority in the IMS at an incident.

Communication systems Radios, computers, printers, and pagers, the numerous hardware and software that goes into a communications network.

Company A team of firefighters with apparatus assigned to perform a specific function in a designated response area.

Compressed air foam system A foam system in which compressed air is injected into the foam solution prior to entering any hose lines.

Conduction A method of heat transfer through a medium, such as a piece of metal.

Consensus standards Standards developed by consensus of industry or subject area experts, which are then published and may or may not be adopted locally; even if not adopted as law, these can often be used as evidence for standard of care.

Convection A method of heat transfer by which the air currents are the means of travel.

Critical incident stress management (CISM) A process for managing the short- and long-term effects of critical incident stress reactions.

Critical incidents Incidents that have a high potential to produce critical incident stress, for example, children, a family member, or a coworker being involved in incidents.

Dampers Devices to prevent smoke travel throughout the HVAC system; often activated by a fire alarm system.

Deflagrate To burn rapidly, which in reality can be considered a low explosive; slower than detonation.

Demobilization plan Process of returning personnel, equipment, and apparatus after an emergency has been terminated.

Department of homeland security Established under the National Strategy for Homeland Security and the Homeland Security Act of 2002, the mission of the DHS is to, "lead the unified national effort to secure America. [To] prevent and deter terrorist attacks and protect against respond to threats and hazards to the nation. [To] ensure safe and secure borders, welcome lawful immigrants and visitors, and promote the free-flow of commerce" (http://www.dhs.gov, 2006).

Detonate To explode with sudden violence.

Direct flame impingement A method of heat transfer by which there is a direct contact to the object by the open flame.

Divisions An IMS designation responsible for operations in an assigned geographical area.

Engine The term for the fire apparatus used for water supply to the incident scene; may also be termed a *pumper.*

Executive fire officer program (EFOP) An initiative of the United States Fire Administration/National Fire Academy designed to provide senior officers and others in key leadership roles with . . . enhanced executive-level knowledge, skills, and abilities necessary to lead . . . transformations, conduct research, and engage in lifelong learning (http://www.usfa.dhs.gov/training/nfa/efop/, 2006).

Exhauster Device designed to speed the operation of a dry pipe valve by bleeding off pressure.

Exothermic reaction A chemical reaction that releases heat.

Facade An artificial face or front to a building.

Finance Part of the general staff of the IMS, responsible for all financial matters.

Fire brigades The use of trained personnel within a business or industrial site for firefighting and emergency response.

Fire control plan A required plan on a ship depicting fire control procedures and resources specific to the ship.

Fire department connection (FDC) A siamese connected to a sprinkler or standpipe system to allow the fire department to augment water volume or pressure.

Fire load All the combustible parts of contents of a building.

Fire pump A stationary pump designed to increase water flow or pressure in a sprinkler or standpipe system.

Fire stops Pieces of material, usually wood or masonry, placed in studs or joist channels to slow the spread of fire.

Fire tetrahedron Four-sided pyramid-like figure showing the heat, fuel, oxygen, and chemical reaction necessary for combustion.

Fire triangle Three-sided figure showing the heat, fuel, and oxygen necessary for combustion.

Flashover An event that occurs when all of the contents of a compartment reach their respective ignition temperatures in a very short period of time.

Garden apartments A two- or three-story apartment building with common entry ways and floor layouts, often with porches, patios, and greenery around the building.

Groups An IMS designation responsible for operations as an assigned function; for example, rescue group.

Gusset plate A connecting plate used in truss construction; in steel trusses, the plates are flat steel stock; in wood trusses, the plates are either light gauge metal or plywood.

Halon An extinguishing agent that works by interrupting the chemical change reaction.

Hydrocarbons Any of numerous organic compounds such as benzene and methane that contain only carbon and hydrogen.

Immediately dangerous to health and life (IDLH) Used by a number of OSHA regulations to describe a process or an event that could produce loss of life or serious injury if a responder is exposed or operates in the environment.

International Association of Firefighters Labor organization that represents the majority of organized firefighters in the United States and Canada.

Interstitial space Opening between the top and bottom chords of a parallel chord truss.

Kevlar A material with abrasion and heat resistive features used for firefighting PPE.

Ladder The fire apparatus with aerial and ground ladders, referred to sometimes as a *truck;* may or may not have pumping capabilities.

Logistics Part of the general staff of the IMS, responsible for all logistical needs and supplies.

Main drain The drain for a sprinkler system that drains the entire system.

Main water control valve The main water supply valve in a sprinkler or standpipe system.

National Emergency Training Center Location in Emmitsburg, Maryland, for the United States Fire Administration, the National Fire Academy, and the Emergency Management Institute.

National Fire Academy (NFA) Agency under the USFA that develops and delivers training and educational programs for firefighters and fire officers.

National Institute of Occupational Safety and Health (NIOSH) A government agency dealing with workplace safety and health issues. NIOSH began a project in 1997 in which it will investigate firefighter line-of-duty deaths.

Nomex A material with abrasion and heat resistive features used for firefighting PPE, particularly for outer shells.

Normal operating pressure The pressure on a water distribution system during normal demand.

Occupational Safety and Health Administration (OSHA) The federal agency tasked with the responsibility for occupational safety of employees.

Operations Part of the general staff of the IMS, responsible for all operational functions.

Outside screw and yoke (OS&Y) One type of main water control valve characterized by the visible screw and yoke.

Panel points Metal plates, sometimes referred to as gusset plates, that have teeth that enter the wood member to hold them together.

Parallel chord truss The top chords and the bottom chords run parallel with each other with the web between them.

Peaked roof truss This truss is found in most of today's homes and commercial buildings. The truss is triangular in shape to provide the peaked roof.

Peer diffusing The concept of using a trained person from the same discipline to talk to emergency responders after a critical incident as a means to allow them to talk about their feelings about the event in a nonthreatening environment.

Personal alert device Device such as the Personal Alert Safety System (PASS) that emits a loud alert or warning that the wearer is motionless.

Personal protective equipment Complete ensemble of clothing worn by firefighters for firefighting; may include different components for different types of emergencies—for example, wildfire PPE differs from structural firefighting PPE.

Personnel accountability reports (PARs) Verbal or visual reports to incident command or to the accountability officer regarding the status of operating crews; should occur at specific time intervals or after certain tasks have been completed.

Planning Part of the general staff of the IMS, responsible for all incident planning functions.

Plenums Part of an HVAC system that provides an avenue for air flow.

Polar solvents Water soluble compounds such as alcohol; require foams that are alcohol-type concentrates (ATC).

Post indicator valve (PIV) One type of main water control valve characterized by the visible window, which indicates the position of the valve.

Postincident analysis A critical review of the incident after it occurs; should focus on improving operational effectiveness and safety.

Pyrolyzing A chemical change in wood resulting from the action of heat.

Radiation A method of heat transfer through light waves, much like the sun warms the Earth.

Rapid intervention crews (RICs) Assignment of a group of rescuers with the sole purpose of rapid deployment to reports of operating personnel in trouble or missing.

RECEOVS A fire incident management goal set: rescue, exposures, confine, extinguish, overhaul, ventilation, salvage.

Regulations Requirements or laws promulgated at the federal, state, or local level with a requirement to comply.

Rehabilitation The group of activities that ensures responders' health and safety at an incident scene; may include rest, medical surveillance, hydration, and nourishment.

Rescue See *squad*.

REVAS A fire incident management goal set: rescue, evacuation, ventilation, attack, salvage.

Riser The vertical piping in a sprinkler or standpipe system.

Rollover The rolling of flame under the ceiling as a fire progresses to the flashover stage.

Row houses Homes attached with common walls and roofs.

Saponification Process that occurs when wet chemicals come into contact with grease and the like, forming a soap-like product.

Scrub area The area or the building that can be reached with an aerial ladder once the apparatus is set up; a consideration for ladder truck placement.

Sector A geographic area or function established and identified within the IMS for operational purposes.

Self-contained breathing apparatus Type of respiratory protection in which a self-contained air supply and related equipment are worn or attached to the user; fire service SCBA are required to be of the positive pressure type.

Shelter in place A form of isolation that provides a level of protection while leaving people in place, usually homes or unaffected areas of large buildings.

Siamese Device to connect two hose lines into one line with either a clapper valve or gate valve to prevent loss of water when only one line is connected.

Size-up A decision-making process that starts before the incident and allows the firefighter or incident command to gather information and develop appropriate strategies.

Spalling The loss of surface concrete, sometimes forcefully, that occurs when the moisture in concrete begins to expand when the concrete is subjected to fire or exposed to heat.

Span of control The ability of one individual to supervise a number of other people—usually three to seven, with five being ideal (the number depends on the complexity of the situation)—or units.

Squad A unit that may carry firefighters, firefighters with specialized tools, or a medical rescue (EMS) unit; may be referred to as a *rescue truck*.

Stack effect As heated air rises then cools it forms a layer, or ceiling, from which more rising smoke banks.

Standard of care The concept of what a reasonable person with similar training and equipment would do in a similar situation.

Standardized apparatus Apparatus that has exactly the same operation and layout of other similar apparatus in a department—for example, all of the department's pumpers would be laid out the same, operate the same, and have the same equipment; useful for situations when crews must use another crew's apparatus.

Standards Often developed through the consensus process; standards are not mandatory unless adopted by a governmental authority.

Static supply A water supply that is always in the same location, such as a lake or pool.

Stratification Heated smoke rises until it establishes equilibrium with the surrounding area and then moves horizontally.

Strip shopping centers Rows of attached mercantile occupancies with a common look and roof line.

Subsurface injection The tactic of introducing foam at the bottom of a bulk storage tank, allowing the foam to float to the top and blanket the product.

Surface area The exposed exterior surface of an object.

Taxpayers A term of building, more common on the East Coast, in which a mercantile occupancy is on the first floor and living areas occupy the floors above.

Test header Is a group of outlets used to test the capacity of a building fire pump system.

Thermal radiation feedback As heat is transferred in a compartment, the walls and furnishings in that compartment heat; this heat then feeds back and further heats the compartment.

Throat The component of a building that connects two wings.

Ties Connecting members, such as gusset plates, that hold the members together.

Trench cut A cut for ventilating a confined fire in which the cut is the entire distance of the roof; useful in confining a fire in a cockloft that is spreading horizontally.

Two-in, two-out rule The procedure of having a minimum of two firefighters standing by completely prepared to immediately enter a structure to rescue the interior crew should a problem develop; to be in place prior to the start of interior fire attack.

Unified command The structure used to manage an incident involving multiple jurisdictions or multiple response agencies that have responsibility for control of the incident.

United States Fire Administration Agency under the Department of Homeland Security that directs and produces fire programs, research, and education.

Wall post indicator valve (WPIV) One type of main water control valve mounted on a wall and characterized by the visible window, which indicates the position of the valve.

Water tender Fire apparatus that is a mobile water supply; may be termed a tanker in some areas of the United States.

Waterflow alarm A mechanical or electrical device attached to a sprinkler system to alert for water flow.

Web Inside members of the truss.

Wildland/urban interface A wildland area with dwellings or other buildings intermixed creating a fire problem during wildfires.

INDEX